Builders

Builders

Herman and George R. Brown

JOSEPH A. PRATT
AND CHRISTOPHER J. CASTANEDA

Texas A&M University Press
College Station

The paper used in this book meets

the minimum requirements of the

American National Standard for Permanence

of Paper for Printed Library Materials, z39.48-1984.

Binding materials have been chosen for durability.

Library of Congress Cataloging-in-Publication Data

Pratt, Joseph A.

Builders : Herman and George R. Brown / Joseph A. Pratt and
Christopher J. Castaneda.

p. cm. —

Includes bibliographical references and index.

ISBN 0-89096-840-3

1. Brown, Herman, 1892–1962.   2. Brown, George R., 1898–1983.
3. Brown & Root (Firm)—History.   4. Offshore oil field equipment
industry—Texas—History.   5. Offshore gas field equipment industry—
Texas—History.   6. Construction industry—Texas—History.
7. Businessmen—Texas—Biography.   I. Castaneda, Christopher J.,
1959– . II. Title.

HD9570.B67P73   1998

338.7′624′09764—dc21   98-29635

                                                                        CIP

For
ALICE (SUZY) PRATT
and
TERRI ALFORD CASTANEDA

CONTENTS

ILLUSTRATIONS

TABLES

PREFACE

The Brown Brothers

The combined talents of Herman and George R. Brown constituted a formidable business team. The Browns grew up in the Central Texas towns of Belton and Temple during the years preceding World War I and entered adulthood with a strong work ethic and ambition. A close personal bond held together their often volatile relationship. As successful business partners, they worked with each other daily while living as neighbors in Houston. Together they built a major construction company while pursuing other business and civic ventures. The Brown brothers formed a type of two-headed businessman, with the energies and vision of two different men possessing complementary talents.

Herman (1892–62) was six years older, and he remained the "big brother" in all aspects of his relationship with George. Not quite six feet tall, stoutly built, and "somewhat military in bearing," Herman had a forceful personality that left a lasting impression on most people he encountered. Lady Bird Johnson captured his essence with a simple statement: "He was the bulldozer."[1] This practical man was a natural builder with a talent for sizing up the requirements necessary to complete a job. Brash yet intelligent, Herman acted decisively, imposed his powerful will on stubborn problems, had a glass of Scotch, and went on to the next job. Herman was the unquestioned boss at Brown & Root, where he demanded and received fierce loyalty from his managers and workers alike. Indeed, he seemed most comfortable in the company of his own workers. Brown & Root reflected Herman's character; it became a creative, risk-taking company that would tackle any project, no matter how large or complex. In this respect, Herman Brown resembled late-nineteenth-century entrepreneurs such as Rockefeller and Carnegie. He believed passionately in efficiency and attention to detail, and he expressed a paternalistic commitment to his employees and community.

Younger brother George (1898–83) was a twentieth-century-style businessman. Slightly taller and thinner than Herman, he had a formal university education in engineering. George was more sophisticated than his older brother; he was less abrasive and confrontational with others as well. A soft-spoken natural salesman who talked with a slight speech impediment, George put all audiences at ease. Lady Bird Johnson remembered him as "slim, an elegant dresser, a genial, lovable man, but if I had just one word, I would say 'creative.'"[2] George spoke often about "the romance of engineering," and he could have applied this phrase to his ability to build coalitions and consensus as well as his ability to build bridges and dams. He seldom backed down from a good argument with Herman, but once the shouting ended, he regularly backed away and allowed big brother to have the final say. However, when the brothers acquired a major interest in Texas Eastern Transmission Company (a large natural gas transmission company), George took the lead managing this new company, becoming chairman of the board. After Herman's death in 1962, George was on his own for the first time in forty years, and he assumed a greater role in Brown & Root and a leading role in building the brothers' philanthropic organization, the Brown Foundation.

The brothers were distinctly different men, but together they possessed the attributes needed for extraordinary success. One potential barrier to their shared success was the personal tension that any two people might experience over a long partnership. Brotherly love overcame this in part. Herman and George felt a strong mutual loyalty, both recognizing the pecking order between a big brother and his younger sibling. On a more practical level, the brothers made an excellent choice early in their business relationship when they agreed to share an equal interest in all business ventures. Their handshake agreement in the early 1920s shaped their business dealings for the rest of their lives. The brothers argued heatedly about politics, business strategies, and personnel matters. But having early on resolved their financial relationship, they did not argue about money.

The Browns' journey began on the primitive roads of early-twentieth-century Central Texas. In 1914 Herman took over a small road construction company active in Central Texas. For the next twenty years, he literally scraped out a living grading and surfacing Texas roads. He formed Brown & Root in 1919 with capital provided by his brother-in-law, Dan Root. George joined his brother at Brown & Root in 1922. Hard times during the Great Depression forced the brothers to search aggressively for new construction jobs, and they hit pay dirt in 1936 when they won the contract for the Marshall Ford Dam west of Austin on the Colorado River. Their successful completion of this project marked the Browns as major builders in their region. They surged into the national spotlight during World War II with several large-scale military projects, including a substantial shipbuilding operation on the Houston Ship Channel. When hostilities ended, Brown & Root was well-positioned to take advantage of the post–World War II boom in Texas, and the company grew steadily into a giant, diversified, international construction firm ready to go almost anywhere and build almost anything.

As Brown & Root prospered, the brothers diversified their business. Immediately after World War II, they took the lead in organizing the Texas Eastern Transmission Company, which purchased the Big Inch and the Little Big Inch pipelines. Originally built by the government during World War II to transport much-needed petroleum from the Southwest to the Northeast, these lines became the backbone of one of the largest natural gas transmission companies in the nation. The Browns remained active in the management of Texas Eastern, whose expanding pipeline system presented a ready market for Brown & Root's contracting services.

In the years after World War II, Brown & Root expanded within the United States and overseas in search of large engineering and construction projects. This led to major jobs for the U.S. military, including military facilities on Guam and air force and naval bases in Spain and France. In a consortium with other large construction companies, Brown & Root later built extensive and at times controversial military facilities in Vietnam. Its nonmilitary construction projects during these boom years included offshore oil drilling facilities, barges, and pipelines for North Sea oil and gas production, as well as several large dams. One of its most highly publicized endeavors was the ill-fated "Project Mohole," a much-discussed proposal to drill through the earth's crust and determine the composition of its core. Although funding for this ambitious undertaking ultimately disappeared, forcing its abandonment, the choice of Brown & Root as the contractor reinforced the firm's image as one of the nation's leading construction companies.[3]

By the late 1950s, Herman and George Brown had earned considerable wealth and a degree of national prominence. Their photograph appeared on the cover of *Business Week* in May of 1957 with a caption noting the Brown brothers' dual image within the business community: "George and Herman Brown of Texas' Brown & Root: In front as builders, backstage as politicians."[4]

As builders, the brothers had certainly moved "in front" by the mid-1950s. Their Houston-based company, Brown & Root, was one of the nation's largest construction firms, and the Texas Eastern Transmission Company gave them a substantial presence in natural gas, one of the nation's fastest growing post–World War II industries. In the arena of politics, insiders recognized the Browns as powerful backstage supporters of then Texas senator Lyndon Johnson, influential lobbyists of the Texas legislature, and prominent civic "movers" in Houston.

Herman Brown's death in 1962 marked the end of an era at Brown & Root. Plagued by serious health problems in the early 1960s, Herman Brown took the lead in negotiating the sale of Brown & Root to the Halliburton Company. After Herman's death, George Brown completed the sale of the company that had carried the Browns to wealth and fame. Suddenly, after forty years of close cooperation with his older brother, George had to adjust to life without his longtime partner. As a Halliburton director and a significant stockholder in the parent company, George also served as president of Brown & Root. But while much of his energy remained focused on business matters, George expanded his commitment to other activities, including philanthropy. Through the Brown Foundation, created in 1951 and greatly strengthened in 1962 with the proceeds from the Brown & Root sale, George became active in the development of educational and cultural institutions, notably Rice University; the Museum of Fine Arts, Houston; and Southwestern University in Georgetown, Texas. After his death in 1983, the Brown Foundation carried forward both the philanthropic work and the name of its founders.

Forty years after the *Business Week* article, few people know the businesses they created or the civic projects they sponsored; fewer still know anything about the brothers' lives. The one feature that remains in the public memory is their political history, as biographers of Lyndon Johnson continue to highlight the Browns' financial support of Johnson.[5] Although these accounts generally depict the Brown brothers as men of great political influence, they include little discussion of their business success, the source of their influence.

This dual biography of Herman and George Brown seeks a more balanced historical portrait of the Brown brothers. Such a portrait must begin with their lifelong involvement in the construction industry, because this is where the Browns earned their fortune and entered the public spotlight. Both their civic leadership and their political influence flowed from their success as builders. The construction industry has seldom attracted the attention of researchers and writers interested in the development of the American economy. Historians focused instead on John D. Rockefeller, Andrew Carnegie, Henry Ford, and others who created the nation's basic manufacturing industries. As a result, we know much about these "captains of industry" but much less about those who built the nation's roads, dams, and even many of its factories. Yet, heavy construction played a critical role in building the basic infrastructure required for the steady expansion of the modern American economy.[6]

Brown & Root and other major construction companies grew with the economy, gradually gaining the capacity to build bigger and more complex projects. Risks confronted them at every turn, and would-be competitors regularly fell by the wayside in the rough-and-tumble world of construction. The Brown brothers had the skills, timing, and luck required to prosper. Their story helps us understand a little-studied, but nonetheless vital, sector of the economy.

This dual biography is an effort to recover a sense of the history embodied in the Browns' work and their name. It strives above all to recreate the Browns' business careers. The focus of the narrative is the process by which the Browns built permanent institutions. Brown & Root and Texas Eastern still exist as organizational monuments to the hard work of the Browns and others. Despite the fact that both companies are now parts of larger concerns, their histories remain instructive about business life in mid-twentieth-century America. An examination of the roles of Herman and George Brown in building these companies sheds light on numerous related issues, including the nature of entrepreneurship in these years, the management of growing firms, and the impulse toward that diversification. Their story also provides a case study of how two successful businessmen used their resources and talents to help build other institutions, notably in this case, Rice University, Southwestern University, and the Brown Foundation.

This book also addresses topics—including antiunionism and political influence—that contemporaries and historians alike have associated with the Brown name. One central reality of the construction business was that it was labor-intensive, making the creation of an effective accommodation between owners and workers a critical ingredient of business success. The Browns' attitudes toward their workforce became an important part of their national reputation. Herman, in particular, developed very strong ideas about labor while working on the roads of Central Texas during the formative years of Brown & Root. He passionately believed that those who worked for him should not be forced to join a union as a condition of labor. He felt that employer and employee owed each other loyalty, and that he—and not an outside union—was the best judge of the needs of his workers. Such attitudes reflected his personal philosophy as a self-made man who identified his own background with those of his workers. They also reflected his belief that nonunion labor was more flexible, more efficient, and at times less expensive than union labor.

Herman Brown held tightly to these views on labor as the world changed around him during and after the New Deal. The expansion of Brown & Root's workforce, new federal government regulations of labor-management relations, and the growth of unions challenged his traditional views on labor. In response, he became an active, effective supporter of right-to-work laws and various other constraints on union activity. He helped shape both state and national labor legislation and then aggressively used these laws to fight unions. In the process, his name became synonymous in Texas with the open shop; it remains so in the minds of many older Texans more than thirty years after his death. We have tried to include a balanced account of the Browns' treatment of workers in the historical context suggested by the rise of organized labor in Texas before, during, and after World War II. We doubt, however, that a reader with strong views on either side of this issue will be convinced that we have succeeded.

A second enduring image of the Browns—that they were "among the most astute, behind-the-scenes politicians the state of Texas has ever known"[7]—also reflected their experiences in the construction industry, which pulled them deeply into politics. Beginning with their earliest road building work, the Browns learned the importance of maintaining good relationships with the government officials who controlled construction contracts. Throughout the mid–twentieth century, public works was an important part of the construction industry, and the Browns became adept at securing government work in this sector of their industry. The Browns understood that the best advertisement for public works projects was a reputation for completing projects on time and on budget, and Brown & Root became known as a hard-driving company that could get things done. But competitive companies also needed good information regarding potential government contracts, and Brown & Root developed excellent networks of information and influence in Washington, as well as Austin.

The most visible member of this network was Lyndon Johnson. From the 1930s forward, the Browns and Johnson maintained a close working relationship firmly grounded in friendship and mutual self-interest. The Browns remained generous and loyal political supporters of Johnson as he moved from Congress to the presidency. Johnson returned the favor by staying on the lookout for opportunities to assist his friends, primarily by keeping them informed about potential government projects. It is our belief that there was nothing sinister or even particularly unusual about this relationship in the context of the history of campaign financing before—and most would agree after—the Watergate scandals of the early 1970s led to the rewriting of campaign finance laws. What made this relationship unique was its duration—and the fact that the Browns supported a young man who subsequently became Senate majority leader, vice president, and president. As Johnson became more prominent in national politics, his dealings with the Browns came under intense scrutiny. While Johnson's biographers stress the Browns' financial contributions to Johnson, the strong personal ties that developed among these men over three decades are equally significant in this context. What began as a marriage of mutual convenience between owners of a construction company and an ambitious young politician from their region ripened into a lifelong friendship, as three small-town Texans moved together onto the world stage in their respective fields.

At the local level, the Browns became influential members of the business and civic elite of Houston. Along with a group of friends drawn from the leading businesses in the city, the Browns sought to encourage what they defined as a "healthy business climate" in their city and state. Political observers christened this group the "8-F crowd" because 8-F was the Browns' hotel suite at the Lamar Hotel in downtown Houston where the brothers and their friends regularly met. This suite is generally remembered by critics as a smoke-filled apartment where big deals and politicians were made, but it was both less and more. It was less in that those who regularly passed through 8-F were not as cohesive or as powerful as their critics would have it. Though the most visible group of "movers and shakers" in this era of the city's history, they were hardly the only group seeking to influence politics in mid-twentieth-century Houston. But it was more, in that 8-F's impact went far beyond politics and economics. The 8-F crowd also guided many other phases of Houston's development from the 1930s into the 1970s, mobilizing resources to help develop the city's educational, cultural, and philanthropic institutions.[8]

The city and the state, in turn, shaped the lives of Herman and George Brown. Perhaps the most significant difference between the Browns and the first generation of industrialists who rose to prominence in late-nineteenth-century America is that the Browns rose to power from a southwestern, rather than a northeastern, base. Born in the 1890s, after the Rockefellers and the Carnegies had already brought industrialization to much of the northeastern United States, Herman and George Brown grew up in the less developed Southwest. They were not flamboyant Texans who wore ten-gallon hats, alligator boots, and denim. Rather, their small-town childhoods and their early road building experiences in Central Texas taught them fundamental values: the importance of hard work, loyalty, and a pragmatic view of politics. They also learned that hard work could be sustained best if combined with camaraderie and pleasure, and they played as hard as they worked. Perhaps the most significant lesson they learned during their formative years in Central Texas was the value of aggressively and single-mindedly pursuing life's opportunities. They never lost the perspective of small-town boys wondering at the marvels of the world in which they lived. They never lost the ambition to succeed in a world they knew to be a harsh and demanding place.

Their lifetimes of hard work helped transform Texas from its nineteenth-century agricultural past to twentieth-century industrialism. They built roads that provided transportation and dams that prevented flooding and generated hydroelectric power. They built some of the early factories along the Houston Ship Channel, as well as many of the cross-country pipelines that connected the region into the nation's industrial economy. They took skilled workers and construction expertise developed in Texas around the world in search of opportunities to build big projects. But more than anywhere else, the city of Houston and the state of Texas still bear these builders' marks. Perhaps sufficient time has now passed that we can look back at their lives as a whole and understand these two brothers in the context of their own time and place.

One fundamental choice made this book possible to write, while limiting its usefulness on several significant issues. We chose to write primarily an account of the business lives of two men, a biography of enterprise. The lack of sources made it quite difficult to write a broader social and cultural account including more on the personal lives and opinions of the Browns. One of the clearest weaknesses of the book is its lack of a fuller treatment of the many roles played by Margarett Root, Herman's wife, and Alice Pratt, George's wife. To call these women "helpmates" of their husbands is to demean them and to miss their importance in their husbands' lives. They were partners in their husbands' careers in the ways that many women of their place and time were partners: they managed the demands of the household and the family while their husbands pursued business opportunities around the world; they provided emotional support to men pummeled by the demands of a very competitive business world; they served as surrogates for their husbands in many cultural, social, and civic endeavors. They supported their husbands' business ventures by serving as sounding boards for new ideas and as skilled hostesses for frequent business-related entertaining.

Just as gender-related issues are raised by the "place" of the Browns' wives, the issues raised by race also touched most facets of the Browns' world in the Jim Crow South. But here again, we have found very little information that would allow us to discuss the impact of race on the Browns' lives except in the most general of ways. The Brown brothers were what might be considered "mainstream" supporters of the Jim Crow system. They were not aggressively racist in their speech or behavior, at least when judged by the prevailing assumptions and attitudes of southern whites in these years. But they accepted the racist traditions and assumptions of the South, and they did little to try to alter this system. To his credit, George Brown took a leading role in desegregating Rice University when he served as chairman of the board of that long-segregated institution in the early 1960s. His primary concern, however, was for the future of Rice, not for the future of race relations. An excellent book remains to be written on the role of Houston's business elite in the use and dismantling of the city's Jim Crow system. We chose to write a different book by focusing on those aspects of the Browns' lives that made them most noteworthy, their business careers. We hope that the reader comes away from our account with the conclusion that this choice offers strengths that offset its weaknesses.

Our work reflects the cooperation of many individuals. Herman and George Brown were as private in life as they were successful in business. Few personal or business documents written by their hands survive. This situation made in-depth research into their lives and careers quite difficult. Without the help of many people who knew them, our dual biography of Herman and George Brown would have been impossible.

An opportunity to gain access to historical records in the form of documents and, more important, the memories of their associates, friends, and family came in 1989. In that year, Texas Eastern Corporation, which George Brown chaired for many years, was acquired by Panhandle Eastern Corporation (now Duke Energy). Subsequently, Brown family members, including Isabel Brown Wilson, George Brown's youngest daughter, contacted the firm and requested that George Brown's memorabilia be returned to the family and not destroyed during the corporate reorganization. At the same time, we were completing a history of Texas Eastern.[9] The firm's management directed the Brown family to us because we were familiar with the company's history and the Browns' legacy there.

During our initial meeting with Wilson, we all agreed that the Brown brothers' lives and careers were worthy of study if adequate research material could be located. Of foremost concern was the historical record locked in the minds of surviving associates, friends, and family. As a first step in creating a historical record, the Brown Foundation agreed to provide funding to enable an existing program at the University of Houston, the Oral History of the Houston Economy, to undertake a comprehensive series of Brown-related interviews. During 1991 and 1992, Christopher Castaneda interviewed more than seventy persons who had known the brothers both personally and professionally. Included were William Akers, Dan Arnold, Harry Austin, Lloyd Bentsen, Mrs. L. T. Bolin, John Bookout, Brown Booth, Searcy Bracewell, Herbert Brownell, Alan Chapman, Ed Clark, John Connally, Howard Counts, Naurice Cummings, George Darneille, Dr. Michael DeBakey, Harold Decker, Ward Dennis, Katherine Dobelman, Charles Duncan, James Elkins, Jr., Thomas Feehan, Durwood Fleming, Lawrence Fouraker, Henry H. Fowler, Evelyn Frensley, Harold Geneen, Lem Goodwin, Norman Hackerman, John P. Harbin, Bill Hobby, Oveta Culp Hobby (telephone), Doris Johnson, Lady Bird Johnson, William T. Kendall, Joe Lochridge, Grogan Lord, Ben Love, H. Malcolm Lovett, Albert Maverick, L. F. McCollum, George McGhee, Robert McKinney, Randall Meyer, Patrick Milton, Walter Mischer, Edgar Monteith, Ralph O'Connor, Frank (Posh) Oltorf, Kenneth Pitzer, H. W. Reeves, Carl Reistle, Jack Robbins, Fayez Sarofim, Louisa Sarofim, Emmett Shelton, Albert Sheppard, James Sims, A. Frank Smith, M. S. Stude, John E. Swearingen, E. H. Thornton, Charles Tillinghast (telephone), William Verity, Dr. E. L. Wagner, Delbert Ward, Don Warfield, Merritt Warner, Louie Welch, Nancy Wellin, Isabel Brown Wilson, and Mike Wright.

After completing the interviews and surveying other primary and secondary sources, we agreed that adequate materials existed for a book on the business lives of the Brown brothers. We continued our work with the assistance of the Brown Foundation, which provided funds to help defray research expenses in the early years of this project. Subsequent support over the long life of the project came from the Cullen Chair in History and Business at the University of Houston. A contract between the authors and the foundation stipulated that we would exercise editorial control of the manuscript, with the understanding that points of disagreement would be discussed and resolved openly. If differences of interpretation could not be resolved, the foundation retained the right to express divergent interpretations in explanatory footnotes. The ultimate goal was the writing of a university press-quality manuscript covering the business careers of Herman and George Brown.

Unlike large private corporations that control corporate archives, the Brown Foundation does not possess significant historical records about the brothers' lives, but family members provided useful historical information and otherwise opened doors to people who knew the Brown brothers well. We gained considerable insight into the Browns from interviews and from a variety of secondary sources. We also benefited from access to the historical records of numerous institutions that the Browns helped to build. We should in particular note the assistance of Billie McMahon, Jay Weidler, Joe Stephens, and Dell Avery; Jim Hart, David Bufkin, Dennis Hendrix, and Tony Turbeville; Jud Custer at Southwestern University; Nancy Boothe at Rice University; and Charles Carrol and Margarett Skidmore at the Museum of Fine Arts, Houston. We also thank Claudia Anderson, Alan Fisher, Tina Houston, Robin Byrne Pegler, and Linda Seelke at the LBJ Library in Austin. The bulk of the material collected from these archives—along with materials and interviews from our work on the history of Texas Eastern and Brown & Root's marine division—will become the Herman and George R. Brown Collection at the Woodson Research Center at Rice University in Houston. To researchers who follow us, we apologize in advance for problems they might have in tracking down our citations once this new collection has absorbed the numerous documents we accumulated from a variety of sources.

We would like to thank Harold Hyman, the William P. Hobby Professor of History and director of the Center for the History of Leadership Institutions at Rice University, and Walter Buenger, associate professor of history at Texas A&M University, for their careful reading of the manuscript. Their comments and advice markedly improved the final manuscript. Also offering suggestions on parts of the manuscript were members of the Browns' families, Isabel Brown Wilson, Maconda Brown O'Connor, Nancy Brown Wellin, M. S. Stude, and Louisa Stude Sarofim. Edgar W. Monteith, who followed in his father's footsteps as the Brown family lawyer, helped us understand the Browns' world. Katherine B. Dobelman and Debbie Marshall of the Brown Foundation tracked down former business associates and friends of the brothers while giving us a long-term perspective on the evolution of the Brown Foundation. Doris Anderson helped Isabel Brown Wilson prepare records and photographs. Others who assisted in this project included Catherine Felsmann, Bernadette Pruitt, and Sethuraman Srinivasan Jr., who served as research assistants; Suzanne Mascola, who transcribed the interviews; and Christine Womack and Leonard Hardgrave, who helped prepare the manuscript.

The public history program at the University of Houston under the direction of Marty Melosi provided a supportive environment in which we could draw on the research and the collegiality of numerous scholars working on related projects. In addition to the research assistants listed above, the following scholars have made the study of Houston-area history an exciting, collaborative endeavor at UH: Jesse Allred,

Amy Bacon, Bruce Beaubouef, Mike Botson, Bill Kellar, Tom Kelly, Elisabeth Lipartito, Ken Lipartito, Dennis McDaniel, Ernest Obadele-Starks, Lidya Osadchey, James Patterson, Amilcar Shabazz, Dwight Watson, and others. John Boles, Harold Hyman, and Allen Matusow at Rice University, Cary Wintz and Merlene Pitre at Texas Southern University, James Olson at Sam Houston State University, Louis Marchiafava at the *Houston Review,* Walter Buenger and Bob Calvert at Texas A&M, Ty Priest (our co-author on *Offshore Pioneers*), and Stephen Fenberg at the Houston Endowment are also a part of our extended family of historians.

We owe a special note of thanks to John Lindsey, a longtime friend of the Browns and the driving force behind the rise of Texas A&M University Press. When Joe Pratt first arrived at Texas A&M in 1981, John met him at the door with the idea of writing a biography of George Brown. Unfortunately, John was not able to convince George and Alice Brown that the time was right to undertake such a project. Both died several years later, and it is a shame that our research did not begin with their input. As we have trudged along with the research and writing of this dual biography for too many years to mention, John Lindsey and the staff of Texas A&M University Press have remained constant in their support.

Over the long haul of this project, our families offered understanding beyond the call of duty. Terri, Courtney, Ramsey, and Suzie Castaneda were patient and supportive as the project began in Houston and continued after a move to California. Alice (Suzy), Kate, and Maggie Pratt assisted by shutting the door to the study and going about their business. Suzy's main contribution was to "borrow" a photograph of a Brown & Root vessel named "Alice Pratt" and use it to dupe her fourth grade class into believing that the vessel was named after her, not after George Brown's wife.

Despite help from those mentioned above and others, the final choices and interpretations are ours.

Builders

Part One

Roots of Success, 1892-32

Highway 16, south of Robstown, Texas. Constructed by Brown & Root, 1924–25.
Courtesy Brown Family Archives.

CHAPTER ONE

Family Background

Born in Belton, Texas, in the 1890s, Herman and George R. Brown entered a world not far removed from the frontier. Rhinehart "Riney" Brown—the father of Herman and George—arrived in 1879 to find a region poised on the brink of far-reaching economic changes. At the time of his arrival, Belton was a regional commercial center with a population of several thousand serving the needs of Bell County's approximately 20,000 residents, most of whom lived on farms in the rural areas surrounding Belton. The town had a sense of dynamism and promise that beckoned migrants such as Brown.

As late as 1879 many Belton residents retained vivid memories of their region's former frontier conditions. Texas had become a state only thirty-four years earlier, in 1845. After Belton's founding in 1850, fear of Indian raids marked its earliest decades. After the dislocations of the Civil War, cattle drives from Texas to the Midwest brought a brief boost to the regional economy, as many residents raised stock on the unfenced range. Life on the range was rugged and often violent. Only five years before Riney Brown arrived in Belton, a group of vigilantes shot to death nine members of an outlaw gang being held in the Bell County Jail on charges of horse-stealing and murder. The region made newspaper headlines again in 1878, when authorities gunned down the notorious train robber Sam Bass during an attempted bank robbery in nearby Round Rock; a member of the Bass gang had tipped off officials in Belton about the plans for the bank robbery.[1] Riney Brown had not picked a sedate, mature community in which to settle down and raise his family.

Yet by the time he arrived in Belton, the town had grown out of its frontier past into a thriving and permanent town. By the 1870s and 1880s, Belton developed many basic services of a commercial center capable of supporting the activities of the farmers in outlying areas. Improvements in transportation and communications came quickly during this time. Telegraph lines extended to the city in 1878; the first railroads reached the county in 1881 and 1882; a local telephone exchange went into service in 1884. The financial needs of the region spawned the chartering of the first independent banks in Bell County in the 1880s, with the creation of the First National Bank of Belton leading the way in 1882. The year 1884 witnessed the completion of Belton's water supply system, drawing water first from the nearby Leon River and soon afterwards from artesian wells. The state of Texas and the county imposed a permanent system for financing public schools in 1884.[2] Such changes forged stronger ties between Belton and the outside world, enabling the once isolated country town to become more tightly integrated into the state and national economies.

Before these rapid social changes, Belton had been something of an "island community," with little effective connection to the national economy.[3] When Riney Brown arrived in Bell County on horseback in 1879, the residents were still largely self-sufficient. They grew their own corn and wheat, made their own flour, and raised livestock and hunted to supply meat. There was a good deal of excitement caused in 1875 when an Austin businessman brought the first ice-making machine to Belton. Although the rich black soil of the region was ideal for cotton, few grew much of this potential money crop because of the lack of reliable transportation to market. Instead, local farmers took their surplus farm produce to trading centers on the banks of the Brazos and Colorado Rivers and traded it for manufactured goods and farming implements. Others rode ox-drawn wagons to the nearest major commercial city, Houston. But before the coming of the railroads, the four-hundred-mile round-trip from Belton to Houston required two to three months of hard travel.[4] The hardship, expense, and uncertainty of such journeys greatly limited the prospects for trade between Bell County and other regions.

Even within the county the lack of good roads limited commerce between Belton and the surrounding countryside. Several decades after their father came to Belton, Herman and George Brown successfully launched their careers by building country roads. Those of their father's generation had tried with little success to confront the problems caused by poor roads. According to Bell County historian, George Tyler, most roads were "little better than mere trails which twisted about over the prairies or around the hills and sought easy crossings on the creeks and ravines." After heavy rains, many of these outlying roads "became almost bottomless quagmires that were impassable for loaded wagons." Even the streets in the towns were not immune to the problems brought by heavy rains. The thick black soil under the Bell County town of Temple "stuck to wheels like glue" after heavy rains, and "it was not unusual for wagons or even carriages to bog down in the streets and remain there for days or weeks until the ground dried."[5] The absence of bridges made crossing rain-swelled streams dangerous, and the lack of any reliable means of flood control meant that periodic floods washed out roads and destroyed property. To maintain or improve these roads, local authorities relied on the traditional system of requiring all able-bodied men living along each section of road to put in several days of labor each year under an overseer's supervision. This was not simply a response to civic duty. These road crews knew that further regional development depended on having better roads that connected the countryside to the town, and, in turn, the town to the outside world.

Despite the hardships of life in Central Texas, a steady flow of migrants found their way there in the late nineteenth century. The magnet pulling settlers was land. "Indeed," noted several Texas historians, "the lure of cheap land in Texas was second only to that of gold in California."[6] America was a nation in motion, and Texas exerted a powerful lure for many in search of new opportunities and a fresh start in life. "Gone to Texas" remained an often-voiced sentiment, particularly in the states of the older South. Bell County's population growth—5,000 in 1860; 20,000 in 1880; and 45,000 in 1900—reflected the steady arrival of new settlers. Among the migrants who arrived in the late 1870s were both parents of Herman and George Brown.

Their father, Riney Brown, had a German heritage, typical of Central Texas. In the decades before the Civil War, Germans comprised the largest grouping of foreign-born Texas immigrants, and many Germans settled in Central Texas. "They constituted a majority of the population in three counties in the San Antonio–Austin area and a substantial part in six other counties."[7] The German communities attracted not only native expatriate Germans but second- and even third-generation Germans such as Riney Brown. His family had left Germany and traveled to the United States years earlier. Riney Brown's father, Louis Baum of Biedenkopf, Hessen-Darnstadt, Germany, arrived in America in the early nineteenth century and settled in Baltimore, Maryland. After moving to the United States, he Americanized his name to Brown. In Baltimore, he met Annie Otelia Ernst, who became his wife in 1826. Annie had only recently traveled to America from Germany; her family settled in Milwaukee, Wisconsin. Riney Brown, the fourth of seven children, was born in Baltimore on August 2, 1855.

In 1859, the Louis Brown family moved to Moorefield, West Virginia. There, Riney attended public school and graduated from Moorefield Academy. After leaving school, he took a job at a dry goods store in Defiance, Ohio. He also worked in a bakery. Riney enjoyed his work but hated the often harshly cold Ohio weather. A journal he kept during those years contains recurring complaints about the bitter cold. On February 16, 1875, a day so cold he stayed home from work, he wrote: "This is my first winter in so cold a clime, & I now think it will be my last, I have a reckolas desire to travel & have dreams of a suney clime & tropical fruits think of visiting some of the sout sea Islands or the West Indies if I can make money there."[8] Although he stayed in Ohio for almost two more years, Riney Brown's dissatisfaction with the climate and his growing interest in having his own business made him restless and increasingly anxious to strike out on his own.

As Riney pondered his future, he was devastated by news about the fate of his older brother George, who had earlier joined the army and was serving as a first corporal in General George A. Custer's 7th Cavalry. In July of 1876 Riney received a letter from John Kimm, a friend of George Brown. The letter related the events of June 25, 1876, when a Sioux Indian war party annihilated Custer and his contingent of 265 men, including George Brown, at the Little Big Horn. Kimm left no doubt that George Brown died in the massacre. Riney copied Kimm's letter into his diary: "his body laid on the battlefield 3 days. horribly mutilated & stripped by the Squaws, of all clothing the details are sad indeed to be thus killed, out in the wilderness miles from home & kindred. Who know nothing of it—till 30 days afterwards. I was in hopes he had escaped in some way. But I suppose now their is no doubt of his death."[9]

The news of his older brother's tragic death heightened Riney Brown's concern about his own prospects in life. His growing dissatisfaction with his situation led him to look for other work. A lengthy entry in his diary on New Year's Eve of 1876 revealed his growing impatience. After recording his desire to buy out his uncle's bakery in the spring of 1877, he noted that he "Wrote to Parents in regard there too but they seemed to think I was too young yet & wanted me to wait Several Years. If I did not wish to get married & settle down—I think I would go West or to Texas[,] perhaps would become a 'rolling Stone' for I have that disposition. So I think the sooner I settle the better."[10]

Soon after this entry, Riney Brown struck out for a warmer climate. When his diary picks up the story again in May of 1877, he had arrived in Central Texas. Riney survived for a while camping out with several other young men as they traveled by horseback throughout the region surrounding Austin and San Antonio. Horse trading provided a temporary income, but he found this a tough way to get by: "More deceitful people in Texas than anywhere I've been—every man for himself, & every body trying to steal from or cheat his neighbor."[11] Keeping his eyes open for opportunities as he explored Central Texas, Brown inquired about purchasing attractive tracts of land and discussed possible jobs ranging from baking to ranching.

During his journey, Riney encountered the ruins of the San Jose mission outside of San Antonio. His description of the architectural and engineering skill manifested in the structure foreshadowed some of his children's interest in grand construction projects. Riney wrote: "the king of Spain sent an architect of rare knowledge & genious to superintend its erection. It is impossible to paint with words the grand effect of this imposing, yet lowish gray structure, rearing itself from the parched lands, with its belfry, its long ranges of walls, with vaulted archways, its richly & quaintly carved windows & doors, its winding stairways, its shaded aisles. The fine carving has been nearly all picked from off the windows & doors, by relick hunters. I too chiped off a piece of an angel's wing."[12]

Riney Brown traveled throughout Texas as a "rolling stone" for about two years.

In 1879 he settled down in Belton, which he later described as the best town in the state.[13] Drawing on his experience in Ohio, he established his own general merchandise store soon after arriving in Belton. He met and courted Lucy Wilson King, a young woman whose family also had only recently migrated to Belton. After a brief courtship, Riney Brown, age twenty-five, and Lucy King, age twenty-one, married on October 10, 1880.

Lucy King Brown was the daughter of Judge Rufus Y. King, who came to Belton in 1878 at the age of fifty. Judge King's family made its way across the South over the course of several generations. His parents, who had roots in North Carolina and Virginia, had lived in Alabama for almost twenty years before migrating to the newly independent Republic of Texas in 1837. There his father, Hugh Bernard King, obtained a large tract of land in what became Burleson County (east of Austin) and settled down as a farmer to raise his family. Hugh King had studied law, and he became a Texas probate judge, a position he had previously held in Alabama. His son Rufus, born in Alabama, grew up primarily in Texas and graduated from Clarkesville College in 1849. Serving as a captain in the 8th Texas Cavalry during the Civil War, he suffered serious wounds at the Battle of Shiloh that forced him to retire from battle. He studied law before becoming a judge in Lee County. From there, Rufus moved to Belton, where he worked as a commercial salesman and later as an adjuster for the Santa Fe Railroad. In 1889, he was elected to the Texas State Legislature. Lucy King was the only one of four children to survive to adulthood from the union of Rufus King and Fannie Martin, a native of Tennessee who died in 1867. When Lucy was a teenager, her father remarried and produced a second family of four children.

Thus, the parents of Herman and George Brown took two quite different routes to Belton. But Riney Brown and Lucy King quickly began to build their own home, business, and family. Over the next twenty years, they had eleven children, seven of whom survived into adulthood. Riney Brown spent these years building his general merchandise store and becoming a mainstay of the local business and civic community.

His business boomed with the regional economy. The coming of the Santa Fe Railroad in 1880 and the Missouri, Kansas & Texas ("Katy") Railroad in 1882 spurred the transformation of the region. The Santa Fe connected Bell County to Houston and Galveston to the southeast and to West Texas cities such as San Angelo. The Katy provided connections to Waco and Fort Worth to the north. As parts of national systems, these railroads gave the region strong new ties to the national economy. The coming of railroads to Bell County was one part of a broader transportation revolution that reshaped the entire Texas economy. Texas had relatively few miles of railroad track in 1870, but in 1904 the state ranked first among all states in total railroad lines. The railroads replaced much slower and less efficient forms of cross-country trans-

Riney Brown and Lucy King on their wedding day.
Courtesy Brown Family Archives.

portation such as wagons and the infamous cattle drives, greatly accelerating the commercial and industrial development of the state.[14]

The coming of the railroads transformed the region in which the Brown brothers came to maturity. More reliable and less expensive transportation encouraged the expansion of cotton, which became the driving force in the regional economy. Farmers began to look outward at broader markets for cotton while relying more heavily on foodstuffs grown in other sections of the nation. Riney Brown's general merchandise store held an increasing array of goods brought in from all over the nation. No longer was an arduous ox-cart trip to Houston required to obtain goods for sale.

Under the impetus of the railroads, the county's population surged from about 20,000 in 1880 to more than 45,000 in 1900, when Belton had some 3,700 residents.

The railroads had an additional impact on the growth of cities within the region. Both roads passed about nine miles to the east of Belton through the newly created town of Temple, which the Santa Fe established in 1880 as a promotional venture. Close ties to the railroads helped Temple grow rapidly into the dominant city in the region. By 1900 it was already twice as large as the older town of Belton.

Belton nonetheless provided ample patronage for the success of the Brown Hardware Store. The store in downtown Belton was a wholesale and retail department store "which stocked every conceivable article used in the home."[15] Advertisements typically included the phrase, "R. L. Brown: 'The Restless and Sleepless.'" The Brown Hardware Company utilized "modern business methods and persistent advertising"; the 9,000-square-foot shop had a widely known reputation as the "Cut Price House" which "enjoyed the largest trade in hardware, stoves, and all house furnishings within a radius of fifty miles." The store specialized in sewing machines and binders. For a time, Brown Hardware claimed to be the second largest distributor of sewing machines in the state.[16]

During his years in business, Riney Brown had several different business partners, including one who brought havoc to Riney Brown's life and nearly ruined him. An experienced traveling salesman joined Riney Brown's business in about 1900. Together, they operated the Brown Hardware Company until 1904, when the partner looted and ransacked the store and fled to South America with the stolen cash. His transgression left Riney Brown broke and his reputation tarnished.

With his business in disarray, Brown reevaluated his situation. Although Belton provided a good location for his hardware store, Temple clearly was now the center of local commerce and a more promising site for future expansion. By this time, the Santa Fe and the Missouri, Kansas & Texas Railroads brought twenty-seven passenger trains through Temple daily. Although both railroads had spur lines to serve Belton, the bulk of the benefits from their operations accrued to Temple, which had attracted new residents, merchants, stores, banks, and hospitals.[17] Riney Brown decided to relocate both his business and his family to the larger town.

As Riney Brown tended to business, first in Belton and then Temple, Lucy King Brown raised a large family. Of the five girls and six boys born to the couple, three daughters and four sons survived beyond early youth. After the birth of three girls, the arrival of the first son, Louis King Brown, brought "excitement and rejoicing in the Brown family [which] was boundless."[18] Riney Brown was so overjoyed at the birth of a son that in the following edition of the local Belton weekly, he placed a large ad for his store with a new name, "R. L. Brown and Son." He explained in the ad that he had taken his son into the partnership. The other surviving children included Fannie Maude, Emma Lena, Mary Louise, Herman, George Rufus, and King.

Some sense of the tone of family life is preserved in the memories of Fannie Brown, the oldest daughter. She remembered Christmas as perhaps the most special occasion in the Brown household. Riney Brown dressed up as Santa Claus and performed what amounted to an annual Christmas skit: "On Christmas morning we would hear much commotion out back of the house, with sleigh bells interspersed with loud "whoas." There would be a scramble for the windows; then in full view around the pasture gate would come Santa himself, bobbling along with a big pack on his back. He would puff into the house explaining that his reindeer were scared of the houses and people in town, and he was forced to tie them out in the lot to keep them from running away. Then would come distribution of the presents in the sack, with a running banter from Santa that would thrill our hearts."[19]

Another daughter, Emma, remembered the attention her father gave his children on other occasions. On Sundays, he spent many hours building carts and playhouses for his children. As he worked and played with his children, he offered them the benefit of his observations on life. One of the lessons he emphasized was the importance of thrift. Once when his children asked him to buy fireworks, he replied that "they were just as pretty to look at when the other fellow was shooting them and cost us nothing."[20] But his actions, more than his words, provided the key example for his children: hard work and diligence were the cornerstones of a productive life.

As Riney Brown provided a role model for his children, he also played an active role in the development of his community. In Belton and later in Temple, years of hard work earned him the reputation of a responsible businessman and active civic worker. A city alderman, director of the Young Men's Building and Loan Association, secretary of the Belton Merchant's Exchange, and president of the Belton Business Men's Organization, Riney also participated in the Masonic Fraternity, Knights of Pythias, Woodmen of the World, and Redman organizations.[21] He provided a model of behavior that his sons would carry forward on a stage much larger than the Belton of their youth.

Herman Brown, the seventh child born to Riney and Lucy and the second oldest surviving son, was born in Belton on November 10, 1892. The family moved from Belton to Temple when he was about twelve, and Herman attended Temple High School. Serious and rather intense in appearance with a heavy brow and probing eyes, Herman played sports, particularly football and track, but seemed much less enthusiastic about academics. The class prophecy for Herman during his senior year came close to foretelling the future of a man who later would own the Driskill Hotel in Austin: "There's another that works in a grand hotel, Herman Brown sure makes a butler swell." Herman expressed quite clearly his attitude about school life in the quote he included under his photograph in his high school yearbook: "Never do to-day what you can put off till tomorrow."[22]

But this quote certainly did not describe his work habits outside the classroom. While attending school, Herman drove a grocery wagon for one year at the rate of thirty-five dollars per month.[23] Working hard doing chores around the house and helping his father at the store, he learned the value of sustained effort and discipline while developing work habits that would later help him build his own construction business from the ground up.

Herman Brown earned enough money from driving the grocery truck to enter the University of Texas at Austin. Once there, he discovered his disinterest in academics extended to higher education. Impatient with life inside the classroom, he was eager to go out and face the world. After less than one year, he left the university. Finding work at the Bell County Engineering Department, he performed a variety of jobs. There he began his career in the construction business.

Herman Brown's choice of engineering was influenced by the experience of floods in his youth. He had lived through a series of serious floods that had paralyzed parts of Central Texas. In April of 1900 the Colorado River spilled over its banks, causing much damage south of Belton. In the winter of 1902 continuous rainfall made numerous roads in and around Belton impassable; "traffic and business requiring transportation over public roads and streets, came almost completely to a stands-still."[24] Then in December of 1913, when Herman Brown was twenty-one and George Brown was fifteen, came the worst flood in Belton's history. A "30 foot wall of water on the Nolan [River]" surged through Belton, killing five people and destroying fifteen homes. By washing out all three wooden bridges, the flood isolated those living on the south side of the river, reminding all residents of the area of their continued reliance on the extremely undependable transportation system of the time.[25]

The impact of extreme weather on a rudimentary transportation system also seems to have influenced Herman Brown's relationship with his younger brother George. According to a story attributed to George Brown in an unpublished manuscript on the history of Brown & Root, the brothers forged a special bond during a severe storm in the winter of 1907. As Herman and George rode home in a buggy from a friend's farmhouse several hours away from Temple, a sudden blizzard lowered the temperature dramatically. The fifteen-year-old Herman took charge of the situation, wrapping his nine-year-old brother in an extra coat and bedding him down in the back of the buggy. But as Herman drove on, the bitter cold began to take its toll on George. Fearing serious consequences, Herman stopped at a farmhouse and asked for shelter for the night. Once out of the cold, he bathed George to quicken his circulation. The next morning the brothers completed their trip home. Although George suffered no permanent damage to his health, the episode permanently altered the brothers' relationship. According to George, "that storm, the blue norther, was the beginning of our closeness. We both knew this. We felt it. It brought us together as nothing before had done, and we never grew apart. There is no way to explain it. We both had suffered, and that is the way close companionships begin."[26]

George Brown, born six years after Herman, grew up, from the age of six, in Temple. He shared his big brother's experience in learning to work hard in the home and the family business, but he seemed more engaged by school than Herman. From early childhood, he showed a gift for salesmanship. His mother later related to Lyndon Johnson a story of how the seven-year-old George had built a business selling rabbits. "He would just go around to all of the neighbors' houses with his little red wagon and he'd pick up all the leftover vegetables . . . take them home, feed his rabbits, and then they would grow up and have more rabbits and he'd sell them."[27] Another early "marketing" job was selling the *Saturday Evening Post* on the streets of Temple. When George Brown looked back on these early years later in his life, he did not focus on the hardships. Rather, he "felt that he was very blessed because he grew up in a very small town in those days."[28]

He was a bit more involved in activities at Temple High School than Herman had been five classes before him. George sang in the glee club, participated in debate for four years, and served as business manager for the school's annual during his senior year. He supplied the following quote for his photograph in the yearbook: "I profess to know how women's hearts are won." The class prophecy looked into the future and found that "George Brown is an orator of great renown. He is styled 'The Modern Demosthenes.'"[29]

After graduating from Temple High School in 1916, George faced the task of choosing a college. Down the road to the south was the University of Texas at Austin; to the east was Texas A&M College at College Station. These two state-supported schools were the common choice for ambitious students in Central Texas and, indeed, throughout the state. But George happened upon an article about a new private university in Houston, and he decided to make his first journey to that booming city on the Texas Gulf Coast to enter the recently opened Rice Institute. Perhaps this decision was prompted in part by the fact that Rice did not charge tuition. His father had died unexpectedly at the age of fifty-nine in 1915, and George was stepping out on his own in the world. This decision proved fateful for George, for it introduced him to a city and a university that became central concerns of his adult life.

The institute George Brown entered in September of 1916 was an ambitious new venture made possible by the generous endowment of William Marsh Rice. This Massachusetts-born businessman and cotton trader made a portion of his fortune in Houston before retiring to New York City in the late nineteenth century. Determined to leave a lasting legacy for Houston in the form of an institution of higher learning similar to Cooper Union in New York City, Rice left much of his substantial estate as an endowment. After his death in 1900, a prolonged legal battle over his will delayed the opening of the new Rice Institute until 1912, when the first class entered.

George Brown selling the Saturday Evening Post.
Courtesy Brown & Root.

With an initial endowment of about $5 million, Rice Institute held great promise of attaining excellence. George Brown and other outstanding high school graduates from across Texas recognized the opportunity to study at a small college modeled after the leading universities in the nation. The early leaders at Rice had high ambitions for their endeavor. They sought to make their new school a sort of southwestern outpost of Ivy League values and excellence. To this end, they recruited highly qualified faculty members from throughout the nation and a president, Edgar Odell Lovett, from Princeton University.

Those in charge of the institute began recruiting students worthy of their high aspirations. They sought to attract exceptional students with the lure of free tuition for all applicants who could meet a rigorous set of admission guidelines.[30] With its hefty endowment and lofty goals, Rice Institute attracted the attention of ambitious young men and women. From the beginning, it drew many students with backgrounds similar to that of George Brown—ambitious, small-town Texans eager to move into a challenging environment in a booming city. Once at Rice, these students tended to forge lasting bonds with each other. The class that entered in the fall of 1916 had some 350 members. One other young man in George Brown's class at Rice was Albert Thomas, who came to Houston from the East Texas town of Nacogdoches. Years later as a United States representative from the Houston area, Thomas worked closely with the Brown brothers on a variety of issues and projects.

When George Brown entered Rice, Houston had only about 100,000 residents, but it was clearly a city on the move. The discovery and development of oil deposits near Houston pushed the city in a new economic direction. The Houston that George Brown encountered after his 170-mile train ride from Temple was a city bustling with energy and ambition, as was the institute he entered on what was then the southern outskirts of the city.[31] Indeed, the story of Rice Institute was little different than the story of the city as a whole in this era. Only the most optimistic boosters could have looked at the barren site chosen for the Rice campus and envisioned the creation of any lasting institution of higher learning, much less a "Harvard or a Princeton of the South." Only the same optimists could look at their rather ragged, rough-edged boomtown and envision the growth of a major metropolis in the hot, humid, and mosquito-infested marshlands of the Texas Gulf Coast.

George Brown originally intended to become a doctor, but his distaste at the sight of blood quickly changed his plans. During his second year, he became an "Academ," the designation at Rice for a student studying the humanities. Sports were not among his favorite activities. "He was not an athlete," recalled George Brown's lifelong friend H. Malcolm Lovett, the son of Rice Institute's first president, Edgar Odell Lovett, "although he was a very good student."[32]

George Brown attended Rice along with three fellow students from Temple High School. One of them, Hugh C. Welsh, later became one of the Browns' family doctors. In a letter to the *Mirror*, a Temple newspaper, Welsh described several hazing incidents involving the Temple boys. He discussed one example of what he termed "class spirit"— the spirit expressed by sophomores who hazed freshman. Once after George and a friend came into their dorm late at night, the sophomores "punished" the latecomers. They forced George to make the following prohibition speech to a small coed audience: "Ladies and gentlemen: This is no occasion of ordinary character. Tomorrow is a big prohibition election day in Fort Worth. We all know it is going dry. Lets do likewise and put all booze out of Houston; and, furthermore, let's put all evil out also."[33]

At Rice, George Brown joined the Student Army Training Corps (SATC), the predecessor to the R.O.T.C. As World War I progressed, this training corps became a strong presence on the Rice campus. It contained a large percentage of the male student body, which dwindled in size as students joined the armed forces. One story in the student newspaper read, "Shadow of War is Thrown on Campus by Men Leaving for Officer's Training Camp . . . 50 seniors leaving . . . And the Class Organization has About Gone to Wreck."[34]

George Brown felt the pull of the war in several ways. When his respected company commander left for the war, the replacement failed to earn the loyalty of some of the members of the SATC. On his first day in charge, the new commander marched his trainees through a hedge. Not long after Thanksgiving, recalled one Rice student, some of the his charges expressed their disdain for him through a creative prank. Captain Reagan had a "nice suite with a fireplace and a toilet . . . and a study and a bedroom. And somebody put the hose down his chimney and flushed him out of his bedroom in his underwear that night. And, of course, he couldn't take over the military unit after that at all."[35] George Brown and others thought to have been involved in the prank immediately left school, and Brown entered the U.S. Marines officers' training school at Georgia Tech.[36] He spent a year in the marines before the war ended and he left the peacetime military.

When George Brown decided to return to civilian life and finish his undergraduate education, he did not reenter Rice. Instead, he spent a summer and fall at the University of Texas. Following this brief stint, he enrolled at the Colorado School of Mines to become a mining engineer. George Brown paid for his education by leasing two Model T Fords he obtained from his older brother, Louis Brown, who had moved from Central Texas to Denver, where he operated a car rental agency. George earned about one hundred dollars per month renting the cars. "I charged them by the mile," he recalled. "Of course, I had to keep the cars in repair. Sometimes I stayed up all night relining the brakes."[37] There is also at least anecdotal evidence that George made money by supplying "spirits" acquired in Denver to fellow students. One personal letter to George from a friend from the Colorado School of Mines in 1922 has the salutation "Dear 'Bootleggah.'" The nickname "Bootlegger Brown" was later confirmed by several of his friends from his undergraduate years.[38]

Brown graduated from the Colorado School of Mines in 1922 with an engineer of mines degree. In his 1922 yearbook, the following inscription appeared beneath his photograph:

George—Brownie
George is one of our most powerful Stray Greeks. He gains his power thru his
ability to make friends. He is some assayer, too. After hearing his stories of former
experiences one can die. Nothing is left to see or hear. Get him to tell you the one
about the snakes or the duck hunting party. One has only to know George to

realize that the interests of the school lie close to his heart, and that he is earnestly striving to make his stay in school worth while, not only for himself, but for others.[39]

After graduation "Brownie" set off to find engineering work, well prepared by both his upbringing and his education to succeed in his chosen field.

Initially, he chose a quite different path than that taken by his brother Herman. This was not too surprising, since despite their similar backgrounds, the two brothers emerged with distinct personalities. During youth, Herman Brown was not interested in an academic education. With a natural attention to detail and efficiency and a passion for the nuts and bolts of building things, he pursued a "practical" career in construction. George Brown found the same profession on a different route. He chose an academic education and received one at a leading engineering school. George, a friend later recalled, believed that "engineers were men who went to far-off places, who built things all over the world."[40] He was trained as an engineer; he was a born salesman and consensus builder. Together, the brothers possessed the qualities essential for success.

They were fortunate to come of age in a time and place with extraordinary opportunities to put to good use their entrepreneurial energies. All around them were obvious social needs waiting to be met. Unreliable roads remained obvious barriers to both commerce and travel. Often devastating floods along rivers uncontrolled by dams stymied commercial development in some areas and plagued Central Texas. As the state's rural past gradually faded, a new era of development presented a host of opportunities to would-be entrepreneurs. Early in life, the Brown brothers recognized these opportunities and learned how to profit from them while participating in the modernization of Texas.

CHAPTER TWO

Road Builders

Herman Brown was almost twenty-one when he left the University of Texas and returned to Temple. Although eager to make his own way in the world, he had no special talents or training. He faced the same question his father, Riney, faced during his early apprenticeship years working in a hardware store and a bakery: what was the best available opportunity?

Herman Brown made a good choice when he decided to commit himself to road construction. He had grown up with firsthand knowledge of the hazards of bad roads, and he understood the need for improvements. By the 1910s and 1920s, this public demand had created a booming market for new roads, and road builders could not construct roads fast enough to please the growing legion of car and truck owners.

The enthusiasm for better roads found voice in the popular media. In an early issue of the *Outlook* magazine, Newton Fuessle's article "Pulling Main Street Out of the Mud" chastised America for not providing enough good roads for its vehicles. Quoting John J. Raskob, chairman of the finance committee of General Motors Corporation, Fuessle suggested why Americans had not built more roads sooner: "And yet we have traveled with only the speed of a snail to the idea of better roads over which to go to church, school, lodges, mass-meetings, and to market. The first puritanical notion that motor cars were pleasure cars may have had something to do with the reluctance to improve our roads."[1] Fuessle's frustration at the lack of good roads was part of a growing sentiment that the traditional neglect of road building could be tolerated no longer.

Historically, the primitive roads of Texas had "contributed to the inefficiency of freight, passenger, and postal services during the period of the republic."[2] During the Civil War, the state's large size and "sparsity of established routes continually frustrated Texas' efforts to serve as the storehouse of the Confederacy."[3] As had been most evident to the Brown brothers in their youth in Bell County, the traditional nineteenth-century system of underfinanced programs and "volunteer" roadworkers had failed miserably to produce better roads.

A new approach to road construction began in the 1880s and 1890s, when the Texas legislature enacted several laws authorizing counties "to lay out, supervise, and maintain roads" (1879), levy road taxes (1883 and 1890), and issue bonds for bridge construction (1887). In this era, individual counties also stepped up their efforts to improve roads and build bridges. In the 1890s, Bell County, for example, purchased two road graders and two road plows for work on county roads while also passing several substantial bond issues targeted for the construction of bridges.[4]

Yet the growing demand for good roads across Texas far outran available state or local funding.[5] The poor political conditions for road building in Texas did not escape the notice of federal officials. During an official visit to Houston in 1895, General Roy Stone, head of the U.S. Office of Road Inquiry, "chastised Texas for making less progress in road development than any other state."[6] Conditions did not improve during the early years of the twentieth century. Legislative attempts to create a state bureau of roads, establish an office of state expert engineer, and appoint a state highway commissioner all failed.[7]

Despite the difficulties of creating a modern road construction program in Texas, the burgeoning demand for better roads by the growing number of car owners could not be ignored. In 1912, Texans registered 35,000 automobiles in their state compared to one million in the United States. By 1925, Texas registrations had increased to 975,000 while total U.S. registrations had reached twenty million.[8] Against this backdrop, the construction of a greatly expanded highway system became a top priority in American society and in American politics.

The establishment of the Texas Good Roads Association in 1911 provided both a forum and a shared sense of purpose for the builders, promoters, and users of roads. Soon, both federal and state legislative action authorized funding and varying levels of oversight for road building projects. In July 1916, the U.S. Congress passed the Federal Road Act. At the same time, the Texas legislature finally passed a bill establishing the Texas Highway Commission. Governor James Ferguson signed the highway bill into law on the same day that the U.S. Senate approved the declaration of war against Germany.

In the new age of the automobile, Herman Brown recognized an opportunity for work and perhaps profit. Herman Brown's first job at the Bell County Engineering Department in Belton introduced him to the emerging road building and paving business. His particular duties with the engineering department were not especially significant. For a salary of two dollars per day, he assisted the county surveyors by checking building materials. Herman did not work every day for the engineering department, and he was never sure how many days of work the county would be able to offer him. But if this work did not provide a permanent career, it did prepare him to enter the road construction business.

Herman soon found a better job with a local contractor, Carl Swinford, working on a road project in Collin County, north of Dallas. Swinford agreed to match the two-dollar-per-day salary Herman received as an employee of Bell County and to promise him full-time employment. Herman accepted the new position, which gave him a more dependable source of income. Working diligently, he earned a promotion to foreman and an increased salary of seventy-five dollars per month. Yet the promotion proved deceptive. Although Herman worked regularly for almost two years, his financially strapped boss rarely paid him on time. Sometime in late 1914, Swinford decided to retire—or escape—from the road building business. Lacking funds to pay Herman for work already completed, he offered his road building equipment in recompense for nine months' unpaid salary.[9]

At the age of twenty-two Herman Brown thus faced a tough choice. He could sell the equipment for whatever he could get, pay off the mortgage, and look for a new job, or he could put the mules and equipment to work by creating his own road building company. Inexperienced, stubborn, and independent, Herman Brown determined to make a go of it as a road builder.

One of his first decisions as an independent businessman was to cultivate an older, more commanding presence by growing a moustache. Like many of the practical lessons he learned early, this moustache stayed with him the rest of his life. It may well have helped disguise his youthful appearance, since he quickly acquired several small jobs. His first job was a subcontract for a larger contractor in Henderson County, about a hundred miles northeast of Temple. There Herman Brown and his crew "cleared, graded, and sand-clayed their roads."[10] He next sought work from the county commissioners in Freestone County, just south of Henderson County, and he received contracts for grading roads. As would be the case time after time over the next twenty years, the move from job to job required packing equipment and supplies into the wagons, the "forced march" to the next job site, and establishing a new base camp. Herman Brown generally lived in these camps with his crews and closely supervised their daily work.

Brown quickly needed help, however, in managing these job sites, and he searched for other trustworthy men with experience in the road building business. One of his first new worker-managers, Dan Dyess, began as a foreman laying asphalt toppings on roads and continued working for Brown & Root for many years. Brown met Dyess in 1914 at a local high school in Holland, a small town in Bell County. On the school grounds, a group of young men challenged some of the high school boys to an impromptu track meet. Dyess won the broad jump competition, and Herman Brown won the high jump. Perhaps remembering the competitive spirit of the day or simply the friendship of the moment, Herman Brown asked the broad jump champion to come to work for his newly formed company.[11]

Building roads in Central Texas. *Courtesy Brown & Root.*

The work they undertook in these early years was difficult; the roads they built were primitive by later standards. Like most of his competitors in these years, Brown's early road building company did not own trucks and tractors. Instead, recalled Brown Booth, Herman Brown's nephew, "they moved dirt from one place to another with mule teams and a slip, they'd call it, a hand-operated thing they dragged behind the mules that was the prime way they moved dirt from one place to another."[12]

Another road builder, Howard Counts, explained that the mules pulled a cutting tool, or scraper, along the surface of the road. They applied varying amounts of pressure to a "Johnson" bar to control the depth of the cut. The scraper collected the dirt and deposited it further down the road, where it was used to fill in holes and smooth out unlevel areas. "We kept mules in the corral along the side of the right-of-way and we would move them forward each time." The crews, who lived in tents pitched near the job site, often had to drill for water for the mules. Sometimes, heavy rains delayed jobs for months at a time.[13]

Contractors built road surfaces composed of a mixture of sand and clay providing what was sometimes referred to as a "squirt top" surface. The typical road was sixteen feet wide. It was slightly higher at the center so that water flowed to the side of the road into primitive drainage structures. Without equipment to pack down the clay and sand mixture, or black dirt, the road crews often borrowed herds of goats and spent days walking them on the road to compact the surface. Although not hard by later standards, such roads sustained the weight of Model T Fords and other lightweight automobiles of this era.

Dan Dyess became crew foreman in January of 1915, freeing Brown to spend more time soliciting new road building jobs. Brown later estimated that he logged as many as seventy thousand miles a year driving the roads of Central Texas to hustle jobs and check on work in progress. Dyess was young, but he was no stranger to road building, having previously worked for another local contractor. He later recalled that his new boss did not have "much of a spread to go into business with." Included were the following: "15 head of stock, of which one horse was a pretty good saddle and buggy horse, but the others were just a bunch of old sore-shouldered worn-out plugs; four old worn-out fresnoes; one old buggy, six worn-out Studebaker wagons and five tents, one of which was the cook tent, one was for me and Herman to sleep in, and the other three for the Negro mule skinners. Oh yes, there were two plows, one a number 2 road plow and the other just a little old farm plow. The harness was all shot and there was a big bank mortgage on the whole outfit."[14]

Manning this old equipment was a crew of rough, tough-talking men. Road building in the hot, rugged, and often wet country of Central Texas was a brutally demanding job, requiring hard, sustained labor. The men in a typical road building crew in these years knew they would sweat through long hours of backbreaking toil in the hot sun; they would live and work out of tents as they scraped roads out of the countryside. The management of such workers required both strict discipline and effort to maintain morale. Herman Brown was the sort of boss known to look after his men, providing them steady work and ample chance to blow off steam. Indeed, Brown recalled that after winning the bid for one construction job, he made an appointment with the county sheriff to arrange bail for any of his construction hands jailed during the weekend to ensure their release in time for work on Monday morning.[15]

Many of the men in Dyess's charge were "mule skinners," a traditional name given to laborers, typically black, who worked with the mules and could use a whip to "skin a mule at twenty feet."[16] To some extent, the relationship between foreman and mule skinner was not unlike that between landowner and sharecropper. As boss of the road gang, the foreman represented the law, judge, and jury. The foreman resolved conflicts in kangaroo courts and often meted out harsh punishments that could even include lashes with a leather harness for such offenses as stealing.[17] Stories of brutal physical punishment may have been embellished in the telling over fifty years, but the difficult roadworking conditions and primitive equipment made the life of the road crews harsh. "The men knew what kind of life it was, hard and rough, before they signed up," recalled one foreman.[18] But the pay came in hard cash, a welcome change for many of the crew members who had done equally difficult work on family farms. These were not men who held high expectations of life: "We were happy," one mule skinner recalled, to have "food, a place to sleep, shooting craps."[19]

These road crews worked very long hours. One of Herman Brown's original mule skinners, Hayward Earls, recalled years later that workers received a wage of two dollars per day. But the mule skinners found it almost impossible to save any money. Of the two-dollar daily wage, the company retained seventy-five cents as compensation for the laborers' food and lodging expenses. A typical breakfast included eggs, biscuits and syrup, and coffee. For dinner, the mule skinner typically ate bacon, salt pork or sow belly syrup, and cabbage or red beans. At night, the entire crew including Herman Brown slept in tents pitched near the work site. Fights, drinking, and gambling characterized many of the evenings after work. On Saturday nights, recalled Earls, "he [Herman] gave us two dollars to shoot craps. When we got put in jail, Mr. Herman got us out." Those who remained in jail on Sunday received a reprieve from the Sunday ritual of repairing equipment or sharpening tools. Herman Brown enjoyed shooting craps with the mule skinners. One night the local police arrested "Mr. Herman" as he shot craps with his men. Brown was fined $21.70, which Dan Dyess paid to get him out of jail.[20]

In the jail or on the road, Herman Brown identified with his employees. He enjoyed the company of laborers—both black and white—and his employees respected him for his interest in their lot. To the extent his limited finances allowed, he took a personal interest in the welfare of his regular workers. At Christmas, he presented them with a gift of money, clothing, or some token. But such gifts alone did not earn the road crews' respect. His men knew that Herman Brown was not afraid to dirty his hands with hard work; they respected the fact that he endured the same living conditions. His laborers understood that his loyalty to them depended upon their dependability as workers. Herman sought year-round work for his crews, and in return he asked for their loyalty and commitment. Herman felt strongly that he and his workers were a kind of family, bound together by their time on the roads. Years later he summarized the importance he placed on loyalty: "I think we have one of the most loyal personnel you can find in any similar firm in the world—and that one factor I am sure accounts for the continued success of Brown & Root in an industry where the mortality rate is appalling."[21]

Attitudes formed by Herman Brown during his early years with his road crews remained with him throughout his life. He felt deeply that he knew what was best for his workers and that he could look after their interests better than anyone, including a union. Looking back at these early years several decades later, he recalled that "I grew up in the days when the relationship between myself and my employees was a personal one. If he had a grievance he could come to me with it. There was no disinterested third party between us."[22] This was by no means unusual in Texas in the 1920s, since unions had very little presence in the early road building industry. Herman Brown's lifelong philosophy of labor-management relations was forged while working with his labor gangs on the roads of Central Texas. He felt deeply that unions subverted the natural harmony between a sympathetic, paternal owner and his loyal employees. Herman never revised this view of labor that he learned in his lean years of roadwork.

A letter to Herman Brown in 1961 from a former employee seeking a loan paints a bleak portrait of the earliest days of Central Texas roadwork for the company that became Brown & Root. The letter recalls "the first little contract you had on the Fredericksburg road in Travis county. You were struggling along with those mules and dump wagons and just barely getting bye." After reminding Herman of an episode in which "I stuck my neck out" to try to "help you pay your feed bill," the letter writer asks for "a Thousand dollars to put me in shape." Hard times forged strong bonds, and Herman's loyalty to those who had worked with him in these lean years at times translated into assistance later in life.[23]

As he moved back and forth between job sites spread over much of Central Texas, Herman Brown located his business headquarters in Georgetown, about twenty miles north of Austin. In Georgetown during 1917, he met Margarett Root, a local schoolteacher. From a family of prosperous cotton farmers in Central Texas, Margarett attended Southwestern University in Georgetown, a small Methodist-affiliated liberal arts school that was Texas's oldest institution of higher learning. She loved art and literature, and spent much of her life in the world of books and ideas.

As a teacher of impressionable young minds, Margarett needed to be a model woman in the community. This meant that she was not supposed to associate with men who were not "respectable." In the eyes of many in the community, Herman Brown qualified as one such man. He first met Margarett when she taught his younger brother, King. The high-minded schoolteacher and the rough-hewn road contractor did not appear to be a good match on the surface, but they found their characters complementary rather than contradictory.

After formally introducing himself to Margarett at a Saturday night dance, Herman courted her. Once on the way home from a dance, Herman impressed Margarett by jumping behind the counter at a cafe and flipping flapjacks. She liked his energy, and she recognized in him a man who was going somewhere.[24] Despite their apparent differences, the intellectual schoolteacher was certain of her feelings for the rugged but quick-thinking contractor. "The first time I saw Herman Brown," she recalled years later, "I said to myself,' there's the man I am going to marry.'"[25] He proposed to her a few months after the dance.

Their wedding and honeymoon must have raised some questions in Margarett's mind about just where this man was going. They obtained a marriage license and located the justice of the peace at a local tavern that he operated. The tavern keeper married the young couple on September 2, 1917. The bride did not want a big wedding; nor did she accept a wedding ring, which she saw as a symbol of bondage. The newlyweds spent their honeymoon in one of Herman's tents, which served as both his home and office while he managed jobs.

In some ways this set the pattern for the rest of their lives. Margarett proved willing to allow Herman to pour his heart and soul into his business. She accepted his devotion to his work and set about building her own life while organizing a good home for her husband. One of her pet sayings became "rise above it," an all-purpose reminder that one was supposed to endure whatever life had to offer. Later, severe emphysema forced Margarett to restrict her activities, but she responded by devoting herself even more fully to art and literature. For his part, Herman did not find much joy in such pursuits, but he respected his wife's intellect and her spirit. He seemed to relish vigorous discussions with her on any and all issues, particularly those on which they disagreed sharply, such as politics. For her part, she respected the practical accomplishment of her husband, understanding that building was his art and his passion.[26]

As he settled down into a married life from which he drew personal strength, Herman felt pride in his growing business. In 1918 after he won the contract to build twenty-three miles of road near Taylor, Texas, the local paper described him as "Herman Brown, well-known road contractor of Temple."[27] With regular work coming in, he gradually acquired more equipment. By the end of World War I, his business included about twenty-five mules. He also took advantage of the availability of war surplus materials at Camp Mabry in Austin to acquire twelve army tents, new harnesses, Studebaker wagons, and Liberty trucks. The man who sold him this war surplus equipment, F. L. "Slim" Dahlstrom, later became a close associate. Herman Brown—contractor, salesman, road crew manager and worker, and sometimes gambler—was earning a reputation as a dependable builder in a wide section of Central Texas.

To continue to grow while meeting his obligations on a job-to-job basis, Brown needed additional capital. Early in 1919, his need for financial assistance compelled him to take on two partners from Temple. Ab Kuykendall, a Bell County commissioner, and Charlie Thompson, a local mule dealer, entered into a partnership with Herman Brown. Thompson supplied him with the larger mules needed to handle major road building jobs. This new backing proved important during one of Brown's first large road construction projects, a thirty-mile stretch between the towns of Hutto and Bartlett (near Brown's headquarters in Georgetown). Kuykendall and Thompson guaranteed Herman's credit and arranged a loan for him.

Almost immediately, tensions grew between the young builder and his somewhat nervous new financial backers. Fearing that Brown might default on the loan, they attempted to exert stronger controls over his spending by withdrawing their guarantee for his credit. Suddenly, Brown found himself in the middle of a major project without the funds needed to purchase the equipment, supplies, and food. Kuykendall and

Margarett Root Brown and Herman Brown. *Courtesy Brown Family Archives.*

Thompson gave Herman a deadline for either paying off his notes or turning the business over to them. Herman retained control of the company during this period only because its road building contracts carried his name and not the names of his partners.

But Brown had one last card to play. Unbeknownst to Kuykendall and Thompson, he had found another source of funds. His brother-in-law, Dan Root, owned a seven-hundred-acre blackland farm outside of Granger, a small town east of Georgetown. Root had excellent credit, and he agreed to use it to aid his sister's husband. Root and Brown visited the bank in Granger and arranged for Brown to borrow $20,000, with the loan backed by Root's credit.[28]

The next day, Thompson and Kuykendall came out to the camp to eat lunch with Herman, as they often did. At lunch, they informed Herman that his deadline had passed and that they were calling in his notes. Dan Dyess recalled that Brown "got real red in the face like he always did when he was mad, and said, 'Goddamn you, I'll just write you out a check!'" Although shocked that Brown had the money, the two men accepted his check, leaving him in control of the business. Having paid off Kuykendall and Thompson, Herman gladly accepted a new and more reliable partner in Dan Root. In deference and appreciation, Herman named his company Brown & Root. Although stepping in at a crucial time to provide much-needed financial support, Root never played a significant role in managing or operating the firm. As a cotton farmer, he had no interest in taking on a new enterprise. He was simply concerned about the fate of his sister and her husband. Dan Root died a bachelor in

1929, but the Browns retained the name Brown & Root in recognition of his timely assistance in the company's formative years.

Another family member soon joined Herman Brown at Brown & Root. After his high school graduation in 1916, George Brown had left Central Texas while his older brother had stayed behind and built his company. A year and a half at the Rice Institute in Houston, a stint in the marines, and a two-year stay at the Colorado School of Mines near Denver had given George a formal education and exposure to the world beyond Bell County. With his degree in mining engineering, George Brown found employment as a geologist with Anaconda Copper Company. One night, George began exploring 2,200 feet underground for a copper vein at a company mine near Butte, Montana. He discovered the vein, and immediately became a victim of a sudden cave-in. He was trapped for many hours "in a jumble of shaking timbers," and he suffered a fractured skull, a broken collarbone, and several smashed ribs. Recalling his first-aid training, George placed the side of his head against a rock, thereby stopping the flow of blood in his brain and reducing the hemorrhage. He was rescued twelve hours after the cave-in. Following the accident, George returned to Texas to recuperate on a ninety-day leave of absence.[29]

George Brown's mining disaster gave him a measure of celebrity back in Central Texas. At the Saint Mark's Episcopal Church in San Marcos, Texas, he delivered a speech on "Mining Life in America" at the "college tea." The newspaper advertisement for the lecture noted that the majority of miners in the United States did not speak English and "came from the ends of the earth. They are so ignorant in fact, that they do not know how to take care of themselves in case of an accident . . . In these camps, the socialists and the IWW's flourish." The topic of the speech, however, was not necessarily the main draw. The announcement noted that "Mr. Brown was a geologist at Butte, Montana, for the largest Copper Mining Company in the world. Once he was buried alive." If that was not enough to attract an audience, the advertisement announced: "Interesting address, buffet supper, Good Music."[30]

After recovering from the accident at home in Temple, George returned to Montana. But his mother did not want him to continue his dangerous work for the Anaconda Mining Company. She asked Herman if there was any chance he would hire George to work for Brown & Root. Herman gladly offered employment to his brother, whose training would be useful to the company. On May 22, 1922, Herman sent a telegram to George in Montana: "YOUR WIRE TODAY HAVE A YEARS WORK UNDER CONTRACT NOW AND CAN USE YOU BE GLAD TO HAVE YOU REPORT AT ONCE AT A HUNDRED AND BOARD WILL BE ABLE TO PAY YOU MORE WITHIN YEAR IF YOU NEED ANY TRANSPORTATION DRAW ON ME."[31]

At this time, Brown & Root was embarking on a new and important project. After severe floods washed out the major bridges on the San Gabriel River, the company won the contract to build more modern replacements. Herman needed some-one to direct the placement of dynamite charges, and George told Herman that he had "some experience in dynamiting rock underwater." In 1922 George Brown joined Brown & Root.[32]

Sharp contrasts between the two brothers were readily apparent. When George entered the company, he was a young man with formal training in both engineering and mining-related work; Herman learned what he knew about construction through practical experience building roads. George had lived in numerous places outside of Central Texas; Herman had not yet ventured far from home. George was comfort-able in the company of all sorts of people, including "college men"; Herman seemed most comfortable around the rough-and-ready men of his road crews. George was taken by "the romance of engineering," while Herman believed in the "romance" of a hard job well done.

As the brothers gradually established the working relationship that was to carry them through the next four decades, they reached an understanding that removed the prospect of haggling over financial matters. They agreed to a simple approach that proved very effective for them: in all of their cooperative ventures, they would share ownership and control fifty-fifty. This meant that Herman ultimately granted George a half interest in Brown & Root at the time of its incorporation in 1929. It was, however, readily apparent who would be in charge of the company. The older brother and the founder would continue to call the shots; the younger brother would help build the company in numerous ways by applying his engineering training and natural salesmanship.

George was the newcomer to Brown & Root, but he was also the boss's brother. Herman did not treat George with overt favoritism, and he placed him under the authority of the current project foreman. George, however, quickly encountered hos-tility from the foreman. On George's first job on the San Gabriel River project, the supervisor, a former army colonel identified in sources only as Colonel Clark, nearly forced George out of the business by antagonizing the younger Brown brother and taking credit for his ideas. The foreman also told Herman that George spent all his nights out carousing.

George was reluctant to discuss the problem with his brother, so he complained to Dyess that "life is too short for me to work for that sonofabitch—I'm going to quit right now!"[33] Dyess, who already had informed Herman that Clark was telling stories about George, encouraged George to talk to his brother about the problems. George did, and Herman sided with his brother. Soon afterwards, Clark left the company, and Herman chose George to replace him as foreman of the San Gabriel Bridge project. Despite his lack of practical experience in bridge construction, George successfully completed the project.

Despite George's promotion to foreman, Herman was not easy on him, and the younger brother worked hard to gain the respect of the man who started the road building business almost a decade earlier. Dyess recalled that Herman "was a lot harder on him [George] than he was on me." Herman was helping George learn how to be a partner and a leader, and he was a good teacher. The nature of construction projects also encouraged young managers to learn quickly. Individual job sites generally were spread out across the countryside, with little effective means of communicating between the different sites. There was no choice except to give each site leader almost total autonomy over his particular project. In this school of hard knocks, those who survived quickly gained valuable experience as independent managers.

Many of Herman and George Brown's most trusted associates at Brown & Root gained their initial management experience in these years of road building. Three such men all came to the company in 1924. Carl Burkhart, W. A. Woolsey, and Jimmy Dellinger began on the road building crews and moved on to hold positions of greater responsibility in other divisions as the company grew. After entering Brown & Root, they helped create an approach to management that continued at the company for years. Throughout his business career, Herman Brown practiced an extremely decentralized form of management, which he characterized as "the looseness of our organization."[34] He set the agenda and chose the men he wanted for important jobs. Then he moved out of their way and allowed them to succeed or fail. Success brought new responsibilities; failure often meant being eased over into less demanding jobs within the company.

All of the men understood that road building was difficult and financially risky. A serious miscalculation or misstep on a major project could place the entire company in jeopardy. The Browns thus had strong incentives to develop strategies to diversify and reduce risks. In these days before economic data was readily available, Herman Brown studied the want ads in small-town newspapers. If he saw many "help wanted" ads, he felt that the county provided good prospects for future road building. He also looked for the presence of Jewish merchants as an indicator that a small town had developed a healthy economy. This helped him concentrate his work in the most promising areas.

Another risk reduction strategy was geographical diversification. In choosing where to bid on jobs George recalled that "we decided we would never bid on a job in southeast Texas where the rainfall is heavy without bidding on one in west Texas where rainfall is light. That was built-in insurance. The weather was always one factor we could not figure on making a bid. And every time we took a job on credit, we took one for cash." The company also steadily expanded its expertise in new areas of road construction, from building bridges to mastering new methods of paving. George explained that "we always had a job going and an opportunity to move our men from one job to another. We kept the men working, kept them on the payroll, by balancing one job against another. This meant a lot to our men. They knew they had a job year round."[35] Diversification thus allowed the Browns to pursue another strategy that proved central to their success: the creation of a loyal workforce that identified its own future with that of the company. (See table 2.1.)

Diversification pulled the company out of Central Texas in late 1925, when George Brown moved to a job in Wharton, a town on the coastal plain southwest of Houston. There he took on a job new to Brown & Root, concrete paving. Dan Dyess recalled that "we bought a Model T truck and mixed the concrete at a railroad siding and hauled it to the job." Brown & Root's concrete paving work on Fulton and Milam Streets in Wharton was thought to be the first such paving done in Texas with a new design of expansion joints originally developed for use on the heavily traveled streets of Chicago.[36]

Brown & Root used a center expansion joint on standard-width streets, and the firm paved wider streets in strips. To compensate for Wharton's spongy soil, the con-

TABLE 2.1. PROJECTS BID BY BROWN & ROOT, INC. (1924–35)

County Road Bids:	576
City Paving Bids:	55
City Bridge Bids:	31
Sewer Bids:	12
Waterwork Bids:	6
Grading Bids:	4
Sewer/Waterwork Bids:	2
Building Bids:	1
Canal Lining Bid:	1
Clean Drain Ditch Bid:	1
County Bridge Bids:	1
Drain Ditch Bids:	1
Fish Hatchery Bids:	1
Gas System Bids:	1
Jetty Repair Bids:	2
Land Dock Bids:	1
Levee Bids:	1
Seawall Bids:	1
Total Number of Projects Bid:	698
Total Number of Projects Awarded:	179
Total Amount Awarded by City/County:	$8,394,671

Source: Recent Contracts Awarded and Recent Contract Bids listed in *Texas General Contractors Association Monthly Bulletin*, 1924–35.

tractor reinforced the concrete with five pounds of ¾-inch steel rod per square yard. Brown & Root cured its paving by flooding it with water for ten days. A Wharton newspaper writer added that while in Chicago concrete pavement was cured under burlap and opened to traffic in fourteen to twenty-one days, Brown & Root's flood curing method and prohibition of street traffic for thirty days "is the better plan." The firm completed the ninety-day contract in seventy-four days.[37]

Brown & Root's innovative paving work in Wharton evoked considerable reaction from the local populace. The concrete paving clearly provided a smoother ride. One response was the demand to issue new bonds to pay for additional paving, and the city held an election on January 19, 1926. In preparation for the election, proponents discussed the advantages of good paved streets as opposed to gravel streets for economic development; they also encouraged the continued employment of Brown & Root: "We should have more pavement, and we should not delay in getting it. . . . If we have a town that . . . can point with pride to block after block of concrete pavement, there is every assurance that more people will come here to spend their money. And instead of going away with a grouch because they have broken a spring or got stuck on the main street, they will go always feeling that the people here have at least made the town better when you once get in it, and thereafter will come to this point to spend their money."[38]

The city of Wharton passed the bond and the city council contracted with Brown & Root for an expanded program of paving without requiring further competitive bidding. The *Wharton Spectator* strongly supported this decision: "It would be foolish in view of the facts to decline to sign it . . . when the work we have received grades above specifications, and when previous experience shows that Brown & Root are conscientiously filling every obligation of the good contract we already have, and in every way cooperating with the engineer and city council to the end that the least inconvenience and loss of time is imposed upon the citizenship."[39]

While completing this job in Wharton, George decided his future at Brown & Root seemed secure enough to consider marriage. He found his match in Alice Pratt, a young woman whose family had lived in Temple off and on from 1898 to 1908. In 1908, when Alice was only six years old, her thirty-four-year-old father, Minot Tully Pratt, drowned while performing his duties as an engineer and division superintendent for the Santa Fe Railroad. As an early engineering graduate of Stanford University, Minot Pratt had advanced rapidly within the Santa Fe system before his untimely death.[40] His daughter subsequently moved to Lometa, Texas, with her mother and stepfather. She later lived in Dallas with an aunt and uncle.

Alice graduated from Southwestern University in 1922. She later joked about her mediocrity as a student there, but while in Dallas and Georgetown, she developed a passion for art. This also reflected her exposure to a paternal uncle, Bela L. Pratt, who was a successful sculptor. This avocation proved useful for her aunt and uncle by providing them a pretext to send her away on a whirlwind tour of the art museums of Europe in the summer of 1925. Part of their motivation appears to have been to convince their daughter to forget about George Brown, a young suitor who did not seem particularly promising to them. Upon her return, Alice and George married on November 25, 1925, at Christ's Church, Episcopal, in Lometa.[41]

Before the wedding, George had asked Alice to make a choice required by economic necessity. She could have a nice engagement ring or an extended honeymoon trip. Alice chose the trip, and the couple made plans for a two-week honeymoon in New Orleans. Realizing that he could not afford such a long stay in an expensive hotel, George arranged for Herman to wait a few days and wire him in New Orleans that a serious emergency at Brown & Root required his immediate return. The ruse worked, and the newlyweds returned early from New Orleans to Wharton, where they moved into a boardinghouse while George got back to work laying pavement.

On the surface, this episode suggests the harsher side of life as a Brown & Root "widow." But read in another way, it suggests the foundation of an enduring relationship. George was concerned enough about Alice's feelings to go through the charade of planning the honeymoon trip; Alice was concerned enough about the work of her husband to allow him his deceit. They were a good match; Alice's energy and interests filled an important spot in George Brown's life. He tried to explain this late in life to a friend: "I could never have done what I did without Alice. I'm not talking about the usual things like running a house or raising children. . . . In all my career that had to do with contacts, with people . . . Alice geared everything to make me look like the intelligent one. . . . If there was anything that didn't fit the pattern, she wanted to assume that her chief goal was to make me look good."[42]

At the time of the Wharton job and George Brown's marriage, Brown & Root appeared to be in strong financial condition. A statement of assets and liabilities filed with the Security Bank and Trust Company in Wharton in July of 1925 suggests how much Brown & Root had grown in the early 1920s. Among total assets of $114,450, the statement lists ninety-two mules and thirty-four wagons; it also lists a variety of mechanized equipment, including six trucks valued at $4,500 and an unspecified number of automobiles valued at $2,500. Brown & Root was gradually moving toward greater reliance on motors in place of mules. Among the assets appears an "undivided ½ interest in 630 acres" located five miles north of Taylor and valued at $73,000. The conservative approach to finance favored by the Browns is reflected in this statement, which lists only $11,300 in liabilities against a total of $187,450 in assets. Brown & Root was no longer a small firm existing job by job. (See table 2.2.)

No amount of planning and hard work could keep Brown & Root prosperous unless it remained successful in gaining contracts to build roads. This demanded careful

Assets

92 head mules and horses	11800.00
Harness	1400.00
Tents and camp equipment	925.00
34 wagons	3200.00
2 grades	750.00
Small grading tools	1300.00
3 concrete mixers	1700.00
2 hoisting machines	1100.00
12 pumps	1800.00
1 water wagon	350.00
1 boiler 35hp	500.00
concrete tools	785.00
small tools in use	1600.00
1 truck	1000.00
5 trucks	3500.00
4 trailers	1650.00
Lumber on hand unused	1000.00
Automobiles	2500.00
Due from Insurance Co. on auto	600.00
Live stock & tools & farm equipment	10100.00
New small tools unused	3800.00
Roller	225.00
2 Asphalt Distributors	12000.00
2 Sweepers	8000.00
Boiler	500.00
Small paving tools	4750.00
Tanks & coils	1200.00
Super heater	390.00
Hay on hand	300.00
Feed on hand	800.00
Money due us (Georgetown)	2100.00
San Marcos paving bills due	7500.00
Estimate Florence due	7500.00
Retainage Florence	3600.00
Real Estate Round Rock	300.00
Real Estate Georgetown	4500.00
American Surety Co.	2400.00
Stocks on hand	2000.00
Est. earned not due State of Texas	3000.00
Total	114,450.00

Liabilities

Farmers State Bank	10000.00
FD Love	1300.00
Total	11300.00
We have following real estate which we have not listed above ———————————	73000.00

Undivided ½ interest in 630 acres located 5 miles North of Taylor

Source: Herman and George R. Brown Archives, Houston, Texas.

attention to government contracting, and the booming road construction business inevitably raised controversial questions about contracts involving public funds. Herman Brown spent much of his time in the formative years of his company hustling contracts; George later became central in this part of the business. Both brothers quickly learned that the political arena was no place for the faint of heart. In a relatively poor state, the large sums of money being spent on roads and bridges spurred intense competition for contracts. This, in turn, produced a measure of corruption and influence peddling.

John Huddleston, a historian of Texas roads, has written about the controversial world of contracting and subcontracting for roads in the 1920s. He noted that Governor James Ferguson and his wife, Miriam, also known as Ma and Pa Ferguson, "plunged the [Texas Highway Department] into scandal in 1925–26, but other politicians such as Governors Dan Moody, Ross Sterling, and Coke Stevenson maintained the political independence of the road agency."[43] According to Huddleston the Ferguson's anti-auto stance changed to one favoring autos because "the construction and maintenance of the Texas road system offered lucrative patronage."[44] Brown & Root learned to take what it could get from a corrupt system that at times did not allow independent companies to compete fairly with favored contractors. Dan Dyess recalled that during Jim Ferguson's time as governor, his company, the American Road Company, controlled most state highway work. "We subbed a lot of work from them, which was about the only way you could get any state work."[45] In Huddleston's opinion, highway scandals "more than any other factor" encouraged the political turmoil that surrounded the Fergusons in the mid-1920s.[46]

Corruption in highway contracting became an increasingly controversial issue in Texas in the 1920s. Brown & Root and other construction companies sought to survive in a competitive environment marked by widespread favoritism in roadwork contracts. In a telegram to a Texas highway official, Thomas MacDonald, head of the U.S. Bureau of Public Roads, complained "that favoritism is being shown contractors in letting work or in requirements after contracts have been awarded. Send your recommendations as to suspension of federal aid participation . . . The commissioners understand that we have regarded the engineering conditions in Texas as serious for a long time."[47]

Large sums of money flowed through the county commissioners' offices as well as through the state highway commission, as county governments financed more and more roads and bridges to accommodate growing automobile and trucking traffic. Contractors sought methods of gaining favor with county commissioners. One common method of lobbying for county road contracts was simply to "wine and dine" the commissioners. Such close contact at times helped Herman Brown persuade the city council or the commissioners' court to advance him funds before the road building work started. This money allowed Herman to "walk around" the town square and pay his debts "to the grocer, druggist, and feed store."[48] At a time when favoritism pervaded the political world in which contracts were let, it paid to find ways to become a favorite.

The most obvious way to gain favor was by submitting the lowest bid and completing the highest quality work on time and under budget. Brown & Root learned this lesson early on in its history, and it established a reputation as a reliable, efficient builder. The firm also had the advantage in Central Texas of being run by "local boys." But as the Browns moved farther out from their home region and entered the competition for more lucrative contracts, they learned to "unstack" the deck from which contracts were dealt. They had to become more adept at playing the game of political influence. This was a natural part of doing business in the world of public works contracts. They accepted this reality and sought ways to master the interworkings of the bidding process. (See table 2.3.)

The necessity for road builders of this time and place to "play politics" is evident in court testimony from these years. In explaining the costs of doing business for Brown & Root, a brief in defense of the company in a case brought by a disgruntled former employee seeking back pay noted that "many others were engaged in like work, and had like organizations. Competition for contracts was keen. No one was successful in getting more than 10% of the jobs bid for." Any possible way to increase the odds of winning bids thus had a far-reaching impact on a company's overall success. Brown & Root "had done much of this kind of work, and had built up a large and efficient organization . . . of regular employees who were experts in construction work and in estimating its costs." As a supplement to well-prepared bids, effective lobbying was an essential part of a business in which the cost of "trying for jobs" added "ten to twelve percent of the estimated field cost."[49]

With success in Wharton, the Brown brothers decided to break into the large market for paved city streets in Houston. As part of an overall expansion plan of 1926, George moved to Houston and created a new office there while Herman relo-

TABLE 2.3. ROAD CONTRACTS BID BY BROWN & ROOT, INC. (1924–35)—
SELECTED COUNTIES

County	No. Bid	No. Awarded	% Awarded of Total Bid
Harris	79	56	71
Travis	18	7	39
Bexar	14	4	29
Pecos	13	6	46
Cherokee	12	0	0
Nueces	12	12	100
San Patricio	12	1	8
Ft. Bend	10	2	20
Jasper	10	1	10
Wharton	9	1	11

Source: Recent Contracts Awarded and Recent Contract Bids listed in *Texas General Contractors Association Monthly Bulletin*, 1924–35.

cated the headquarters of Brown & Root from Georgetown to Austin. This configuration lasted for the next twenty years.

As one of the fastest-growing cities in the nation, Houston was an obvious target for expansion. Its population more than doubled from 1920 to 1930, when it numbered almost 300,000.[50] The opening of a ship channel to the Gulf of Mexico just before World War I made Houston an important port for cotton and oil products.[51] The massive oil discoveries on the Gulf Coast during the early twentieth century spawned a growing industrial base centered around oil refining and oil tool manufacturing. Capital, as well as people, then poured into Houston. This young and dynamic economic center beckoned to Brown & Root, which sought to take advantage of some of its extraordinary opportunities.[52]

The Browns entered the city by bidding on numerous contracts in the Houston area. George Brown later recalled that "the contractors in the city didn't want us to come in. They fought us and made things as hard for us as they could. They told the city council and the city engineers that we didn't have the experience necessary for big city paving."[53] To gain a foothold in this new market, Brown & Root needed to complete a high-profile job.

This opportunity came when George Brown secured the contract to pave the streets of River Oaks, a new master-planned subdivision that sported some of the city's most fashionable homes.[54] Mrs. L. T. Bolin, whose husband later became a high-ranking executive at Brown & Root, recalled the work Jimmy Dellinger performed on the River Oaks job: "he lived out here in River Oaks when they started paving it. Right up here on Lazy Lane . . . where he had his corral with his mules and all. He just started paving." These paving crews consisted of mule skinners who lived in a tent city on the outskirts of River Oaks and waited at a specified location every morning for a company vehicle to pick them up and transport them to the job site.[55]

Brown & Root's next major project in Houston involved paving Market Street road from Houston to Baytown under a contract dated July 12, 1927. This was the company's first concrete slab paving job. It involved placing riprap along the slopes of the drainage ditches adjoining the road, paving with asphalt, and constructing a concrete bridge over Fresh Water Bayou. Brown & Root did not build the bridge, however, because the Harris County commissioners' court decided to replace the proposed concrete treble bridge for which Brown & Root bid with a double culvert.[56]

Dan Dyess recalled that Brown & Root made its best money on small paving jobs in Houston and other cities and towns. Herman guaranteed his paving work for five years, and this guarantee occasionally won his company jobs even when Brown & Root was not the low bidder. Throughout the 1920s and early 1930s, the overwhelming majority of the firm's road projects were in Harris County. Its largest road projects occurred after 1928. One of the largest was a 1934 county road job in the Central Texas counties of Bexar and Kendall. It consisted of building and laying an asphalt top on twenty-five miles of Highway 9, and the company received more than $400,000 for its work. This very large road project, among others, marked Brown & Root as one of the state's leading road contractors. (See table 2.4.)

During the early 1930s, Brown & Root began to diversify its operations into other areas. The firm followed the practice of some of its competitors in "taking paper," or accepting the notes of property owners for their share of a city's paving costs. In city paving work, the city paid the cost of building the road through the issuance of as-

TABLE 2.4. TOP TEN ROAD CONTRACTS AWARDED TO
BROWN & ROOT, INC. (1924–35)

Year	County/City	Description	Amount
1934	Bexar	County Road Bid	418,974
1928	Harris	County Road Bid	373,579
1929	Yoakum	City Paving Bid	236,371
1935	Jefferson	County Road Bid	188,956
1930	Houston	City Paving Bid	169,208
1933	Harris	County Road Bid	161,194
1934	Tarrant	County Road Bid	157,379
1930	Pecos	County Road Bid	142,217
1928	Westlaco	City Paving Bid	135,200
1928	Odessa	City Paving Bid	132,256

Source: Recent Contracts Awarded listed in *Texas General Contractors Association Monthly Bulletin*, 1924–35.

sessment certificates. The certificates, explained attorney Edgar Monteith, "were payable by property owners on an installment basis, usually over about six or seven years . . . the legal connection was very important because you had to get a valid assessment against these abutting properties and that was a very sensitive area and it had to be done right."[57]

To accept and process paving notes from such jobs, Herman and George Brown formed a new company called Brown Securities. The "paving paper"—warrants or promissory notes—typically carried a five-year term at 8 percent. In 1928, Roy Montgomery Farrar, the president of Union National Bank in Houston, urged the Browns to liquidate their paving notes.[58] Although Farrar was a well-established banker with ties to Jesse Jones,[59] the Browns did not initially heed his advice to sell the paving notes.

The Browns believed that the continually expanding Texas road building business did not require the sale of their paving notes or a refusal to accept additional ones. Dan Root's death in 1929, however, gave the Browns the opportunity to protect the firm from possible financial difficulties by reorganizing it. On July 2, 1929, the Brown brothers incorporated Brown & Root. Its original capitalization consisted of $200,000 in capital stock of which Herman contributed $155,000 followed by his wife's $30,000 interest, and George Brown's $10,000 contribution. W. A. Woolsey, Herman's longtime right-hand man, received a $5,000 interest in the firm. The original directors included Herman, Margarett, and George Brown; Woolsey later replaced Margarett as a director.[60]

Soon after Brown & Root's incorporation, Farrar again suggested that the brothers sell their paving paper. Farrar said "you'd better liquidate. You're broke and don't know it. All you've got is paper. Credit's running wild; it's out of hand. I tell you again, sell your paper and sell it quick!" "I felt then," George Brown recalled, "that he knew what he was talking about. I went up to Chicago and sold almost all the paving paper we held. I had to take a ten percent discount on eight percent paper, and I didn't like that, but Farrar had convinced me that it was the thing to do." The stock market crashed the following October. Many road contractors who retained their paving paper found themselves broke and out of business. Later, Farrar told George Brown to go down to the safe-deposit box room in the bank and write a letter to him apologizing for "the dirty old son-of-a-bitch you called me when I told you to sell your paving paper!"[61] Fortunately for the Brown brothers, the company they had created proved strong enough to survive the devastation of the Great Crash and the hardships of the Great Depression.

In the years before the depression, the Browns built Brown & Root from a local contractor in Central Texas into a major regional construction company active over much of the state. In the process, they learned valuable lessons that guided the future expansion of their company. Herman Brown's determination, ambition, and deep involvement in the company's early work set a pattern that continued to shape the values and tone of Brown & Root throughout his life. He inspired loyalty and commitment, which he returned in the form of an intensely personal interest in the welfare of his workers, a commitment to provide steady work for loyal employees, and a willingness to give his managers extraordinary leeway to accomplish their work as they saw fit. He and George developed a smooth working relationship that made good use of the strengths of each. Their strategy of diversification entailed much risk, but it spread these risks so that the company became less vulnerable to the uncertainties of roadwork.

Herman Brown turned forty in 1932. By that time his approach to business had been shaped by twenty hard, all-consuming years on the roads. Margarett Brown acknowledged his obsession with building Brown & Root when she joked to her friend Lady Bird Johnson that "if she ever was called on to produce a family crest, it would consist of a mule and a fresno."[62] George Brown shared his brother's intense commitment to Brown & Root and a pride in the company's engineering achievements. His oldest daughter, Nancy, recalled, "Early on in our childhood when Papa would build a bridge, he would take me out there and we would look at it."[63] Both brothers had a passion for their work, and this sustained them through the early years of backbreaking labor required to survive and prosper building roads in Central Texas.

This very demanding business was at once highly competitive and rife with uncertainties. It created hard men. A longtime colleague of Herman Brown at Brown & Root gave a vivid description of his boss: "He was pretty tough. . . . He was very generous with the people that worked for him, the hands and so forth. . . . But if somebody crossed him up, he'd come down on them so hard, you could hear their bones crack. He came up in that kind of country. That was the way of the world in those days in the construction business."[64] Two decades of success in that world had given the Brown brothers the experience and the hard competitive edge needed to surmount the severe challenges presented by the depressed economy of the 1930s.

Part Two

*Laying Foundations
in Depression
and War,
1932-45*

A young LBJ sends a signed photograph to George Brown.
Courtesy Brown Family Archives.

CHAPTER THREE

Public Works and the Marshall Ford Dam

Brown & Root's steady growth brought the brothers financial independence and even a measure of recognition, but the Great Depression raised serious questions about their future. By 1932, Herman and George faced a dual challenge: how could they protect Brown & Root's road building business from the depression, and where could they find new avenues for diversification? As Herman Brown sought projects from his office in Austin, George scoured the city of Houston for any work he could find to keep the company's men and equipment employed.

With some five to six hundred workers at the onset of the depression, Brown & Root needed to adapt quickly as its traditional road building jobs declined dramatically.[1] In the early 1930s, the company's survival required detours into Houston garbage collection and the scrap business. Although Brown & Root had not yet completed any large construction projects outside of road building, the Browns began looking for a "breakthrough" project that would place their struggling company on a firmer footing.

The most obvious place to look for such projects in the 1930s was in the growing market for public works. As the federal government sought to put people back to work, a wave of new programs poured millions of dollars into construction projects as varied as dams, sewage plants, parks, and courthouses. Although politically justified as jobs programs, the unprecedented public works initiatives of the 1930s addressed many long-neglected societal needs.[2] In the process, they offered salvation for hard-hit construction companies that had been fortunate enough to survive the chaos of the early 1930s. Brown & Root joined other firms in aggressively pursuing the new federally funded public works opportunities.

Throughout its history, Brown & Root pursued a strategy of building employee loyalty and boosting company morale by providing full-time work in an industry traditionally characterized by part-time, seasonal labor. To maintain this strategy in the depression of the early 1930s, Brown & Root—like individual citizens from all walks of life in this era—took on virtually any job it could find. In 1932 the company won a contract to pick up garbage in Houston, using workmen and trucks idled by the downturn in road building. Exploring every possible opportunity for profit, Herman Brown ordered his men to "take the organic garbage out of the garbage, feed it to the pigs, and [then] he sold the pigs. He said he knew people had to eat, so for a short while there . . . that's how Brown & Root made it through the Great Depression."[3]

The garbage contract embroiled Brown & Root, and particularly George Brown, in controversy. The *Houstonian,* a local Houston newspaper, charged the firm with hauling the garbage at exorbitant prices. "The way things were handled in awarding the garbage contract," wrote Nat Terence, editor and business manager of the *Houstonian,* "nobody else had a chance except Brown & Root."[4] The controversy arose over the renewal of the company's contract with the city. The city's original agreement for garbage pickup was with the South Texas Transportation Company, owned by Brown & Root. Later, the firm reorganized, still under Brown & Root control, as the Houston Public Service Corporation. With Mayor Oscar Holcombe's support, the city council agreed to award a three-year contract to Houston Public Service for $1,254,000. Prior to officially awarding the contract, a second firm bid $750,000 for the same job. Houston Public Service then lowered its bid by $462,000 to $792,000, and the city council awarded the contract to the Brown & Root subsidiary. Terence played this episode to its fullest and suggested the possibility of corruption, although no official charges were ever made.[5] Houston Public Service continued to pick up the garbage in Houston throughout the 1930s.

Herman Brown called another new venture in these years "the junk business." He entered this business in the mid-1930s, after he and Dan Dyess went to see a man named Slim Dahlstrom for assistance in finding an upright boiler. Years earlier Herman Brown had purchased World War I surplus equipment from Dahlstrom, who specialized in the sale of secondhand industrial and surplus property equipment. Dahlstrom found for Brown and Dyess an upright boiler in good shape; in so doing, he attracted Brown's interest. In 1935, Dahlstrom and the Browns formed the Texas Railway Equipment Company to enter the surplus and used industrial equipment business. Dahlstrom soon purchased for $50,000 a crushing mill originally valued at $600,000. The "junk" partners put the mill into immediate operation and sold $100,000 worth of crushed stone before having to pay for the mill. They later put the mill to work at the Marshall Ford dam site. Thus began a long and profitable sideline for the Browns.

Brown & Root also entered the transportation industry. In 1932, the firm began roadwork in a major oil field near Conroe, some forty miles north of Houston. A large oil discovery there during the summer had attracted a horde of oil seekers whose efforts were hampered by mud and the lack of roads or railroad tracks leading into the area. After winning a contract to build roads serving the oil field, Brown & Root encountered problems hauling dirt and clay by wagon through the muddy field. Joe D. Hughes, a prominent local businessman who owned thousands of mules as well as a fleet of trucks, assisted Brown & Root by using his mule teams to haul construction materials onto the oil field. Impressed with the firm's "mule power," the Browns bought out the company, which continued to grow as a transportation company while retaining the Joe D. Hughes name. In combination with Brown & Root's garbage collection work and the junk business, this foray into transportation enabled the company to hold its own in a severely depressed economy.

Herman Brown had never been a man prone to extravagance, but the struggle for survival in the early 1930s made him even more committed to cost cutting. He later recalled that the experience of these depression years shaped his fundamental approach to the construction business. "We have never forgotten the lessons we learned in the depression. . . . The aversion to wasting money has stuck with us—whether it is our money or our client's money. . . . I don't think any contractor who has grown up in the dog-eat-dog days of competitive bidding could ever live with his conscience if he needlessly wasted his customers' funds."[6] Of course, one of the key problems for Brown & Root in the 1930s was that so few of its traditional customers had any money to spend, much less to waste.

The search for additional opportunities for growth ultimately pulled the company into still another new direction—dam construction. In searching for politically feasible public works and jobs programs, New Deal officials developed a passion for "multiple use" river basin projects that promised thousands of construction jobs in the short term and flood control and power in the long term. The most visible such project was the Tennessee Valley Authority, a federal government agency that harnessed the power of a region's rivers to spur the development of a vast, previously impoverished section of the South. Other high-profile New Deal–era dam construction included the Boulder Dam on the Colorado River and the Bonneville and Grand Coulee Dams on the Columbia River in Washington and Oregon. Each of these projects involved tens of millions of dollars and plunged the federal government deeply into the business of financing major construction on a level unknown in earlier periods. In the process, these giant projects enabled several regional construction companies to grow much larger. Companies such as Kaiser and Bechtel parlayed involvement in large-scale dam construction near their traditional markets on the West Coast to emerge from the 1930s as much larger companies prepared to compete nationally for major projects.

Brown & Root followed a similar path in the late 1930s. The Brown brothers' opportunity to step into the world of dam construction came in Central Texas, where they had a hard-earned reputation as a major regional builder. Despite prolonged dry spells for much of the year, extremely heavy rains and flash flooding periodically devastated the region. The Browns personally experienced numerous floods near Belton and Temple. Indeed, they had moved into bridge building after winning contracts to rebuild several bridges washed out by severe flooding in 1921. They also had seen and read about the devastating major floods south of Bell County, on the Colorado River near Austin.

The Colorado is one of the major rivers flowing through the center of Texas and its frequent flooding plagued all who lived in the region. The Colorado received its name from Spanish explorers describing the "red" (or "colorado" in Spanish) color of the river during its flood stage. These floods at times occurred north of Austin and in Austin itself, but the worst ones hit south of the city. Between 1900 and 1936, the lower Colorado valley suffered an estimated $80 million in property damage and about one hundred lost lives due to flooding. Seven serious floods occurred during 1935 and 1936 alone, when the sight of the Texas capital under water heightened public calls for flood control. "These floods," wrote one historian, "caused merchants and civic organizations along the river to seek means of controlling the flow of the river through the building of dams and reservoirs."[7]

Various attempts to control the Colorado River failed before the 1930s. The first major dam on the Colorado River near Austin had been built in 1893, only to be washed out by the flood of 1900. Just as repairs on the dam neared completion, flooding again swept it away in 1915. In the late 1920s, the Insull interests, operating under the name of the Middle West Utilities Company, proposed construction of five to six dams on the Colorado River. Construction went forward for a time on several of these dams but came to a halt because of strident political opposition and the collapse of the Insull empire after the Great Crash.[8]

The state of Texas then sought unsuccessfully to carry out a publicly financed and managed plan to control the Colorado River. In 1929, the Texas legislature created a Brazos River Corporation and Reclamation District, and beginning in 1933, the Texas legislature made several attempts to create a similar authority for the Colorado River. After a heated political battle, the Texas Senate finally created the Lower Colorado River Authority (LCRA), which Governor James V. Allred approved in February of 1935. The new agency had broad powers to conserve the waters of the river for any useful purpose, to prevent floods, to develop power plants and sell electricity, and to acquire and sell property. The LCRA oversaw forty river districts or authorities for Texas; it also acquired the assets of the Colorado River Company, which previously completed about half of the Hamilton Dam above Austin. The LCRA received a $4.5 million loan from the Public Works Administration (PWA), a New Deal jobs program specializing in heavy construction, to finish the Hamilton Dam. Upon completion, this dam was renamed the Buchanan Dam in honor of Texas congressman James P. Buchanan, who had been chairman of the House Appropriations Committee.[9]

Not long after the creation of the LCRA, another disastrous flood hit Central Texas. The floodwaters from the Pedernales and Llano Rivers entered the Colorado River below the Buchanan Dam, which could not prevent flooding along the river. The flood caused $16 million in damages in Austin and the lower Colorado valley. In response, Austin mayor Tom Miller initiated discussions with LCRA officials and civic leaders to consider the possibility of rebuilding the old Austin dam.

The mayor appointed a twenty-one-member Citizens Advisory Committee to work directly with the LCRA on the idea of rebuilding the dam. Herman Brown was appointed to the board, probably because of his standing as a major local builder. During the early discussions about dam building, Herman no doubt realized that his company might have an opportunity to build one of the dams needed so desperately to control the floodwaters in Central Texas.[10] In 1936, Brown & Root supplemented its experience building several bridges and pouring massive amounts of concrete with clearing trees from a river basin in the northernmost lake on the Colorado River and building a concrete dam wall.[11] The company badly needed a major construction project, and Herman Brown was well placed to compete for the contract to build a large dam on the Colorado River.

In the Flood Control Act of August 26, 1937, Congress recognized the need for an additional dam on the Colorado River.[12] Next, the Interior Department authorized $10 million for construction. The Bureau of Reclamation took charge of the project, making this the first job in which Brown & Root dealt directly with the federal government.

Created in 1902, the Bureau of Reclamation's original mandate included transforming marginal arid land, particularly in the western states, into viable farmland.[13] The bureau's greatest successes before the 1930s involved the construction of various dams, including the Roosevelt Dam on the Salt River in Arizona, the Elephant Butte Dam of the Rio Grande, and the Pathfinder Dam in Wyoming on the Platte River. During the New Deal, the bureau's Boulder-Canyon Project resulted in the construction of Hoover Dam, Parker Dam, the All-American Canal, and power lines and other related water and electric power projects that benefited millions of people.[14] Although not as large or as visible as the Boulder Dam on the Colorado River in Nevada and Arizona, the proposed Marshall Ford Dam held the same sort of promise for those living along the Colorado River in Texas.

Three firms submitted bids for the new dam. Herman Brown proposed a joint venture between Brown & Root and the McKenzie Construction Company of San Antonio. George Brown later explained why his company undertook the project with a partner: "When we took on a big one," he said, "like the Marshall Ford Dam, we wanted another company up in that dark alley with us."[15] The partnership of Brown & Root–McKenzie Construction Company submitted the low bid of $5,781,235. Utah Construction came in second with a bid of $5,909,049, followed by W. E. Callahan at $7,332,496.[16] On December 3, 1936, the LCRA awarded Brown & Root–McKenzie Construction Company the contract for the Marshall Ford Dam.

Work began almost immediately, on January 14, 1937. The official groundbreaking occurred on February 19, 1937, with Secretary of the Interior Harold Ickes attending. Brown & Root worked on the project through May, 1942. The Bureau of Reclamation directed construction of the dam and furnished steel, concrete, and other building materials while the LCRA acquired the land.[17] For more than five years, Brown & Root focused much of its resources and personnel on this one giant project.

Building the Marshall Ford Dam required Brown & Root to expand. For the first time, the company built a major flood control project and a hydroelectric power plant. But these departures did not much worry the Browns, who were confident they could meet these new challenges. George Brown later reflected on the basic similarities between building roads and dams: "We originally were road builders. To be road builders, you have to know about concrete and asphalt. You have to learn something about bridges. Once you learn these things, it's only a step, if you're not afraid, to pour concrete for a dam. And if you get into the dam business you'll pick up a lot of information about power plants. . . . Each component of a new job involves things you've done before." But whatever the similarities, the Browns also clearly recognized that the Marshall Ford Dam represented an entirely new level of activity. George Brown recalled that "it was a new plateau for us. It was a new challenge. It gave us a new set of conditions to meet. We were doing things we had never done before. And we were working against a stiff deadline."[18]

At every step of the way in securing appropriations and meeting this deadline, the Brown brothers relied heavily on the advice and assistance of Alvin J. Wirtz, a prominent Central Texas lawyer and political insider. Born in Columbus, Texas, in 1888, Wirtz was only four years older than Herman Brown. He graduated from the University of Texas Law School in 1910. While building a strong regional reputation as a specialist in corporate law with an emphasis on oil and riparian rights, Wirtz became deeply involved in Democratic party politics. His tenure in the Texas Senate from 1922 until 1930 included a term as president of the senate; he also served as a delegate to the Democratic Convention in 1928 and 1932. He was active in efforts to harness the Colorado River long before the Browns entered the picture. Wirtz had represented the Insull interests in the late 1920s as they sought to build dams on the Colorado River. He then served as the receiver of their property before becoming a backstage force behind the passage of the legislation that created the Lower Colorado River

Authority. As a member of the prominent, well-connected Austin-based law firm of Powell, Wirtz, Rauhut & Gideon, Wirtz exerted influence because of his long experience in Texas politics and law, his precise legal intellect, and his passion for progress in controlling the rivers of Texas. For a time Wirtz served both as an attorney for Brown & Root and as general counsel for the LCRA. In 1940, President Roosevelt appointed him undersecretary in the Department of the Interior. Wirtz proved a most valuable ally for the Brown brothers in their quest to build the Marshall Ford Dam. In the process, he became a valued associate of the Browns and Herman's special friend.[19] According to a mutual friend, Brown "had a tremendous trust in Wirtz's discretion and his integrity and his mind. . . . Of all the people Herman knew, the one whose mind he respected most was Senator Alvin Wirtz."[20] Without Wirtz's assistance and expert advice, the Browns might never have had the chance to build the Marshall Ford Dam.

The original design for the Marshall Ford Dam called for a 190-foot-high structure, 4,000 feet in length. This edifice required two million cubic yards of concrete and one million yards of earth fill. Excavation work uncovered about ninety graves of Native Americans, which were removed from the Marshall Ford Basin.[21] Since the Bureau of Reclamation was in the process of planning the Grand Coulee Dam on the Columbia River in Washington State, the bureau applied much of that design work to the plans for the Marshall Ford Dam. This hastened the entire design and building process.[22]

Early work on the Marshall Ford Dam encountered severe problems with funding, design, and labor. The funding problems emerged even before construction began. President Roosevelt had personally authorized construction of Marshall Ford Dam without waiting for congressional approval. Under the rules of the Emergency Relief Appropriation Act, however, each public works project required congressional approval, usually after hearings and reports before the committee of federal funds. Noting the lack of such approval, the comptroller general's office refused to authorize the first $5 million payment for construction costs. Texas representative James P. Buchanan, chairman of the House Appropriations Committee, finally "persuaded it [the comptroller general's office] to allow work to begin." In return, Buchanan assured the comptroller general's office that Congress would authorize the dam during the next year.[23]

After Brown & Root won the bid, the firm moved forward with its construction plans, unaware of the potential problem in funding. But when the brothers learned of the appropriations complication, they paused to rethink their predicament. Building the dam required a large initial capital investment, and the Brown brothers could not be certain that the second $5 million payment would be made. George Brown later recalled that "Wirtz was telling us all along that the money was wrong, and that if someone in Congress raised a question they would stop it [would stop paying out money under the contract]."[24] This was not the first time that a Bureau of Reclamation project had begun before acquisition of the proper authorization from Congress, but Brown & Root found little comfort in precedents as it pushed forward with its largest—and riskiest—project ever. Given the large initial capital investment required to begin construction on the dam, Brown & Root assumed considerable financial and political risk in undertaking the Marshall Ford Dam project.

The company first had to construct support equipment at the dam site. To purchase some $1.5 million worth of heavy equipment, Brown & Root borrowed funds. The single most costly piece of equipment was a cableway, with two metal towers erected on opposite sides of the canyon connected by heavy gauge cables. The cableway supported a trolley system for transporting buckets of concrete to the laborers who poured the dam's foundation. The cableway, the trolley, and other smaller items required in the start-up construction phase of the project cost Brown & Root about $500,000. The investment in this equipment would be lost without a second appropriation. The comptroller general's office told Buchanan that there would not be a second appropriation unless Congress approved the dam in 1937. The Brown brothers nonetheless decided to go forward with the project. They were willing to take the risk in pursuit of the considerable rewards that might come with completion of the project.

By September, 1936, Brown & Root began pouring concrete from the installed cableway. At this point, a new complication emerged as federal officials raised serious questions involving land ownership. When Congress had created the Bureau of Reclamation in 1902, it had empowered the bureau to build projects only on land owned by the federal government. The Board of Land Appeals had reaffirmed this requirement in subsequent decisions under its auspices. After Brown & Root began construction of the dam, an attorney in the comptroller general's office checked the land title at the dam site. He discovered that the federal government did not own the land. Not only did the project lack congressional authorization, it violated federal law governing the bureau's activities.[25]

This problem was unique to Texas. In all of the western states besides Texas, the federal government had acquired ownership of all public lands as the territories accepted statehood. Texas, however, had been an independent republic prior to its admission to the union in 1845. When the United States admitted Texas, the state retained ownership of its public lands, including riverbeds. Since the Marshall Ford Dam was the first dam to be built by the Bureau of Reclamation in Texas, this unique issue had not arisen before.

George Brown related later to biographer Robert Caro, "We had put in a million and a half dollars in that dam, and then we found out it wasn't legal. We found out the appropriation wasn't legal, but we had already built the cableway. That cost several hundred thousands of dollars, which we owed the banks. And we had to set up a quarry for the stone, and build a conveyor belt from the quarry to the dam site. And we had to buy all sorts of equipment—big, heavy equipment. Heavy cranes. We had put in a million and a half dollars. And the appropriation wasn't legal!"[26]

Within a few days of the discovery, Wirtz told the Browns that one solution was a federal law specifically authorizing Brown & Root's contract. The comptroller general's office indicated that it would be able to continue to fund the project under such a law, and Congressman Buchanan agreed to work on gaining congressional approval of such a law. Buchanan's sudden death from a heart attack in February of 1937, however, clouded the future prospects of the Marshall Ford Dam in a fog of uncertainty.

Alvin Wirtz peered through this fog and saw a practical plan of action. The goal remained the same: to persuade Congress to pass a law approving Brown & Root's construction of the federally funded Marshall Ford dam on land owned by the state of Texas. The person most likely to submit this legislation to Congress would be Buchanan's replacement as the representative from the congressional district in which the dam was located. Two individuals emerged as the most likely successors to Buchanan. These were C. N. Avery, Buchanan's campaign manager, and an up-and-coming young politician named Lyndon B. Johnson.

Before his entry into electoral politics, Johnson had been the state director of the National Youth Administration (NYA), where Alvin Wirtz worked with him as chairman of the Texas State Advisory Board of the NYA. Twenty years younger than Wirtz, Johnson consulted often with his fellow Central Texan on NYA matters, as well as on political issues. Wirtz came to respect the ideas and the ambition of his young friend, and he supported Johnson's decision to campaign for Buchanan's congressional seat. Herman and George Brown had worked closely with Buchanan, and they supported his friend C. N. Avery in the election. But Johnson traveled the region to build popular support, and he won the election. Herman Brown had met Johnson before the election, when the young head of the NYA in Texas visited him to discuss jobs programs. Soon after Johnson's election, the Brown brothers and their wives met the new congressman and his wife. Lady Bird Johnson recalled, "We had a mutual friend, his name was Jim Nash—big, jovial, well-to-do Roman Catholic. A man who helped make Austin run . . . And Jim Nash said, 'Well, I'll just give a dinner party and I'll invite you and Lady Bird and both the Browns.' Well, we actually got together."[27] The Browns had a very practical incentive to get to know the new representative; they needed Johnson's aggressive support in protecting their hard-won contract to build the Marshall Ford Dam.

Upon arriving in Washington, Johnson gave this matter his close attention, and his success in protecting the Browns' interest in the dam quickly cemented a close relationship between him and the Brown brothers. He reached the capital on May 13, 1937, and the Rivers and Harbors Committee's report on the Marshall Ford Dam was scheduled to be issued on May 24. Johnson obtained the necessary authorization from the committee: "The project known as Marshall Ford Dam, Colorado River project, in Texas, is hereby authorized . . . and all contracts and agreements which have been executed in connection therewith are hereby validated and ratified."[28]

A series of Senate, House, and conference committee approvals were also necessary before the second $5 million appropriation would be made available to Brown & Root. To guide the bill through this congressional quagmire, Johnson sought help from President Roosevelt. Johnson discussed the problem with a close associate of Roosevelt, Tommy Corcoran. Johnson asked Corcoran to bring to Roosevelt's attention at the right moment the issue of the Marshall Ford Dam. Corcoran did, and Roosevelt reportedly replied to Corcoran, "Give the kid the dam." Corcoran later recalled that he interpreted Roosevelt's decision to imply that Johnson needed the dam to secure a firm financial base for his political career: "When Roosevelt told me to take care of the boy, that meant to watch out for his financial backers too. In Lyndon's case there was just this little road building firm, Brown & Root, run by a pair of Germans." Corcoran subsequently "made a hell of a lot of calls on that dam." Opposition disappeared, and Brown & Root continued to receive funding for the dam.[29]

With appropriations secured for the early stages of the project, Herman and George Brown faced the practical problem of building the dam. Finding good employees and maintaining a productive and reasonably satisfied workforce became a significant problem. The U.S. Employment Service and Works Progress Administration (WPA) set certain employment standards for Brown & Root and McKenzie Construction Company. H. P. Bunger, Construction Engineer for the U.S. Department of the Interior, wrote the construction team and informed them that the contractors were permitted to hire highly skilled workers—those individuals responsible for each piece of major equipment—without predesignation, that is, without prior approval of their employment. All employees below the level of highly skilled workers had to be predesignated by the government for employment.[30] Predesignation in this case referred to workers obtained through the U.S. Employment Service or WPA.

C. E. Blakeway, employment manager for the construction team, replied that "in order that our job function to the highest degree of efficiency, it will be necessary to have better cooperation, a better class of labor and more prompt referrals from the Employment Service and Works Progress Administration."[31] Difficulty in finding semiskilled as well as skilled workers prompted the construction firms to request more flexibility in hiring workers. The existing pool of unemployed workers on the Texas relief rolls did not contain enough men with the necessary skills.

George R. Brown, Lyndon Johnson, unidentified, Tommy Corcoran, Herman Brown, 1939. *Courtesy Brown & Root.*

Herman Brown complained to Lyndon Johnson about the difficulty of finding properly skilled employees. "They have just called me from the Dam," Herman wrote, "saying that they are unable to get rough carpenters, riveters, or, in fact, all of the skilled type from the relief rolls and the office here is rather slow in making exceptions; however, I think we will be able to get this adjusted in the next day or two."[32] The difficulty of finding qualified workers was only the first of even more difficult labor-related problems to emerge later in the construction work. The winter weather only made matters worse. "We have been having too much cold weather to suit us at the Dam," wrote George Brown to Lyndon Johnson, "as we have had to close down three times during the last ten days because of freezing weather."[33]

Labor conditions continued to deteriorate. The Browns' partner, A. J. McKenzie of the McKenzie Construction Company, met twice with Lyndon Johnson in Washington in early 1938. During those meetings, Johnson informed McKenzie that he received a letter from a Texas labor organization charging the contractors with "irregularities." Johnson, however, did not furnish McKenzie a copy of the letter or describe the specific charges. Upon returning to Texas, McKenzie learned that the Texas State Federation of Labor had filed the charges and a federal investigator was investigating them. McKenzie complained that the nature of the charges was being unfairly kept from him and he was unaware of what labor irregularities might exist. The letter indicated that despite the joint venture relationship, McKenzie did not have a close rapport with Herman Brown. "Herman Brown told me in effect," he wrote to Johnson, " . . . that you felt my request for a copy of such charges was out of place, and perhaps, impertinent."[34]

Friction in the working relationship between Brown & Root and McKenzie became more apparent in later correspondence. Herman Brown wrote Johnson and informed him that A. J. McKenzie was traveling to Washington. Brown suggested he see Johnson and tell him what he was doing "so that you would be informed and that there would be no static. He re-assured me," wrote Herman, "that he was leaving everything else, except the engineering problems, to George and me and I think he really means it this time."[35] Herman felt that negotiating in Washington was the Browns' business, not McKenzie's. With a growing controversy over the future of the dam's construction, there was no room for a divided front in Washington.

A continuing flood problem soon created a public outcry. The combination of the Buchanan Dam and the Marshall Ford Dam as designed under construction could not prevent major flooding along the Colorado River. This was confirmed in July, 1938, when still another flood caused much destruction and financial loss south of Austin. Those responsible for the Marshall Ford Dam now faced serious questions about the future effectiveness of flood control on the Colorado.

In addressing such questions, Bureau of Reclamation engineers and engineers from other agencies determined that the height of the Marshall Ford Dam needed to be increased to prevent losses from major floods.[36] The bureau conducted studies comparing the flood control abilities of the existing 190-foot dam with a proposed 270-foot dam. The Department of the Interior's evaluation was that "substantially full control of floods of a maximum runoff of 480,000 second feet, being well above the highest of record, can be accomplished with the dam height of 270 feet." Without this higher dam, the Interior Department estimated major flooding above the 190-foot dam's capacity could cause up to $32 million in damage. The Interior Department also required a new bid for awarding the contract to raise the dam.[37]

Convinced of the necessity for a higher dam, authorities began planning almost immediately to appropriate $17 million for the additional work. An additional irony now entered the picture. The bureau had legislative authority to fund dams only for flood control. If the high dam was to be built for the purpose of flood control, what had been the purpose of the low dam? Although the original justification for the low dam had been flood control, it had in fact been used also for hydroelectric power generation. Wirtz decided to remove this potential problem simply by rewriting the stated purpose of the dam. The low dam would indeed be officially a power produc-

Marshall Ford Dam, December, 1938. *Courtesy Brown & Root.*

tion facility while the high dam, the additional eighty feet, would be for flood control. Using accepted formulas for determining the cost allocation of a dam between flood control and power production, the flood control portion of the dam could cost no more than $14,850,000 and the LCRA could fund no more than $9,515,000. The two amounts totaled $24,365,000, or $2,635,000 less than $27 million. The $2,635,000 portion of the dam built for power production could not receive federal funds.[38]

Washington attorney Abe Fortas devised a scheme whereby the first thirty-three feet of the dam to go on top of the original low dam would be for both flood control and power production. This section would be used for power production only when the floodgates were opened. Under this scenario, the PWA agreed to fund the construction of this portion of the dam for approximately the amount needed to bring the total funding available to $27 million. Fortas convinced Harold Ickes, the man with the power to approve the scheme, of its legality. Although some legal sleight-of-hand had been required to push through the appropriations needed to complete the high dam, all of those involved seemed to agree that the project was both socially and politically justified. Ickes agreed and issued his final approval.[39]

The plan also required congressional approval, and not all congressmen liked the idea. The new funding plan was put forward as a part of an amendment to the Rivers and Harbors Bill that exempted the Marshall Ford Dam project from reimbursing the bureau for the power production portion of the dam. When the amendment's opponents spoke forcefully against it, House Majority Leader Sam Rayburn of Texas answered, voicing strong support for the amendment. Rayburn also read a supportive statement from Congressman Joseph Jefferson Mansfield of Texas, a member of the Rivers and Harbor Committee and key supporter of the dam. After Rayburn's speech, the House approved the amendment.[40]

Immediately after the vote, George Brown wrote Mansfield, "I just heard good news on the Marshall Ford Bill. Lyndon tells me the letter you wrote to Speaker Sam Rayburn really had the desired effect." George also professed his desire to return the favor: "if there are any of your friends I can at any time favor, please do not fail to let me know."[41] Though still relatively new to the world of Washington politics, the Browns were quickly learning the realities of practical lobbying. They had their hands on a project that was good for them and good for their region, and they enlisted the strong support of those who represented the people who would benefit directly from the building of the dam. Fortunately for the Browns, their political allies had sufficient clout to push for an enlarged dam on the Colorado River.

Since Brown & Root already had in place much of the heavy equipment including the cableway needed to build the high dam, it could bid low on the high dam project, leaving ample margin for profit. Subsequent changes in various specifications on the project gave Brown & Root opportunities for additional profits. But the firm did not always get the terms it wanted. George Brown described to Johnson a meeting involving the Browns, government engineers, and McKenzie to negotiate new prices for construction on the high dam. George wrote, "They would only give us $20,000 for flood hazards—we asked for $50,000, and they cut our price on concrete $.50 a yard—we asked that it only be cut $.25." He also noted that the government demanded its terms be met in this case since Brown & Root already had the necessary equipment on hand. He conceded the government's point. "Our part of the $5,000,000 [the second appropriation for the low dam] is something less than $2,000,000, which is a nice piece of work." Johnson oversaw the progress of the change order. He sent a telegram to George Brown in August informing him of its imminent approval. He reported to Brown: "For your confidential information until announced by Department, Comptroller-General signed decision this morning . . . I got decision sent by special messenger to Interior and will check Reclamation later this afternoon. Will stay here until everything is cleared out and go order given."[42]

Johnson had come through with flying colors on the Marshall Ford Dam, and the Brown brothers appreciated his efforts and understood their obligation for his strong support. Soon after the approval of the final appropriations for the high dam, George Brown wrote a heartfelt note to "Dear Lyndon." Its conclusion suggests the depth of the Browns' commitment to their newly proven Washington ally: "In the past I have not been very timid about asking you to do favors for me and hope that you will not get any timidity if you have anything at all that you think I can or should do." Brown then emphasized the extent of his support of Johnson: "Remember that I am for you, right or wrong and it makes no difference whether I think you are right or wrong. If you want it, I am for it 100%."[43] Herman Brown had earlier written to Johnson to express his gratitude: "I appreciate your wiring me last week, and, also, appreciate your help and advice regarding Marshall Ford Dam."[44]

During the course of work on the high dam, funds ran low as Brown & Root waited for payments. To acquire the remaining $3 million necessary to complete work on the high dam, Brown & Root needed to meet its schedule by avoiding work stoppages. One strong political selling point for the project was its employment of up to 2,500 workers. As Lyndon Johnson pushed the next appropriation through Congress, the carpenters union began agitating for a wage increase from $1.00 to $1.25 per hour.[45] Prolonged wage disputes among one classification of workers threatened to disrupt working conditions throughout the project. Brown & Root's managers began working with the LCRA, which set the rates, to equalize all wages. In the meantime, Johnson wrote Harry Acreman, secretary of the Texas State Federation of Labor: "I am using my best efforts to eliminate the grievances [of] which you spoke." Most important, Johnson advised "that any strike on the job should be avoided if possible."[46]

The continuing labor problems threatened to delay future funding needed to finish the dam. Costs continued to climb while equipment remained idle.[47] Wirtz informed Harry Acreman of the Texas State Federation of Labor that he was "terribly disappointed" about the labor problems. Any strike, he wrote, would affect the LCRA much more severely than the contractor. In any case, Wirtz reported that he "took the [labor] matter up with the Bureau of Reclamation and was advised that the matter would be reopened and reconsidered."[48] The labor problems subsided, and funding for completion came through.

This entire episode was a rude awakening for Herman Brown and Brown & Root, which had never before used such large numbers of newly recruited workers. The temporary workers who staffed many of the jobs on the Marshall Ford Dam did not have the same loyalty to the company or its founder as did the road crews that Brown & Root had held together for long periods in the past. The Browns also had never before encountered a determined effort by organized workers to demand better conditions. Coming at a time when a successful strike might have threatened the continued existence of the company, this tension between the Browns and some of their workers no doubt underscored Herman Brown's belief that unions were the curse of the construction industry.

In addition to labor problems, Brown & Root faced challenging technical problems in completing its dam across the Colorado River. The site of the Marshall Ford Dam was not readily accessible by existing transportation. The nearest railroad was the Southern Pacific line some thirteen miles to the east; a connecting line had to be built to haul in the hundreds of thousands of tons of aggregate, cement, steel, and machinery needed to construct a major dam. Roads also had to be extended, and a highway connecting the dam site to Austin was built. Given the history of past efforts to build dams on the Colorado, great care had to be taken to guard against floods during construction, when the dam was vulnerable to high waters. To protect the site, the contractors built two 26-foot temporary tunnels to divert the river's flow during construction. These tunnels proved their worth during a flood in July of 1938. Upon completion of the dam, these tunnels were closed off and filled with concrete.[49]

The low and high dam together consisted of a 2,710-foot reinforced concrete section extending across the main channel. Pouring such a massive volume of concrete created extreme cooling problems. As concrete hardens it generates heat; large volumes of freshly poured concrete require constant cooling in order to harden quickly and properly. At the Marshall Ford Dam, Brown & Root developed an artificial cooling system consisting of many miles of pipe to provide the cooling equivalent of a thirty-ton ice plant.

Another concrete section was 4,100 feet long, filled with one million cubic yards of dirt, clay, and rock, and faced with hand-laid rock. The concrete section contained three miles of inspection tunnels ventilated by an air conditioning system that removed excess moisture, preventing condensation. Twenty-four pairs of floodgates controlled from within two large galleries inside the dam allowed up to 120,000 cubic feet of water per second to flow from Lake Travis. The dam's power plant contained three 25,000-KVA turbogenerators.[50] (See table 3.1.)

After its completion in 1941, the Marshall Ford Dam played an important role in controlling floods along the Colorado River. Before its completion, annual property losses averaged $2.2 million, but there were no "flood losses of any consequence in

TABLE 3.1. MARSHALL FORD DAM

Height of Dam:	270 feet
Length of Concrete:	2,710 feet
Length of earth filled portion:	4,100 feet
Total length of dam:	6,810 feet
Surface area of completed dam:	41,940 feet
Length of lake:	65 miles
Maximum width of lake:	8.5 miles
Install capacity:	75,000 KVA
Water impounded:	2,000,000 acre feet, or 652,000,000,000 gallons
Maximum depth of lake:	225 feet
Elevation of dam at sea level:	750 feet
Lake shoreline:	270 miles
Cost of dam and power light:	$29,620,000

Source: Compiled from the Official Program for the dedication of Mansfield Dam, Herman and George R. Brown Archives, Houston, Texas.

the valley downstream" after the dam's completion.[51] It generated low-cost power for regional cooperatives, and it created Lake Travis, which quickly became an important recreational center for Central Texans. The dam clearly proved its value to its region. The dam originally had been named after the Marshall Ranch, through which cattle traditionally had been driven across the Colorado. After the completion of the dam, the Texas legislature renamed it in honor of Texas congressman Joseph Jefferson Mansfield, who had been a member of every conference committee on rivers and harbors work since 1921.

The LCRA ultimately built or acquired four other dams on the Colorado in addition to the Marshall Ford and Buchanan Dams, and reservoirs north of Austin. These included the Austin, Inks, Marble Falls, and Wirtz Dams. Brown & Root also built the Marble Falls Dam as well as the Wirtz Dam, which had originally been named the Granite Shoals. The Marshall Ford Dam was the "key unit of the system operated by the Lower Colorado River Authority."[52] It was a significant engineering accomplishment and a highly visible project that marked Brown & Root as a major regional construction company.

The Marshall Ford Dam project led Brown & Root to another related venture. In 1939, the Pedernales Electric Cooperative (PEC) awarded Brown & Root the contract to install 1,830 miles of electric line. Lyndon Johnson and E. Babe Smith, a graduate of Southwestern University and a rancher from northeastern Burnet County who desperately wanted electricity, organized the PEC with headquarters in Johnson City on May 8, 1938. Johnson and Smith encouraged local farmers and rural residents to join with a five-dollar membership fee and agreed to grant easements across their land for electric lines and retire any loans made by the federal government.[53]

To receive a loan from the Rural Electrification Administration (REA) to install the electric lines, the PEC needed to sign up three customers per mile. Despite a diligent campaign, Johnson and Smith signed an average of only two customers per mile. Johnson knew that he had one avenue of recourse, a meeting with Roosevelt. Once again, Johnson approached Tommy Corcoran, who gained him access to the president. Johnson later recalled that his meeting with Roosevelt went well. During the meeting, Roosevelt reportedly telephoned REA administrator John Carmody and expressed his support for the project. The REA notified the PEC on September 27, 1938, that it was making a $4,322,000 loan, later increased by $1.8 million, for the extension of 1,830 miles of electric lines to 2,892 families previously without electricity. Brown & Root got the job.[54]

This project also meant jobs for local workers. The money spent on construction, Herman Brown explained in a letter to J. E. Van Hooke of the REA, would directly benefit the people in the region served by the project: "Our reason for not moving off any faster was that we promised the Board of Directors of the Pedernales Electric Cooperative that we would meet their request to employ local labor on the project, as the cash money derived from their work was the only means some of them had to pay for connecting up with the electric service. This we have done and have used the men they recommended to us in each locality. We have had to train their men to do the work and it has necessarily made progress slow at first."[55]

Brown & Root encountered some of the same labor problems on the REA project as it had on the Marshall Ford Dam. Herman Brown continued to fend off attempts by both labor organizations and the REA to force Brown & Root to alter its traditional, nonunion employment practices. In response to a telegram from R. J. Beamish of the REA suggesting that Brown & Root pay higher wages to attract better-quality workers to get the job done, Herman replied, "We have operated in this section of the state for twenty-five years without labor troubles . . . we do not expect to pay any higher rates than set out in [the] proposal."[56]

Herman Brown then sent the correspondence to Johnson with a note: "I do not know what this man has in his mind nor do I know who is furnishing him with this kind of thunder. We have discussed it with the local board and we expect and intend using all the local people that it is possible for us to use and we do not expect to let Mr. Beamish or anyone else make us move in tramp linemen from all over the southwest to do this job."[57] Herman Brown resisted when he felt that "outsiders" sought to dictate to him the terms under which he would hire workers, and his experiences in the Great Depression only hardened his resolve to keep control over the management of labor at Brown & Root.

The Marshall Ford Dam was Brown & Root's breakthrough project, allowing the company to move up into the ranks of the major construction companies in its region. In subsequent years, Brown & Root built numerous other giant dams, with major projects throughout the United States and in foreign sites as distant and diverse as Thailand and Haiti. Although several of these dams were much larger than the Marshall Ford Dam, none was more important to the company. It pushed the company to a new level of size, profitability, and visibility. After completing this giant project on budget and in a reasonable time, Brown & Root found itself in a strong position from which to compete for other large projects that would have been beyond its capacity only five years before. In 1937 the company had won the initial contract in part on its reputation as an efficient, dependable builder active in the region in which the dam was located. In 1941 it had established credentials that qualified it to bid on much larger projects both inside and outside of its traditional region.

Brown & Root was not alone in its use of federally funded public works to prosper in the New Deal era. On the West Coast, both Henry Kaiser's company and the Bechtel company used the Boulder Dam project to transform themselves into major national concerns. The so-called Six Companies (which actually numbered eight) that cooperated to build the much-publicized Boulder Dam shared a budget of about $49 million; Brown & Root was one of two companies that built the much-less-discussed $29 million Marshall Ford project. In both cases, road and bridge building prepared the companies involved to take the leap into the construction on a more massive scale. Both cases also required the companies to master new technologies and to train workers in new skills. And both pulled the companies deeply into the world of national politics, which before the 1930s had not been an important consideration for construction companies because there were so few significant national government contracts.[58] Brown & Root, Kaiser, and Bechtel parlayed major New Deal public works projects into platforms for becoming larger, nationally active construction firms. All three played a prominent role in the construction industry during and after World War II.

Brown & Root emerged from the Marshall Ford project with far greater technical, financial, and managerial resources than it had before the mid-1930s. The Brown brothers had gained valuable experience managing large labor forces and in the politics of contracting with the federal government. In Alvin Wirtz, they had found a lawyer/lobbyist whose practical mastery of corporate law proved invaluable. In Lyndon Johnson, they found an eager, ambitious young Washington politician with strong personal motives to assist the Brown brothers, who were prominent builders active in Johnson's home district.

Much has been written about the political connection between the Browns and Lyndon Johnson.[59] It was forged in the late 1930s, and it remained strong throughout the lives of the three men. Obviously, there was a direct, practical reason for their earliest cooperation. At a pivotal time in the history of Brown & Root, Johnson proved to be a valuable ally in Washington, paying close attention to the needs of Brown & Root. Under extreme pressure to defend the giant public works project in his district, the freshman congressman demonstrated that he could deliver the goods. The Browns reciprocated by paying close attention to the needs of Johnson as he campaigned for offices in an electoral system that demanded substantial resources for success. The new congressman felt that he was a man of destiny who needed support in his home district to build a political base for future success. He did this through support of a major regional contractor in the completion of a project that provided jobs, flood control, and electricity to the citizens of his district. For their part, the Browns knew that they were as logical a choice as any other contractor to undertake this break-through project in their home region, and they quickly realized that Johnson was a much-needed and dependable ally in what was to them the strange new world of national politics.

From the Browns' point of view, their support of Johnson was simply American politics as they had experienced it since their dealings with county commissioners. "Wining and dining"—providing favors needed by individuals facing election campaigns—had deep roots in American politics. The rules remained ill-defined because no one involved had any real incentive to clarify them. Those involved in the construction of public works knew that the politics of contracting were very competitive and potentially corrupt. The bottom line for the Browns remained: can we do the job once we receive the contract? Their answer at Marshall Ford was yes, and nothing that has happened in more than fifty years since its completion has proven them wrong.

The practical ties between the Browns, Lyndon Johnson, and Alvin Wirtz were cemented by bonds of friendship and a shared sense of place. All four had grown up in Central Texas, and they shared a strong desire to see their region prosper. Wirtz and Herman Brown were near the same age; Johnson, though ten years younger than George Brown, shared much in common with him. All had witnessed poverty. All believed they could improve the region's living conditions. One person with a front-row seat from which to observe these men, Lady Bird Johnson, later recalled their common passion: "They were a good deal alike in that they shared a vision of a new Texas and they were going to be part of it. By gosh, they were going to make things happen—bring Texas whatever industry and whatever had made the eastern part of the United States the so-called elite and rich part. . . . They were all builders, strong, young, aggressive, determined."[60] They saw their personal ambitions—wealth and power—as a natural part of this broader process of creating a "modern" Texas.

Their shared victory in the extended congressional battles over the Marshall Ford

Dam appropriations built lasting political ties among them. But equally significant were the lasting personal ties forged in these years. As they worked on the Marshall Ford Dam and other related issues, the Browns, Johnson, and Wirtz spent considerable time in one another's company, and they became friends as well as associates. While the men separated to work in Washington or at the dam site or in Austin or Houston, their wives also became close friends. Johnson had only recently married, and his wife greatly enjoyed the company of Margarett and Alice Brown: "It was impossible . . . to know the two of them [George and Herman Brown] without cranking their wives into the equation right off." During the 1930s, Lady Bird spent more time in Austin than Houston, and Margarett made a particularly strong and lasting impression on her: "I considered her wide-eyed as a real intellectual, and she was the first such person that entered my sphere of influence."[61]

Visits with Margarett and Herman Brown were not generally purely social calls. The Browns entertained regularly in these years before Margarett's health deteriorated. Their home near the governor's mansion and the state capitol building became a meeting place for people engaged in Austin's leading industry, politics. Many evenings the Browns hosted a sort of open cocktail hour, when various people "just came and had a drink and got caught up on the news, and it was very, very animated."[62] They also had well-attended dinner parties, at which Margarett "would intentionally mix up the group—some Texas legislators, some businessmen, some people from the academic community. . . . And she would throw out a question and there would be good conversation, substantive conversation. . . . And if the ladies wanted to talk about the servants and the children, they'd better congregate in another room."[63] Years later Lyndon Johnson fondly recalled how "the minds of two of us—Lady Bird and I—went back to those evenings of good talk a long time ago at No. 4 Niles Road [Herman Brown's home in Austin], which have a very special place in our hearts and memories."[64]

George Brown later recounted his own memories of these rambling debates about key issues of the day. He remembered a meeting of the minds between himself and Johnson, particularly on the need for better education for the mass of Americans. Herman treated Johnson somewhat like a younger brother, and George and Johnson would at times gang up on Herman to force their points on him. All were supporters of Roosevelt, although Herman had more qualms about FDR's social programs than did George and Johnson. They shared the excitement of the New Deal era as well as of their common journey into a broader world beyond their roots in Central Texas. From these earliest days, Johnson seemed to relish matching wits and wills with Herman, a crusty man sixteen years older who never pulled punches in his arguments. "Lyndon and Herman would have some knock-down, drag-outs, but they would always get back together because they all appreciated each other as a worthy opponent." As they grew older and Johnson took on the trappings of higher and higher office, he continued to respect Herman's opinions. Herman was a man incapable of giving anything but a forceful, direct answer to any question. "George and Lyndon were more alike and more affectionate as time went on," and they became closer friends than did Herman and Johnson.[65] But all of the men and their wives made close personal ties in the 1930s that held them together for the rest of their lives.

As Herman and Margarett built a life centered around work and politics in Austin, George and Alice put down deep family roots in Houston during the 1920s and 1930s. They had three daughters in rapid succession, Nancy in 1927, Maconda in 1930, and Isabel in 1931, and they moved their growing family into a house in River Oaks, a relatively new subdivision of upscale homes a short commute from downtown Houston. There they built a comfortable family life. George relished his role as father, and Alice continued to pursue her interest in the arts while raising her children.

George Brown family: George, Alice, Isabel, Nancy, and Maconda.
Courtesy Brown Family Archives.

The press of work did become heavy in these demanding years at Brown & Root, and the brothers invested more and more of their vital energies in their business affairs. This was perhaps a predictable price of success, but it could be a heavy one. In a letter to her friend and sister-in-law Alice Brown in 1936, Margarett pointedly discussed their husbands' obsession with work: "I have a certain sympathy with these men who work very hard until they are old and then wake up to the idea that perhaps they missed something. . . . There is a lonely lesson there for George and Herman. They'd better gather their rose buds as they go along or soon winter is upon them."[66] Both men began suffering health problems. George Brown developed a severe case of ulcers that plagued him for much of the rest of his life. In 1940, at the age of forty-two, a flare-up of his ulcers forced him to take an extended rest period from work to allow his stomach to heal. But long periods of relaxation away from the job of building Brown & Root were not in the long-term picture for Herman or George. Their company had survived the Great Depression, emerging from the 1930s more prosperous than it had entered this tumultuous decade. Before the brothers had sufficient time to heal their ulcers or to step back and admire what they had built, World War II would bring even more dramatic opportunity and change to their lives.

CHAPTER FOUR

Defense Contractor

Brown & Root spent its formative "childhood" years as a Texas road builder. It spent its "adolescence" learning about federal government contracting while constructing the Marshall Ford Dam. The company came of age building defense-related projects in World War II. Like most Americans of this era, Herman and George Brown turned away from peacetime pursuits for the duration of the war, dedicating their energies to the national effort to win military victory on two fronts. The brothers spent the war years building a major naval base in Corpus Christi, Texas, and operating a massive shipyard near Houston. While contributing to the war effort, these projects also pushed Brown & Root to a much larger size and to a new level of engineering expertise, prestige, and profitability. This change in size was the most visible impact of the war. In the late 1930s, the two- to three-thousand-person workforce at the Marshall Ford Dam seemed large; but only five years later Brown Shipbuilding Company employed as many as ten times that number. A $29 million contract to build a dam during the depression no doubt seemed to be "the big time"; yet during the war years the Brown brothers completed contracts valued at almost $600 million.

In the late 1930s, the United States government gradually built up the nation's military capacities in response to the threat of war in Europe. Recognizing that this build-up inevitably required substantial construction work, the Browns kept a close eye on developments in Washington. After several years of stops and starts in efforts to gain a foothold in this new market for defense-related construction, Brown & Root succeeded in winning a contract to build a major new cadet training base for the navy at Corpus Christi. Winning this contract was difficult, since Brown & Root had to convince officials that it could quickly and effectively complete a vital job for which it had no previous experience. But under the mounting military pressures brought by the growing certainty of war, Brown & Root proved up to the job. Ground breaking took place in July of 1940, and the first cadets began training at the partially completed base in April of 1941.

The company's movement into defense-related projects reflected the growing importance of military construction as the nation slowly came to grips with the prospects for American involvement in a world war. As the major European nations began to move toward open conflict, political support grew in the United States for a defense buildup, including the creation of a navy capable of operating in both the Pacific and Atlantic Oceans. President Roosevelt supported the Congressional Act of May 17, 1938, which authorized the secretary of the navy to appoint a committee to report on the need for additional submarine, destroyer, mine, and naval air bases in the United States. In response, the committee investigated possible sites for naval bases and training facilities. To finance the expansion, Congress passed the president's request for a $1 billion appropriation. Congress targeted a portion of these funds for the construction of naval bases and training facilities.

To assist them in identifying promising opportunities in this burgeoning new market for defense-related construction, the Browns drew on their ties with Congressman Lyndon Johnson, who sat on the Naval Affairs Committee. The Browns first bid on a base to be built in San Juan, Puerto Rico, but the navy rejected their bid. The Browns discussed the project with Johnson, but he was unable to convince the navy to reconsider their offer. As a relatively new congressman, Johnson had only limited influence in the procurement process, especially for projects outside of Texas. After discussing Brown & Root's bid for the Puerto Rican base with Admiral Ben Morrell, chief of the navy's shore construction program, Johnson reported to George Brown that "I talked with Admiral Morrell, and he said there is nothing further that we can do at this time."[1]

Johnson nonetheless kept George Brown up-to-date on the project, presumably in case the navy decided to reopen bidding. In mid-May, Johnson once again contacted Brown. "I'll do all I can to get you any information on the Puerto Rico project," Johnson informed him, "and will let you know when anything breaks. You know how hard it is to get any dope in advance, but I'll have my eyes open. I'll probably wire you if I run into anything which seems likely."[2] Although the Puerto Rico project slipped away from the Browns, the brothers remained interested in building military facilities.

They realized that they could pursue this interest most successfully within their home region, where they had a number of valuable allies. During the late 1930s, several of the most influential politicians in Washington were Texans, and they had strong incentives to assist the Brown brothers as well as other Texas contractors seeking federal projects. In addition to Johnson, Texans John Nance Garner, a former speaker of the House, was Roosevelt's vice president (1933–41); Sam Rayburn was a prominent congressman; and Albert Thomas represented the Houston area in the House. Like most other representatives, Thomas lobbied hard for federal projects in his district and state. In this case, what was good for Brown & Root would also be good for Houston.[3] The Browns knew that military projects would be built in their region, and they wanted to be involved in building them. Their friends in Washington aggressively sought funds for regional projects, knowing that Brown & Root was a likely choice as a contractor.

Herman and George Brown followed with interest the navy's consideration of locating a naval air training center on the Texas Gulf Coast. The U.S. Navy recommended construction of a base in 1938, and in September of that year the navy sent a six-member team to investigate possible sites for bases in Texas.[4] Four locations attracted their attention: Houston, Corpus Christi, Brownsville, and Port Arthur/Lake Sabine.[5] Both Houston and Corpus Christi represented large metropolitan and commercial areas with influential political representatives. Congressman Albert Thomas, member of the House Naval Subcommittee, actively promoted a Houston location; Corpus Christi area congressman Richard Kleberg and the city's former mayor, Roy Miller, lobbied for their city.

Thomas had been elected in 1937 to represent the 8th Congressional District, which included all of Houston until redistricting added another congressman in 1958. Obtaining a major naval base for the Houston area would be a coup for the young congressman. With the zeal of a man who would continuously represent Houston from 1937 to 1966, Thomas set about the task of convincing the navy to take a hard look at his district. Along with the Houston Chamber of Commerce, he devised a plan to bring the proposed naval training station to a site near Galveston Bay.[6] Thomas appealed to the navy's site inspection team to give the Houston area serious consideration. In a letter to team leader Admiral Hepburn, Thomas encouraged the group to use his offices during their visit to Houston to survey possible sites. "If there is anything I can do to assist you and your command in any way," Thomas wrote, "I am yours to command." The navy's reply was not particularly encouraging: "we will really be in Houston for such short periods only that it will be impractical for us to make use of your kind offer."[7]

After examining the information collected during the site inspection team's visit to the Gulf Coast, in December of 1938 the navy recommended the establishment of a naval air training station in Corpus Christi.[8] The following May, a second committee selected the Flour Bluff area in Corpus Christi as the specific location of the base. George Brown followed these developments and wrote Roy Miller to express his excitement for the Naval Air Station's prospects. "I spent Saturday in Corpus," he wrote, "and they are all pepped up over the prospects of the naval base. Lyndon tells me there is a very good chance of getting $5,000,000 for it this year."[9]

The following day, George Brown wrote Lyndon Johnson to express his appreciation for his assistance on the Marshall Ford Dam and to inquire about the prospects for the Naval Air Station. Brown wrote, "I hope you know, Lyndon, how I feel reverence to what you have done for me and I am going to try to show my appreciation through the years to come with actions rather than words if I can find out when and where I can return at least a portion of the favors."[10] Johnson replied, "Your letter of May the second made me feel good all over. I felt pretty sure that the good news would put a spark of life into you and so I got it off just as fast as I possibly could."[11]

Despite the initial excitement, plans for the base moved forward slowly. During November, a board of naval officers confirmed the suitability of the Corpus Christi site. The Naval Subcommittee of the House endorsed the Corpus Christi location on January 23, 1940. This recommendation only confirmed the Corpus Christi site as one suitable for the base but neglected to address the issue of when or even if the government would actually build the base.[12]

Amid rumors that the navy had rejected the Corpus Christi site, Albert Thomas continued lobbying for some type of naval or marine base in the Houston area. Thomas kept in close contact with the Houston Chamber of Commerce in locating suitable areas on which to place such a base. During the spring of 1940, Thomas wrote Charles Edison, secretary of the navy, and noted that "this area is wide open to possible attacks through Mexico . . . one-fourth of the entire supply [of petroleum] of the United States is refined within one hundred miles." Thomas continued, "I am authorized to offer the Navy Department, free of cost, whatever lands are needed for such a base, up to three thousand acres."[13] Even this generous offer of land did not bring a positive reply from the navy.

In response to Thomas's inquiry, Assistant Secretary of the Navy Lewis Compton noted the findings of the 1938 navy report: "The Naval Defense of the continental coast bordering the Gulf of Mexico is primarily a matter of controlling the entrances to the Gulf; that is the Florida Straits and the Yucatan Channel." Compton also stated that if expansion of the navy's Air Arm became necessary on the Gulf Coast, "Galveston, Texas, will undoubtedly be given consideration."[14] When the navy finally chose Corpus Christi as the site for a Gulf Coast Naval Air Training base, Compton explained to President Roosevelt that "the Corpus Christi site afforded the most suitable terrain, both land and water, for the establishment of a naval air station for training . . . [however, the navy] is not actively planning to establish the station at Corpus Christi."[15]

As the military situation in Europe worsened in 1940, however, the Corpus Christi base became a high-priority project. Johnson was a member of the House Naval Affairs Committee, and a White House message to the navy suggested that he "was to be consulted—and his advice taken—on the awarding of Navy contracts in Texas."[16] By May of 1940, events in Europe prompted Congress to begin allocating funds for military projects, including the base in Corpus Christi. The *Corpus Christi Caller* reported that Kleberg and Lyndon Johnson actively collaborated to secure authorization of the base.

Albert Thomas and Herman Brown. *Courtesy Brown & Root.*

The House Naval Affairs Committee unveiled plans for the base on May 15 during hearings regarding funding for twelve new military installations. At that time, the navy opened a round of bidding for "test piles and test holes" at the Flour Bluff Point as a preliminary project to prepare for large-scale construction. As low bidder, Brown & Root received the contract for the test piles. The firm agreed to start work within four days of the contract date and complete the job within thirty days. Without the public's knowledge, the navy had already contracted for the design of the base and commissioned architects in Florida to draw the plans.[17]

While Brown & Root employees worked on the earliest phase of the Naval Air Station, Congress continued deliberating on the appropriation request. Realizing that the project would soon receive presidential approval, Johnson told reporters that Texas might well secure more military contracts in the near future. "When you consider that we are now training only 500 pilots annually and need to increase the figure immediately to 10,000," he said, "you can realize the possibilities for Texas."[18] Herman and George Brown were well aware of the potential this project held for their own future in the contracting business.

On June 11, 1940, the navy approved the contract supporting a naval air station at Corpus Christi, and two days later Franklin Roosevelt signed the cost plus fixed fee contract for $23,318,000 with Brown & Root, W. S. Bellows Construction Company, and Columbia Construction Company. Roosevelt also approved funding for ten thousand planes, eighteen thousand pilots, and twenty-two new combat vessels. The contract designated that construction commence almost immediately, on June 24. Brown & Root had entered defense contracting in a big way.

When questioned years later about Lyndon Johnson's assistance to Brown & Root in acquiring government contracts, George Brown recalled that "the only contract he [Johnson] ever played any role in was the Corpus Christi Naval Station."[19] Brown remembered how Johnson put in a good word about Brown & Root to Admiral Ben Morrell, chief of the Bureau of Yards and Docks, who granted the contract. Asked if Roosevelt assisted Brown & Root in this matter, George Brown replied, "Not that I know of—no more than he would any other contract that was let through the government."[20] In early February, 1941, President Roosevelt petitioned Congress for an additional $16 million appropriation for the Naval Air Station. The final cost of the entire base eventually rose to $125 million.[21]

George Brown cemented his reputation as Brown & Root's proficient salesman by negotiating long and hard with the navy to gain the contract. He later remembered that "getting the Corpus Christi contract was the hardest selling work I ever did. The Navy told us it just didn't think we could do the job."[22] Despite George Brown's best efforts to convince the navy that Brown & Root could do the job alone, Brown encountered a firm navy position: secure the participation of Henry Kaiser or face losing the contract.

The development of Kaiser's construction firm from its California base paralleled that of Brown & Root in Texas. By 1940 Kaiser had achieved a sufficient reputation and clout in Washington to be considered the type of company whose participation in a project assured its success. Kaiser had entered the construction business in 1914, the same year Herman Brown took his first job in the industry. After a decade of street building in Washington State, Oregon, and California, Kaiser's firm diversified into pipeline construction and dam building. During the 1930s, sometimes alone and sometimes as part of the famed Six Companies, Inc., Kaiser helped build the Hoover (Boulder), Bonneville, and Grand Coulee Dams.[23] These massive and much-publicized projects gave Kaiser a higher profile than Brown & Root, and Kaiser's participation in the Corpus Christi base made the project more palatable to the navy.

In the early summer of 1940, George Brown arrived in Washington to negotiate an agreement with Kaiser for his firm's participation in the construction project. The navy had instructed Brown to solicit Kaiser's participation at 50 percent or less. George Brown met Kaiser at the Shoreham Hotel in Washington. During what evolved into a tough late-night negotiating session, Brown offered Kaiser only 25 percent of the contract along with the added incentive that he would not require Kaiser to provide men or materials, just his name. Kaiser countered with a proposal of 75 percent for him and 25 percent for Brown & Root. When it appeared that Kaiser's offer was unchangeable, Brown prepared to leave. Falling for the bluff, Kaiser finally agreed to George Brown's terms.[24] Kaiser's participation appeared under the name of the Columbia Construction Company.

The consortium poured the first concrete for a permanent building on July 26, 1940. Construction proceeded rapidly on the 2,050 acres required for the air naval station. The three contractors established an operating committee consisting of one representative from each firm, W. A. Woolsey of Brown & Root, W. S. Bellows of Bellows, and S. H. Wilde of Columbia Construction. This committee met frequently to direct construction efforts and resolve problems.

The contractors first had to prepare the site for construction. They built a small, twenty-mile-long railroad line to transport heavy materials into the base area. Since they could not use local water because of its high salt content, the contractors brought in water through a distributor. A joint effort by Southwestern Bell Telephone Company, Western Union, and the Postal Telegraph Companies established an elaborate communications system for both public and military use. In addition to constructing roads and numerous runways both at the main station and at nearby locations, the consortium built a number of buildings including hangars for storing and repairing planes, utility shops, recreation and supply buildings, and an administration building.

Brown & Root also acquired the contract to build barracks at the Corpus Christi base. As the prime contractor, Brown & Root controlled work there and subcontractors solicited work from it. For example, Alvin Wirtz, who was still undersecretary of the interior, wrote George Brown with a favor regarding the naval housing job. "I have been trying to get the architectural work on this project for my brother's firm," Wirtz wrote, "but so far have not been able to find the key to the situation. Can you tell me anything about it?"[25] It is not certain if Wirtz's brother got the job, but his letter suggests how politics shaped federal contracting and jobs.

The consortium's employment peaked at more than 9,300 in January, 1941, and union membership was not encouraged. Labor disputes on wartime government projects were not uncommon, and Brown & Root—imbued with Herman Brown's staunch antiunionism—soon confronted serious charges of unfair labor practices as defined by the National Labor Relations Act (1935). In 1942, two employees filed a complaint with the National Labor Relations Board (NLRB) claiming that company officials "urged, persuaded and warned their employers from becoming or remaining members of a union or engaging in concerted activities." After a trial examiner found in favor of the employees, the NLRB dismissed the case on a technicality. (In the future, Brown & Root faced similar challenges to its policies toward labor, and most would be decided in labor's favor.)[26]

On March 12, 1941, only eight months after the navy contracted for the base, Secretary of the Navy Frank Knox dedicated the Corpus Christi Naval Air Station.[27] By then, the contractors had completed approximately 60 percent of the base; the navy scheduled training for the first cadets to extend from April 1 to July 1, 1941. Designed to turn out 300 qualified pilots per month with a total of 2,500 cadets in training, the Corpus Christi Naval Air Station employed a staff of 12,000, including 800 officers.[28]

The high point for Brown & Root's management came on January 8, 1943, when the government awarded the Army-Navy Production Award, also referred to as the Army-Navy "E," to the joint venture. This award went to "those plants and organizations which showed excellence in producing ships, weapons, and equipment for the Navy." The Corpus Christi base was one of sixteen projects that received the "E" that year out of six hundred public works projects under consideration.[29] The navy presented the award only twenty-eight months after work had begun. Brown & Root, Bellows, and Columbia completed the primary phase of construction nearly eight months ahead of schedule.[30] At the presentation ceremony, Captain L. N. Moeller, USN, noted that the station's cost reached $75 million (it escalated another $50 million during the next several years).

With the completion of a major defense project ahead of schedule and at a reasonable price with no significant interruption or problems, Brown & Root now had an important notch on its belt. The Corpus Christi project brought Brown & Root's name to the attention of federal officials interested in identifying competent contractors capable of meeting pressing deadlines for a variety of military production jobs. No doubt, the Browns' close ties to Lyndon Johnson and Albert Thomas helped them gain initial consideration for the project. But Brown & Root proved to the federal government what it had shown county road commissioners throughout the 1920s; it could deliver the goods. In the future, government officials would not require that Brown & Root bring in a more prominent partner to win a major contract; its work on the Corpus Christi Naval Station earned the company a place on the government's list of trusted defense contractors.

After Albert Thomas failed to secure the naval air station for the Houston area, he continued lobbying the navy and other military departments for projects in his district. Corpus Christi won the naval air base despite Thomas's efforts to bring the base to Houston, but he later assisted the Browns in acquiring a small military project from an overburdened local contractor. This minor project quickly evolved into a shipbuilding operation that dwarfed in size any work the Browns had ever undertaken.

During March of 1941, the same month as the dedication of the Corpus Christi Naval Air Station, Platzer Boat Works of Houston received a $2,552,000 contract from the navy to build four Patrol Craft (PC), or subchasers.[31] The navy originally designed the experimental 173-foot PCs for light escort duty as well as for submarine search-and-destroy missions; they were used later for mine sweeping. Ultimately, Platzer's ambitions outran its capabilities. Local bankers initially agreed to finance the operation if the navy awarded Platzer the PC contracts. But after Platzer had secured a contract to construct four PCs, the bankers withdrew their financial commitment, leaving the company with an uncertain future.[32]

Thomas informed Admiral Sam Robertson, head of the navy's Bureau of Ships, that Herman and George Brown of Houston's Brown & Root could assist Platzer. A navy official telephoned George Brown and asked him whether or not Brown & Root had an interest in building ships. Brown recalled that the navy "didn't want to cancel him [Platzer] out, didn't want to hurt him. They wanted to know if we would take the contract if they would get him to assign it to us. And after much conversation with my brother and L. T. Bolin and others—we'd never built a ship in our life—we knew what the contract price was, so we decided that we could do it."[33] The Browns agreed to work with Platzer only if they financed the project and retained a controlling interest. Without much choice, Platzer agreed. The Brown brothers entered the joint venture on their own and not as representatives of Brown & Root.

Platzer's plan had been problematical from the start, and the Browns made significant changes early in their involvement. Harry Austin, a nephew of Herman and George who worked for them at the Platzer shipyard, recalled that "he [Platzer] was going to build [the ships] at his little place in Harrisburg on the inside of the bridge. . . . And they would make them [the PCs] in pieces and float them underneath that and then put them back together . . . because the bridge was so low that that was the only way they could do them."[34] The Browns decided to move the shipyard operations to another location farther down the Ship Channel, outside of the bridge.[35] They asked L. T. Bolin, the Brown Shipbuilding foreman, to help locate a better site for the yard. Bolin traveled by motorboat farther down the Ship Channel—past cow pastures, cotton warehouses, and oil refineries—to the point where the channel met Greens Bayou. This site had five thousand feet of usable land on Greens Bayou and three thousand feet of deep water frontage on the Ship Channel side. The Greens Bayou frontage was shallow but could be dredged. From that isolated, undeveloped, triangular-shaped piece of land, the Browns could build the ships and launch them directly into the Houston Ship Channel unobstructed by bridges.

The Brown brothers agreed that Bolin had found the best location, and they learned that the land had a single owner. The brothers contacted the owner, who agreed to sell the property so long as the brothers purchased all his 156 acres, which was more than was needed. As negotiations for the purchase went forward, workers began clearing the land.[36] On August 7, 1941, the Browns purchased 156 acres for one thousand dollars per acre on the triangular-shaped site, and workers immediately began constructing the shipyard with navy representatives overseeing construction.[37]

The new location ultimately proved excellent for building and launching vessels, but its remoteness initially posed serious problems. The nearest highway was three miles away, and the only road passable in wet weather was a mile and a half away. Brown & Root built new roads in order to transport equipment to the site. To build the shipyard as inexpensively as possible, the crew used local pine trees for pilings on the Greens Bayou side of the yard in order to support the weight of the vessels under construction.

Much of the heavy equipment and machine tools needed at the site was salvaged from a variety of other ventures. The shipyard made use of "what other people would call junk," recalled one shipyard worker, "and it was a fast and cheap operation at first."[38] Railroad tracks needed for transporting heavy materials and equipment into the isolated area came from an abandoned Dallas-Corsicana interurban line, the rails of which were ripped up and used to connect the shipyard with the Houston Beltway. The Browns purchased an idle sawmill from Conroe, fifty miles from the location, and rebuilt it as a warehouse and machine shop on the site. They bought machine tools previously used in a Wichita Falls oil field equipment factory, and shipped in a large lathe from Seattle. They also obtained a used plate roller from the East Coast.

The organization of the shipbuilding endeavor had to be catch-as-catch-can. In the midst of a national emergency, with industrial goods in short supply, the Browns hustled for resources and personnel. George Brown later captured the tone of the times: "We had never seen a ship built, and we decided we could build one, even though we did not know a bow from a stern."[39] Even their technical specialists on ship design had to learn as they went along. After Albert Sheppard had completed his work in helping design the shipbuilding facilities at Greens Bayou, the chief engineer, Pat Anderson, asked him if he would stay with the company and help build ships. Sheppard replied that "I didn't know much about it. He answered 'Neither do I, we can learn together.' So we became ship builders."[40] The considerable risk involved in such a situation helped convince the Browns to organize this activity outside of Brown & Root, so Brown Shipbuilding came into being to own and manage this new venture.

Constructing the first four ships proved difficult, but the early start on building the shipyard allowed the first keel to be laid on August 14. The inexperienced partners

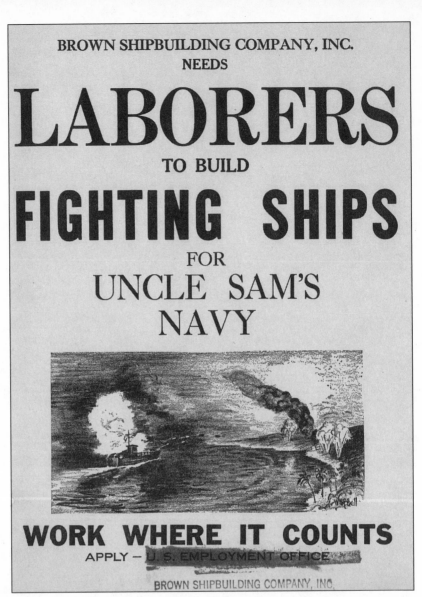

Help Wanted poster for Brown Shipbuilding. *Courtesy Brown & Root.*

found a dearth of experienced shipbuilders in the Houston area. Although Houston was the third largest port in the United States, it did not have a significant shipyard. The Brown brothers employed some experienced shipbuilding foremen from shipyards in the Beaumont–Port Arthur area. They also relied on workers from oil fields, farms, and various shops to learn to weld and otherwise assemble ships; only 5 percent of their original employees had shipbuilding experience.

Work began almost as soon as the Browns purchased the land. With the first four PC subchasers under construction, the Browns decided that George should travel to

Washington and attempt to obtain additional naval ship contracts. In September, 1941, the navy awarded the firm, still operating under the name Platzer Shipbuilding Company, a $5 million contract for an additional eight PC subchasers. At this point Platzer became uneasy about the rapid growth of the shipyard and decided it was time to leave the business. The Browns bought out the rest of Platzer's contract and began operating as the Brown Shipbuilding Company on October 1, 1941 (the firm incorporated in May, 1944), with L. T. Bolin as the general manager.[41]

But as the first two hulls neared completion in January, 1942, an unexpected problem threatened the work. A dredge deepening Greens Bayou uncovered a strata of quicksand near the bank of the bayou. The sand flowed into the bayou, causing the foundation underneath the two hulls to drop out of place. One hull fell eighteen inches and its bow swung three feet out of line. "It looked as if the first two hulls were headed for a premature launching," the company reported later, "which would cause them to wind up as a twisted wreck of unsalvageable steel on the bottom of the bayou."[42] Herman Brown rushed to the scene to view the predicament. He and his brother had invested nearly $1 million of their own funds in the business instead of applying for government financing. Now the dangers of using their own money became quite clear. Herman reacted philosophically: "This is just one of the risks you have to expect in the contracting business." The shipbuilding company called in workers from other Brown & Root jobs in South Texas to drive new piling and shore up the bank.[43]

Recognizing the growing need for ships and always ready to promote government expenditures in his district, Albert Thomas encouraged the navy to look with favor on the expansion of the Browns' shipbuilding operation. "We are hopeful that additional orders will be forthcoming," he said, "[and] we have good reasons for believing the shipyard will be a permanent industry, continuing its operations after the war."[44] Another major shipbuilder on the Houston Ship Channel, the Houston Shipbuilding Company, also received substantial navy contracts to build cargo vessels rather than defensive or offensive ships.

World War II made shipbuilding Houston's largest industry by 1943. Of the approximately $1 billion in war contracts let in the Houston area, about $600 million went to shipbuilding. In 1943 the Houston Shipbuilding Corporation was Houston's largest shipbuilder with forty thousand workers and an estimated weekly payroll of $2 million. At that time, Brown Shipbuilding employed more than fifteen thousand and had a weekly payroll approaching $700,000. Other shipyards in the Beaumont and Port Arthur area also received large contracts for navy vessels.[45]

The Browns hired many of their former electrical engineers, electrical mechanics, and craftsmen from the Marshall Ford Dam to work at the shipyard, but they needed more workers. Regarding new employees, Brown Shipbuilding's Operating Committees agreed unanimously that in view of the war emergency, the company would hire both nonunion and union workers. The committee agreed further that the company would "meet, or exceed where practicable, wages and working conditions prescribed by the Gulf Coast Stabilization Committee and recognize that the primary principal for retention and advancement of employees was the individual merit of each."[46] Brown Shipbuilding organized an Industrial Relations Department to deal with all employee-related issues. For the first two years of operation, it faced few problems or opposition from labor organizations.[47]

On February 27, 1942, the company finished its first ship, the 165-foot PC-565. This ship was christened by Jacqueline Perry, the fourteen-year-old daughter of Commander E. B. Perry, supervisor of naval shipbuilding on the Gulf Coast. Reporters and photographers yelled for her to smash the bottle, but in the confusion, she did not hit the ship hard enough to break the bottle. As the ship slid down the ways, a workman grabbed the bottle and smashed it against the ship.

The shipyard conducted sideways launchings because Greens Bayou was too narrow for traditional launchings. Platzer's most unique contribution to the shipbuilding effort was using bananas instead of grease to lubricate the "ways," the large pieces of timber on which the finished ships slid down into the water. The bananas did cause some problems for the shipyard. "It got a little smelly after a while," recalled one worker, " . . . [and] we decided we couldn't go on with this banana stuff, we had to have some grease."[48]

When reporter Lloyd Gregory asked George Brown what he was thinking as he watched the PC-565 slide down the way, Brown grinned and replied, "I was praying she wouldn't turn wrong side up!" A worker voiced similar apprehensions, but in a more positive light. After the launching sent a virtual tidal wave high over the ship's side, he noted that "if she can stand that much it'll take more than a torpedo to sink'er."[49] Despite this somewhat inauspicious beginning, PC-565 became the first PC of any yard to sink a German submarine.[50] In some ways this episode epitomized the frantic, early rush to turn out ships vital to the war effort. Brown Shipbuilding improvised as it went along, relying heavily on the energies and commitment of its employees. From this point on the key was simple: keep the ships sliding down into the Ship Channel and out to the Atlantic and Pacific fronts.

The shipyard held a brief fifteen-minute ceremony for the christening of the third subchaser, but it was a particularly meaningful moment for the Browns. Their mother, Lucy King Brown, attended, and fourteen-year-old Nancy Brown, George Brown's oldest daughter, christened the ship.[51] The *Houston Post* reported the event: "Tall, lean Mr. Brown beamed while his pretty pink-clad, yellow-haired 14-year-old daughter, Nancy Nelson cried, 'I christen thee U.S. PC 567,' and smashed a bottle of champagne on the sharp bow of the sleek gray craft. The instant she did this the props were released, and to the music of a band, the scream of whistles, and the cheers of about 2,000 people, the ship plunged broadside down the skids and into the water with a loud booming splash."[52]

L. T. Bolin, chairman of Brown Shipbuilding's operating committee, captured the prevailing sentiment: "The men who have built this ship have worked hard and would enjoy a little time off, but the fall of Bataan has once more made us realize the inadequacy of our efforts, and try as we may, we cannot equal the sacrifice made by the men who defended Bataan."[53] Within half an hour of the launching, the crowd dispersed and the nine-month-old shipyard appeared as if no special event had occurred at all. Such ceremonies gave only the briefest of respite to the shipyard's workers, who knew that the nation needed ships to be produced as quickly as possible. They worked hard, driven by a sense of national purpose and shared sacrifice.

Brown Shipbuilding's importance to the navy increased substantially early in 1942. In January, the navy launched its Destroyer Escort (DE) program and immediately contracted with Brown Shipbuilding for eighteen DEs at $3.3 million each. The navy designed the DEs to battle the U-boats that plagued Allied ships in the Atlantic. At about three hundred feet long, DEs averaged seventy to eighty feet shorter in length than destroyers (DDs), but the navy intended DEs to function as DDs in terms of escort duties and firepower. Their highly effective, secret radar systems made DEs vital components of naval battalions.[54] (See table 4.1.)

The shipyard expanded to build destroyer escorts quickly and efficiently. Along with the new contract for the more sophisticated DEs, the navy also authorized the Browns to build a new $6 million shipyard, to be named "Yard B" adjoining the existing "Yard A." J. E. Walters, former superintendent of the Marshall Ford Dam project, supervised construction of the new yard. Upon its completion, Walters became manager in charge of all ship construction at Yard B. M. P. Anderson, Brown & Root's chief engineer at Marshall Ford, designed the new yard and became the assistant works

TABLE 4.1. SHIPS BUILT AT BROWN SHIPBUILDING

Type	Number Constructed	First Keel Laid	Last Completed
PC	12	Aug. 14, 1941	Apr. 13, 1943
LCI	32	July 27, 1942	Feb. 13, 1943
DE	61	July 15, 1942	July 31, 1944
LSM	214	Feb. 15, 1944	Nov. 1, 1945
LSM(R)	36	Mar. 24, 1945	Oct. 20, 1945
ARS(D)	4	Aug. 25, 1945	Dec. 12, 1945
Total	359		

Source: Compiled from Brown Shipbuilding folder, Herman and George R. Brown Archives, Houston, Texas.

Herman, Lucy King, and George Brown surrounded by the Brown Band, Dec. 21, 1942.
Courtesy Brown Family Archives.

manager.[55] Unlike Yard A, which the Brown brothers owned, Yard B was owned by the U.S. government and was leased to Brown Shipbuilding for one dollar per year.

The navy soon contracted with Brown Shipbuilding to construct a third type of ship. During the spring of 1942, the Browns received orders for thirty-two LCIs, landing craft infantries, at a total cost of $17 million. The navy needed LCIs to land infantry on the coast of North Africa, an action to be followed by the storming of Europe. The works laid the keel for the first LCI on July 27, 1942, and workers constructed these ships along with the PCs at the first yard while workers at Yard B built DEs.[56]

The DE was the most complicated ship built by Brown Shipbuilding, and the shipyard had some difficulty adopting the new construction methods necessary to build it. According to the shipyard magazine, building the first DE became a "nightmare." Construction work for the ship fell four and one-half months behind schedule. Eventually, DE-238 required ten and one-half months of work from keel laying to completion.

Brown Shipbuilding launched its first destroyer escort, the USS *Stewart,* on November 22, 1942. One month later, on December 21, and a little more than a year after the Browns took over the shipping operation, Frank Knox, secretary of the navy, presented the "E" pennant to Brown Shipbuilding. Asa Platzer, of the original Platzer Shipbuilding Company, accepted the award. Those in attendance included Secretary of Commerce and fellow Houstonian Jesse H. Jones, House Speaker Sam Rayburn, Texas governor Coke Stevenson, and Lyndon Johnson, along with numerous other political and military figures. This "E," like the one awarded to Brown, Bellows, and Columbia for the Corpus Christi Naval Air Station, signified the determined effort of the Brown brothers and their employees to develop rapidly a shipyard capable of producing ships vital to the Allied war effort. To commemorate the "E," Brownship launched seven ships the day Frank Knox presented the award; five LCIs went down the ways simultaneously. Later in the day workers launched two DEs.[57]

Prior to receiving the award, Herman Brown, who rarely spoke to the press, discussed the success of the shipyard, which had opened only eighteen months earlier with few experienced shipbuilders. Herman said, "Both my brother George and I are proud of the award, but we feel that most of the credit goes to our loyal employees."[58] To another reporter, Herman Brown said: "[The employees'] achievement proves green, inexperienced Americans can do any job if they want to. Less than 5 per cent of the shipyard employees knew anything about ship construction at the outset. Car salesmen, undertakers, school teachers, waitresses, farmers and those from almost every line of work have built these ships. They regard the army-navy E not only as a reward but as a symbol of greater incentive to even greater production."[59]

Thirty thousand persons, including most of the eighteen thousand employees of the shipyards, attended the ceremony. Secretary of the Navy Knox told the onlookers, "I wish to tender my congratulations as Secretary of the Navy to the Brown brothers and to the men who work for them. You may not know it, but you are performing a miracle here. The like of it has never been seen before." One month after this ceremony, *Time* magazine noted that Brown Shipbuilding was "one of the most remarkable shipyards in the whole U.S.—yet most citizens have never heard of it."[60]

The navy renewed the "E" six months later. Admiral C. C. Bloch, chairman of the Navy Board for Production Awards, characterized the renewal: "In the first instance it was difficult to win the Army-Navy 'E' and by meriting a renewal, the management and employees have indicated their solid determination and ability to support our fighting forces by supplying the equipment which is necessary for ultimate victory."[61] The navy signified renewals by sending a new "E" Pennant to Brown Shipbuilding with one star per renewal.[62] Brown Shipbuilding won three more renewal stars.

Despite some difficulties in learning to build its first DE, the company's workers built the following ships much more quickly.[63] The shipyard's success prompted the navy to increase its DE order to fifty-six for a total contract amount of $200 million. Albert Thomas announced the award in Washington on June 28, 1943: "I am told by

"E" Ceremony: Frank Knox, Sam Rayburn, Jesse Jones, George R. Brown,
Maj. Gen. Richard Donovan. *Courtesy Brown & Root.*

navy officials that they regard the Brown Shipbuilding Company as one of the best
and most efficient yards in the country."[64] During October, 1943, Brown Shipbuilding received positive publicity when it delivered eight DEs in a single month, a feat
that tied the highest monthly output by any yard. Navy Secretary Knox recognized
the company's performance during December, 1943, when he noted that because of
its efforts in the DE program, Brown Shipbuilding would be included on his "honor
roll" of five privately owned shipyards that constructed DEs. The Consolidated Shipbuilding Company of Orange was the only other Texas yard on the honor roll.

The company's success in building LCIs brought it a contract for another landing
craft vessel, the landing ships medium. LSMs transported heavy equipment such as
tanks over long distances. The navy used LSMs built at Brown Shipbuilding in land-
ings at Leyte, Luzon, Iwo Jima, and Okinawa. In announcing the LSM contract in
late September, 1943, Congressman Thomas noted that the initial contract called for
130 LSMs for $65 million, or $500,000 each. The navy simultaneously canceled a
recent order for thirty-two additional Destroyer Escorts. This cancellation reflected a
change in war strategy. In the Pacific, military planners determined that landing crafts
were in much higher demand than destroyer escorts.[65]

Workers turned out the LSMs at an incredible rate. The navy first requested ten LSMs in May before raising its request to fourteen. The shipyard reverted to a seven-day-a-week schedule and delivered twenty LSMs in May. Altogether, the shipyard built 202 LSMs in fifteen months. The construction of each vessel required approximately 288,000 worker-hours and an average of fifty-two days to complete.

By late 1944, the navy decided that it needed a ship capable of launching multiple rockets at enemy positions in the Pacific. Initially, the navy requested that shipbuilders convert existing LCIs into "rocket ships." In early January, 1945, Brown Shipbuilding's LCI-345 became the first LCI in the Central Pacific to be outfitted with rocket launchers. While these vessels performed reasonably well, they were essentially makeshift. On February 3, 1945, the navy requested that Brown Shipbuilding begin building a new generation of ships "built from the drawing board up."[66] The company launched the first of these new ships on April 21, 1945. Later in the year, it launched salvage ships, its sixth type of ship, to clear harbor wreckage along the coasts all over the globe. By mid-July, 1945, Brown Shipbuilding had built 327 warships, with more on the way.[67]

Brown Shipbuilding's work constructing ships offered a chronicle of major developments during the war. The company began building PCs needed to destroy and drive away enemy submarines lurking off the Atlantic coast. Next, Brownship constructed infantry landing craft for the planned invasion of North Africa and other invasions of Europe. Then came the DEs, which escorted larger warships and hunted submarines moving farther eastward into the mid-Atlantic that remained a threat to oceangoing commerce and warships. Next came LSMs to transport tanks into the South Pacific for the war against Japan. Near the end of the war, the company built ships designed specifically for firing rockets to "soften up" land-based enemy positions in the South Pacific and in Japan preceding planned Allied invasions.[68] The company and its workers thus had a ringside seat from which to observe the evolution of naval warfare. They took great pride in contributing the many types of ships needed to secure victory on two fronts.

At its peak in October, 1943, Brown Shipbuilding employed twenty-three thousand workers at its two shipyards. With such numbers, labor problems inevitably emerged after the waning of the patriotic fervor of the early years of the war. Officials in Washington voiced concerns about possible labor problems in Houston early in the war. The general director of shipping of the Office of Production Management, Harris Robson, told Albert Thomas in early December, 1941, that he had heard that Houston had unsatisfactory labor conditions that might inhibit delivery of war-related supplies. Thomas sought to soothe Robson's worries: "You can be assured there will be no further labor strikes or stoppages of work in the Houston area. The boys are one hundred percent behind the President and the war program. We have thousands of water-front workers who can expertly handle ships sent to the Port of Houston."[69]

As Brown Shipbuilding grew, labor problems increased as well. After the opening of Yard B, a variety of labor-related issues demanded an increasing amount of management's time. The increased workload associated with the DE program centered at Yard B and required Brownship to add a substantial number of new workers. By early 1943, Yard A employed 3,300 workers while Yard B employed 7,511; the number of Yard B increased substantially over the next few months to almost 15,000. When the lack of local housing created a severe housing shortage, George Brown enlisted the help of Albert Thomas to prod the government into providing additional housing. In early March, 1942, Brown wrote Harry D. Knowlton, regional defense chief, noting the "shortage of housing for the shipbuilders in the lower wage bracket." Despite plans to increase the shipbuilding work force by between 4,000 and 5,000 within the next ninety days, the navy planned to build only 800 new houses in the same time period. George Brown personally traveled to Washington in the summer of 1943 and received assurances from the navy that materials would be made available for the construction of 3,800 family housing units to be built by private builders in Houston.[70]

As Albert Thomas worked to increase the available housing for the thousands of defense workers employed at the Dickson Gun Plant, Reed Transmission Company, the Houston Shipbuilding Corporation, and the Brown Shipbuilding Company, the navy studied labor problems at Brown Shipbuilding and other production facilities. The War Manpower Commission produced a report titled "Labor Market Survey Report on the Houston, Texas Area." The report noted that the Browns' company suffered from a high employee turnover rate and a high absenteeism rate. "Turnover," the report stated, "has lately become a serious problem at both yards." The high turnover and absenteeism rate resulted from losing workers to active military service, "long hours (10 hours per day), excessive commuting distances, lack of nearby housing, and poor lunchroom facilities." The report also noted "an unnecessarily severe policy of refusing to rehire any worker who [was] discharged by any foreman or official of the company" and stated that a "failure of the firm to attempt to hold workers who express a desire to quit also is a contributing factor."[71]

Brown Shipbuilding had a straightforward explanation for the poor absentee and separation rates. An article titled "Slugger or Slacker" in the company newspaper reflected Herman Brown's views: "'Slacker' is a rough word . . . we still think it's exactly the right word to apply to some of the men working here in the yards. No, they aren't shirking service in the armed forces, and they aren't draft dodgers; they're WORK-dodgers. And any man who imagines that our war effort doesn't call for the best he's got, every hour of the day he spends at the yards, is a slacker. . . . Look yourself over carefully—are you a producer, or a slacker?"[72] Herman Brown had always demanded intense loyalty from his workers in return for secure employment, and it was not surprising that he held strong views on the obligations of his work-ers—and himself—during war.

Labor was a particularly vital commodity during the war, and employers suspended some traditional employment practices in order to utilize as many persons as possible. At Brown Shipbuilding, Yard A followed traditional employment policies by restricting women to clerical and technical occupations and nonwhites to unskilled jobs. According to the War Manpower Commission, Yard B had "an acute shortage of electricians" in 1943 and "a good many of the other skilled occupations might be characterized as inadequate, but the inadequacy of such workers has not yet interfered materially with production." Women at Yard B held a variety of positions previously filled only by men, including expediters, burners, welders, and loftsmen. Blacks, in particular, did not achieve the same level of assimilation into skilled positions as women, although the report noted that "tentative plans envision special crews of nonwhites in selected semi-skilled occupations."[73]

When in need of additional workers, Brown Shipbuilding often hired persons waiting outside the plant gate or employed workers from other shipyards or from Brown & Root. In some instances, a friend, businessman, or politician suggested that the firm hire a particular person. In early October of 1942, Lyndon Johnson wrote George Brown and told him that Thomas Corcoran had asked him recently if he could find employment for the brother-in-law of a mutual friend. Johnson wrote to George Brown, "I am writing you this long letter, and the substance of it is that I have already employed [this] brother-in-law for you. I am not your employment manager. You did not authorize me to do it, but I did it anyway, and I hope you will confirm my action."[74]

George Brown later asked Lyndon Johnson for personnel help of a sort as well. Lieutenant Robert McKinney worked in the navy's Office of Procurement of Material, and his official involvement in matters relating to expediting ship production brought him into contact with George Brown. After the war ended, McKinney applied for an early dismissal in order to resume his work at a New York brokerage firm. Meeting resistance from his immediate superior, McKinney sought a dismissal from a higher official, who rerouted the request back to McKinney's immediate supervisor. Angered at McKinney's insubordination, the supervisor demoted McKinney and refused to dismiss him from the navy. After George Brown heard this story, he introduced McKinney to Johnson, and McKinney told Johnson his story. Two days later, McKinney received a call at 3:00 A.M. from Johnson, who said, "this is your old friend Lyndon. I just wanted you to know that you're now out of the Navy."[75]

The special treatment a handful of wartime workers received was not representative of business-labor relations during the war. When nine labor unions separately representing machinists, iron workers, operating engineers, plumbers, electrical workers, and other groups either telephoned or wrote letters to Brown Shipbuilding requesting recognition as the bargaining representative for their workers, Brown Shipbuilding either declined or refused to accept the letters. Subsequently, these and other unions filed complaints with the NLRB, which consolidated them into one hearing held in May, 1944. At this hearing, Alvin Wirtz represented the Browns. He argued that the NLRB had no jursidiction in the matter. Company management took the view that the NLRB did not have jurisdiction over Brown Shipbuilding because it was "building combat ships for the United States Navy exclusively, with funds provided by the United States Navy, and the finished ships delivered directly to the fighting forces of that Service."[76] Wirtz cited the NLRB's previous decision regarding Brown & Root's Corpus Christi project. In that labor case the NLRB agreed that it did not have jurisdiction. At Brown Shipbuilding, the NLRB replied, its jurisdiction was certain. The NLRB ordered elections.[77]

In August, 1944, workers at Brown Shipbuilding voted in an election mandated by the NLRB the previous June to determine if various unions should be designated as the exclusive bargaining agencies for workers of six crafts—machinists, electricians, sheet metal workers, structural iron workers, plumbers and steam fitters, and operating engineers. Soon thereafter, the NLRB ordered Brown Shipbuilding to hold similar elections for sign painters, railroad trainmen, welders, boilermakers, and painters.[78] In a message to employees regarding elections, Herman Brown stated that "you have a right to express your opinion in this important matter, and should do so by all means." Never one to withhold his opinion on a subject for which he held strong views, he also told his employees: "It has never been necessary for any employee in any enterprise with which I have been associated to designate a union as his exclusive bargaining agency in order to get a square deal. I have never required an employee to join a union, or stay out of a union, in order to get a job, hold a job or get a promotion."[79] Herman Brown would not readily retreat from his long-held commitment to an open shop.

The newly elected union representatives submitted contracts to Brown Shipbuilding providing for a closed-shop agreement. The agreement, if signed, would have required the company to compel its nonunion workers to join an appropriate union or face termination. Brown Shipbuilding responded that under the demanding shipbuilding schedule contracted by the U.S. Navy, it could not comply with these agreements and continue to produce ships on schedule. The union representatives then took their battle to an additional forum, receiving certification from the War Labor Board "that a labor dispute existed on our job that would seriously affect the war effort."[80]

Before the War Labor Board issued any directives regarding the labor problems, the unions set up pickets at the Greens Bayou site. For the most part, workers crossed the picket lines. To quiet labor discontent, Brown Shipbuilding's management later led a well-publicized drive to raise salaries, which were dictated by government scales covering all employees.[81] George Brown recalled that "the Navy saw what we were doing, it could not have cared less. We never had any trouble. We listened to what the unions had to say, but we went on building ships. At one time we turned out a destroyer escort a week—four a month—and the Navy was satisfied."[82]

In comments made after the war, Herman Brown discussed what he considered the lessons learned in managing the workforce at Brown Shipbuilding. This was a far larger workforce than the Browns had ever managed on a Brown & Root project. Not surprisingly, the lessons Herman Brown took away from this experience reinforced those he had learned previously as head of road building gangs in Central Texas. He noted that "during the war my brother and I took 25,000 housewives, farm boys, and four F's and built 350 fighting ships for the Navy. And we had Army and Navy 'E' pennants all over the place."[83] He felt that one key to Brown Shipbuilding's success in meeting its strict deadlines was the flexibility provided by a nonunion workforce. Unrestrained by union rules concerning craft divisions or seniority, the company could easily shift workers around as needed: "Considerable myth has grown up about the years of training required to become a construction industry craftsman. . . . Many, many of our men can qualify in several different crafts on anybody's job."[84] In Brown's opinion, unions had not been needed, since the workers and their bosses constituted a patriotic "family" that shared the common goal of winning the war.

Many of his workers disagreed, and they voiced their preference for union representation in NLRB-mandated elections. But Brown Shipbuilding blunted this impulse to organize for the duration of the war, as complaints to the NLRB disclosed. Workers who had voted to establish a union no doubt saw this as evidence of the company's willingness to ignore the law in its resistance to unionization. But Herman Brown read a much different lesson in all of this. The fact that his shipyards had taken thousands of untrained workers and built ships efficiently underscored the value of the open shop and the correctness of his own paternal view of labor. The tension between this view and that of workers seeking to organize was somewhat muted by the patriotic fervor of both owner and worker.

A big part of the high morale at Brown Shipbuilding was the knowledge that the ships launched into the Houston Ship Channel at Greens Bayou played crucial roles in the war effort around the world. A company newsletter, the *Brown Victory Dispatch,* regularly featured the exploits of Brownship's vessels to remind workers of their important part in the war effort and to motivate them to work even harder. Brown-built ships took part in well-publicized naval actions that brought a sense of pride to the workers who had labored over them months or years earlier.

Ships from Greens Bayou first saw action in the Atlantic. Five DEs built there— the USS *Haverfield,* USS *Swenning,* USS *Willis,* USS *Janssen,* and USS *Wilhoite*—

Destroyer escort number 410 is launched into Green's Bayou,
January 11, 1944. *Courtesy Brown & Root.*

received presidential citations. The citation commended those and other ships operating with the USS *Bogue,* an escort carrier, in "forcing the complete withdrawal of enemy submarines from supply routes [in the Atlantic] essential to the maintenance of established military supremacy."[85] These successful missions against German submarines took place between 1943 and 1944 and helped keep supply lines open between North America and Great Britain during the war.

Brown Shipbuilding played an equally important role in building ships for naval operations in the Pacific. In early June, 1945, the U.S. Navy granted the company a fifty-eight-hour workweek in order to produce rocket ships for attacks on Japanese beaches and shore emplacements to "soften up" the enemy in preparation for the much-publicized impending Allied invasion of Japan. A Brown Shipbuilding Company newsletter described the purposes of a rocket ship: "After the aerial pounding and shelling they lead the landing ships in and keep punching, and the Japs who try to crawl out of their caves or emplacements to oppose the landing are blown to pieces".[86]

This newsletter regularly included comments from commanders of various vessels constructed at Brown Shipbuilding on the superior quality of the ships. Other articles chronicled the "lives" of Brown-built ships that sank German submarines or landed infantry on Iwo Jima. Yet not all articles described a victorious Brown-built ship. One article followed in detail the destruction of the USS *Samuel B. Roberts,* which the Japanese sank along with two destroyers on October 25, 1944, during a major sea battle in the Philippines. In language typically reserved for the exploits of courageous fighting men, the *Brown Victory Dispatch* described the scene: "She died game, charging an overwhelmingly superior force of Japanese warships which had for the moment eluded the main forces of our fleet and was threatening the whole beach-head of Leyte."[87]

Perhaps Brown Shipbuilding's proudest moment came when management and employees learned that two of its destroyer escorts, the USS *Charles S. Tabberer* and the USS *Melvin R. Nawman,* survived a 130-mile gale that sank three other ships, including two destroyers, off the coast of Luzon. One DE later rescued fifty-five survivors from the destroyers. The U.S. Bureau of Ships commended Brown Shipbuilding workers after the event.[88]

In one of the last naval battles in the Atlantic, the USS *Moberly PF-63,* built by Brown Shipbuilding and placed in commission on December 11, 1944, played a decisive role. On May 5, 1945, only four miles off Point Judith, Rhode Island, a German U-boat sank an American collier. Four of the ship's forty-six crew members perished while all others were rescued by another American ship in the area that heard the collier's SOS. The PF *Moberly,* operating as a Coast Guard vessel, heard the SOS and hurried to the scene along with several other subchasers. After making radio contact, the American ships launched two sets of depth charge attacks, which were followed by wreckage and dead fish floating to the surface. Shortly, the *Moberly* sonar indicated the sub was again on the move. After a series of additional attacks by the *Moberly* and other ships, the sub was destroyed and "flotsam, jetsam, oil and water geysers," as well as material of Germanic origin came to the surface. The *Brown Victory Dispatch* proudly reported this action, noted as the "last Atlantic battle," as a patriotic reminder that Brown Shipbuilding made a difference.[89]

All of the workers at Brown Shipbuilding, including the Browns, sacrificed for the war effort. Herman Brown paid a heavy toll in travel. He regularly lived in Houston at the Lamar Hotel and then traveled home to Austin for weekends with his family. Early in the war, the Browns acquired a surplus DC-3, the first plane owned by Brown & Root, to ease the strain of the travel required by their war work.

The war years brought a more permanent change in the lives of Margarett and Herman Brown. In 1943 they decided to adopt a young girl who had been separated from her family and was living temporarily with a foster family. At the time, Louisa Stude was seven years old, and she entered a new world complete with a substantial home in Austin, air travel for the first time in her life, and new parents, who were then in their fifties. Once she understood that the change was permanent, she spoke

Herman Brown family: Herman, Margarett, Louisa, "Senorita," Mike, James Head and Louise Head, Christmas Day, 1945. *Courtesy Brown Family Archives.*

up and informed the Browns that she also had a younger brother, Mike Stude, who had been living in an orphanage in Austin. The Browns arranged for Mike to join Louisa, and they quite suddenly had a new family. The young Louisa quickly became accustomed to the realities of her new life with her new parents, whom she called Uncle Herman and Aunt Maggie. She found "the excitement of a beautiful home and being given so much freedom, rollerskating around on these beautiful parquet floors." She also was taken by "the warmth of Uncle Herman. . . . Aunt Maggie was giving us our lessons in manners and so forth and he was just sort of having fun." All in all, it was a "wonderful atmosphere of stimulation. One of the things I remember about Uncle Herman was his conviction. I mean everything he believed in, he just pounded on the table. . . . He never had any doubts about the view, the fact that he was right, and it was always a very adult conversation, very much in the area of politics most of the time." Though three years younger, her brother, Mike, also recalls quickly developing a sort of hero worship of Uncle Herman, whose heavy travel schedule limited his time at home, making his presence there a time of celebration.[90] Thus in many ways, World War II was a time of far-reaching changes for the Browns.

In early February, 1946, the U.S. Navy announced it would retain only 42 of the

123 wartime shipyards. Initially, the navy intended to keep Brown Shipbuilding in service as a maintenance and ship repair yard.[91] The general manager of the Port of Houston at the time predicted that "Brown Shipyard is destined to be highly successful in peacetime operation and will prove of great benefit in development of the port." With the third highest volume of tonnage carried in the United States, Houston seemed poised to become a shipbuilding center as well. Yet despite such optimistic views of the industry's future, Brown Shipbuilding's future was by no means secure. George Brown noted that "the Navy wants to keep the yard open. It has expressed a preference to keep it open as a private industry and negotiations are now going toward this end."[92] But as the navy began canceling contracts for new ships even prior to the war's end, Brown Shipbuilding found new work more difficult to acquire. The shipyard did purchase dry dock facilities used in the process of repairing navy cruisers, a large Ford Motor Company automobile freighter, and various drilling barges.

One of the company's last official construction projects for the navy was the construction of four specialized salvage ships designed to pluck radioactive material from the ocean after proposed atomic demolitions of test ships in the Pacific.[93] Despite this project and some repair and maintenance work, shipbuilding in peacetime did not offer the excitement or opportunities of the wartime business. After building more than $500 million worth of naval vessels during World War II, Brown Shipbuilding did not continue after the war.

The Browns examined several options for the shipyard. George Brown came up with one of the most imaginative possibilities after reading an article written by George Darneille, an air force officer, on the tuna fishing industry. Brown contacted Darneille about the article and asked him about the possibility of using at least a part of the shipyard as a Gulf Coast tuna facility. Although Brown quickly discarded this idea, he subsequently hired Darneille, who became a valued associate of the Browns.[94]

The brothers did utilize some of the shipyard's facilities for other purposes. On February 15, 1946, Brown & Root announced that it would convert part of the shipyard facilities into offices for Brown & Root's new Petroleum and Chemical Division. These facilities, intended to help the construction firm diversify into the oil and chemical business, also complemented the continuing shipyard work.

The shipyard's newsletter reported that the new Petroleum and Chemical division was designed to "perform every step in the creation of process plants, from plant location studies, through engineering design, construction and operation, to completion. It will be able to perform all or any phases in completion of natural gasoline plants, recycling plants, compressor stations, dehydration plants, gas treating plants, vapor recovery plants, installations for fractionating and testing of refinery vapors, refinery units and chemical plants."[95] This new division put the Browns in an excellent position to become an important force in the construction of oil, gas, and petrochemical facilities. These industries had grown steadily after the opening of the Houston Ship Channel in 1914 and were destined to boom in postwar Houston.

Despite these efforts to adapt the shipyard to other pursuits, these facilities remained little used. The Browns decided to sell the facilities to another shipbuilder, if such a buyer could be found. In early 1949, the brothers began negotiations with officials from Todd Shipyards, a well-established firm with roots deep in the mid–nineteenth century. During World War II, Todd controlled shipyards on the East, West, and Gulf Coasts, including the Houston Shipbuilding Company, the largest such firm located in Texas, during much of World War II.

On April 28, 1949, John D. Reilly, president and chairman of Todd Shipyards, announced at the company's annual meeting that it had "purchased the major portion of the facilities of what was formerly Brown Shipyards . . . and leased [from the Browns] the underlying land for a period of twenty-years."[96] The war was over, many of the ships remained in service, but the shipyard, like much of the remaining World War II era equipment and property, took on a new peacetime role. Herman and George Brown realized that the shipyard required experienced and dedicated shipbuilders to manage it and make it prosper. Eager to return to the construction business and to rebuild Brown & Root, they sold the facility.

The Browns must have felt a touch of sadness as they sold Brown Shipbuilding. Its success had marked a turning point in their careers. Their previous work on roads and dams had prepared them for the larger scale required to build the Corpus Christi Naval Air Station. The experience from all of this work had helped them succeed at building and managing a massive shipyard whose workforce and cost far exceeded anything in their past. Along with a new scale came a new level of expertise and a new standing in the public eye. The years of shipbuilding also firmly established Houston as the headquarters of the Browns' operations. By the end of the war, the Browns were well positioned to emerge as one of the most significant builders based in the Southwest.

Part Three

The Browns in the Postwar Boom, 1945-62

Herman Brown and Mr. and Mrs. William A. "Bill" Smith in front of DC-3.
Courtesy Brown Family Archives.

CHAPTER FIVE

Diversifying into Texas Eastern

At the end of World War II Herman Brown was fifty-three; George, forty-seven. The brothers had a solid reputation in the construction industry, a well-established presence in Houston, and impressive resources from both Brown & Root and Brown Shipbuilding. As a major defense contractor, the Browns faced the distressing prospect of a sharp decline in the demand for their primary product, naval vessels. Yet at the same time, their knowledge of the defense industries alerted them to numerous promising ventures involving war surplus properties and demobilization. The Browns also entered new ventures in response to the postwar economic boom in Houston and throughout the nation. Although much of their activity involved construction-related projects undertaken within Brown & Root, the brothers also moved into a variety of related ventures organized and managed outside of the company.

The key to postwar expansion for the Browns—both at Brown & Root and in their other personal holdings—was diversification. Brown & Root expanded and diversified its construction of public works while completing an array of ambitious and profitable private sector projects, including extensive construction in the oil, gas, petrochemical, electric, and paper industries. Revenue from Brown & Root allowed the Browns to diversify their personal interests with a series of profitable investments outside of the construction industry. One of these, Texas Eastern Transmission Company, put them on the ground floor of one of the fastest-growing industries in postwar America, the shipment through long-distance pipelines of natural gas from giant fields in the Southwest to seemingly insatiable markets in the Northeast. In addition, the Browns acquired a salvage company, an independent oil producing company, and a portfolio of valuable properties. Several of these ventures took on increasing importance to the Browns in the 1950s and 1960s.

In all of these endeavors, the brothers continued their long-standing tradition of sharing equally in all of their business investments. At times the two brothers exhibited different levels of interest in the management of the various enterprises, but they always held to this principle of equal ownership. In overseeing the development of their concerns, they were greatly aided by the mobility of the DC-3 airplanes they acquired in 1942. These planes allowed the Browns to move quickly and comfortably within their far-flung empire. The brothers relied heavily on numerous loyal and effective managers who took broad responsibilities under the Browns' decentralized management system.

The end of World War II brought many unusual business opportunities for investors and entrepreneurs across the country. As the government sought to return to a peacetime economy as quickly as possible, it became, in one sense, the largest "junk" dealer in the nation's history. It owned approximately $25 billion worth of "war surplus property," which ranged from oil refineries to rubber life rafts, from petroleum pipelines to aircraft. The federal government sought to divest itself of war materials by selling as much surplus property as possible to private industry. Entrepreneurs with ready access to capital, a good eye for the resale market, and an inclination to take risks could find valuable equipment and property at bargain-basement prices.

The rush to sell war surplus properties created a giant new market where potential purchasers bid for properties in much the same way bidders traditionally sought contracts for public works. The rules guiding the bidding process for war surplus properties varied depending on which government agencies controlled the process. As in the Browns' previous bids for public works, an effective proposal was important, as was knowing the persons who occupied positions of responsibility at the appropriate government agencies. Frenzied lobbying efforts in Washington on behalf of businesses seeking to acquire particular properties marked divestment proceedings, as bidders sought an edge in a somewhat chaotic process. The Brown brothers had become adept at bidding for public works contracts in Texas before the war, and their work at the Marshall Ford Dam and on defense projects had introduced them to the world of federal contracting. As they began an interesting journey into the intensely political world of war surplus property acquisitions, they were neither novices nor hardened veterans.

In anticipation of the war's end, Congress passed the Surplus Property Act of 1944 and created the Surplus Property Administration (SPA), which later evolved into several successor agencies. The Surplus Property Act mandated the sale of the government's war surplus property for the best price to companies or individuals capable of using the equipment in a competitive rather than monopolistic fashion. Soon after the war ended, the United States began auctioning property, usually to the highest bidder. The Browns bid on a wide assortment of this surplus property and made good profits from several of their purchases.[1]

Herman and George Brown had learned earlier in their careers that used and even discarded industrial equipment could prove valuable for either continued use or resale. The Texas Railway and Equipment Company, a joint venture with Slim Dahlstrum, had always operated independently of Brown & Root. Like similar firms associated with other large contracting firms, it purchased Brown & Root's used equipment, which then would not appear as an asset on the contractor's financial statements. The salvage company could either repair or sell the equipment. Throughout the late 1920s and into the 1940s, Texas Railway and Equipment operated as a used equipment company, or a "junk" firm in the common parlance. Thus when the Browns looked to purchase surplus properties after World War II, they already had an experienced associate, Slim Dahlstrom, and a well-established company, Texas Railway Equipment.

Along with Dahlstrom, Brown & Root managers (W. A. Woolsey, J. M. Dellinger, and L. T. Bolin), and Robert Thomas, the Browns also formed a partnership known as the Dahlstrom Company. This operated in Vernon, California, in the mid-1940s as a construction equipment repair depot, shop, and storage facility for the U.S. Navy.[2]

The Brown brothers made excellent short-term investments in several war surplus properties. For the most part, the government disposed these properties quickly by offering bargains to those willing to assume the risk of finding peacetime uses for them. The Browns purchased either privately or with other businessmen an old U.S. military fort, thousands of surplus military airplanes, and two major government-financed petroleum pipelines. All proved profitable.

One unusual prospect that attracted the brothers' interest was a historic U.S. Cavalry frontier post, Fort Clark. Located in southwest Texas near the Mexican border and the small town of Bracketville, Fort Clark sat on a rocky limestone ridge at a curve of Las Moras Creek, which emptied into the Rio Grande River thirty miles to the south. Historically, the fort's garrison watched over the U.S.-Mexican border as well as the pioneers pushing westward across the southwestern United States. Originally established in 1852, it took its name from Major John B. Clark, who died in the Mexican War. The fort's garrison and other soldiers lived in temporary barracks until 1857 when the military built a permanent post headquarters. During the Civil War, Texas volunteers occupied the fort after federal troops left. Many notable military figures were stationed at Fort Clark during its history, including Generals Jonathan Wainwright, Ulysses S. Grant, Dwight D. Eisenhower, and George Patton. During World War II, the fort served as a training ground for U.S. Army troops and as a prisoner-of-war camp.[3]

The army deactivated Fort Clark on February 9, 1946, and the SPA declared it surplus property on February 21, 1946. The SPA then transferred all surplus properties to the War Assets Administration (WAA), which had the responsibility for auctioning or otherwise selling them. Hoping that a private buyer would not purchase the fort, a group of prominent Texans, including former vice president John Nance Garner and Fort Worth publisher Amon G. Carter, attempted to persuade the government to transfer it to a nonprofit military school to be called the Fort Clark Military Institute. Despite widespread support for this plan, it was not accepted by the War Assets Administration, which decided instead to auction the fort. The Texas Railway Equipment Company submitted the winning bid of $411,250, which was some $300,000 higher than the next highest bid. The high bid reflected the value of the large tract of land underlying the fort. As soon as it took possession of the fort, Texas Railway Equipment sold the lumber and fixtures from the recently constructed barracks. The buyers of the barracks razed them, saving the salvage firm that work, and within only a few weeks Texas Railway Equipment made back most of the purchase price of the Fort.[4] The Browns kept approximately seventy-five buildings, mostly original structures of sturdy stone construction.

The Browns used Fort Clark primarily as a vacation and holiday retreat, and as Margarett Brown's health worsened, her family spent long summers in the dry climate of West Texas. Several of their friends leased old officers' quarters on the property. Herman Brown particularly enjoyed Fort Clark, which became one of his favorite residences away from Houston. According to his daughter, Herman "loved being there. It was back to the land, for whatever reasons, he loved it. . . . He liked to drive the land. He liked to shoot there. He took his friends, and they had weekend gatherings. He probably was as happy as ever, relaxed." Part of the attraction was to be far away from his work: "It gave him a sense of awayness, and you got in a plane to go there and you were out of the pressures of Houston."[5] Herman chose for his home at Fort Clark the house where General Patton had lived when he had been commander of the fort. George and his family rarely came, in part because during the war he had purchased property near Concan, on the Frio River in Uvalde County.

George Darneille, who assisted Herman and George Brown in the purchase of other war surplus property, recalled that "the reason they really held onto it was Herman loved to go out there and shoot dove and turkey and they used it as a hunting place and as a place for entertaining guests and business people."[6] These dove hunts often became elaborate events, with a Brown & Root plane leaving with guests from Houston, picking up additional guests in Austin, and then taking the planeload on to Fort Clark for a weekend of hunting, talking, and aggressive relaxation. Prominent among the frequent guests at Fort Clark were Lyndon Johnson, John Connally, Lloyd Bentsen, and other state and national politicians, as well as business colleagues of the Browns.

When Hollywood moviemakers approached the Browns and asked the brothers if they could use Fort Clark to shoot scenes from the movie *The Alamo* starring John Wayne, the Browns agreed. But they stipulated that while making the movie the pro-

Margarett Brown at Fort Clark with General Wainwright and his aide, Colonel Arnold, 1948. *Courtesy Brown Family Archives.*

ducers would be required to refurbish some of the older buildings, which the moviemakers could then use themselves during the filming. Fort Clark was never a money-making venture for the Browns, but it gave them much enjoyment.[7] For a relatively small investment, the Browns had obtained a property that gave them a distinctive retreat for entertaining others and for personal relaxation.

Other surplus property ventures proved more profitable, if less exotic. At the war's end, the U.S. government had some twenty-six thousand surplus airplanes. The air force considered several alternative postwar uses for the planes, including scrapping them. George Darneille, who had been in the air force before going to work for the Texas Railway Equipment Company, informed Herman and George Brown of the ongoing deliberations on the fate of these surplus aircraft. The Browns discussed the idea of buying the aircraft, melting them down, and selling the metal to industry. Darneille recalled that Herman told him to "go ahead and see if you can sell them [the government] on the idea."[8] Darneille returned to Washington and "spent six or eight months in Washington working with the War Assets Administration, and finally convinced them that they would be much better off to sell the airplanes and get rid of the overhead and get the metals returned to industry."[9] The cost of storing and guarding the planes alone was exorbitant, and the government wanted to get rid of them as soon as possible. They contained so much metal that President Truman wanted them scrapped quickly, within eighteen months of their sale.

Almost twenty-one thousand of the surplus planes sat on five airfields at Walnut Ridge, Arkansas; Clinton, Oklahoma; Altus, Oklahoma; Ontario, California; and Kingman, Arizona. The air force estimated the original cost for these planes at $3.9 billion. Included were army and navy fighters, bombers, and tactical aircraft; each contained a wide variety of metals. Darneille learned from air force records "where to find the metals in the planes—magnesium, copper, aluminum, silver, platinum. . . . We studied those planes day and night," he recalled. "Before we were through, we knew the location of every ounce of metal in every type of plane the government wanted to sell, how long it would take to get the metal out, and what the salvage cost would be. We knew as much about those planes as the men who had designed them."[10] Darneille and Dahlstrum calculated that one melted-down bomber could provide seventy thousand medium-sized saucepans.

The WAA held an auction for the airplanes, and 106 companies throughout the United States bid on them by airfield. The WAA stipulated that each company could bid for planes located at only one airfield. Herman Brown, however, figured a way to get around that rule. According to Darneille, "Herman got a hold of companies with whom he had done joint venture work and convinced them to put in bids for the planes at the other fields . . . and that group of companies bought all the fields."[11]

Texas Railway Equipment submitted the high bid of $1,817,738 for the 4,890 planes located at the Walnut Ridge, Arkansas, airfield. The salvage firm later purchased 5,000 additional planes located on another airfield. After the auction, the companies that won the bids for the other fields formed a single salvage firm, the Aircraft Conversion Company, which was managed by three associates of the Browns. Slim Dahlstrum managed the entire operation; J. E. Walters, who had managed Yard B at Brown Shipbuilding, oversaw production efforts at each field; and Darneille managed sales and government liaison work. For the 20,960 planes at the five original fields, the five firms bid a total of $6,582,157.

The Aircraft Conversion Company worked efficiently. Material not manually stripped from the planes felt the cold steel of a specially made guillotine, which chopped the planes into pieces small enough to be fed into a smelter. The Aircraft Conversion Company built ten smelting furnaces, each with a yield of 100,000 pounds of aluminum per day. George Brown recalled later that "we had a corner on the aluminum market and didn't know it. The price rose to twenty cents a pound, but we sold from six to eleven; we thought that was good enough and we made a great profit."[12]

The Aircraft Conversion Company recovered more than 200 million pounds of aluminum, 1,000 ounces of platinum, and 80,000 ounces of silver from the planes. Additionally, most of the planes still contained fuel, which was drained and sold. Bulletproof fuel tanks taken from the planes had no obvious salvage purpose until an Arizona rancher bought one for a water trough for his cattle. Other ranchers soon bought tanks for the same purpose. The tires from the planes were sold to docks for use as shock absorbers on piers.

Some planes were virtually brand new. To prevent the salvage company from selling intact aircraft and thereby entering the used plane business, the WAA stipulated that if the firm sold any operating plane, it had to pay the government the amount of money it received above its estimated salvage value. The Aircraft Conversion Company did sell some planes at the request of the U.S. government. Several A-26 attack bombers went to the French government for use in the French-Algerian war. During the early years of the Korean war, the company sold engines and spare parts to the U.S. Air Force. The spare parts gleaned from the 26,000 planes transformed the Aircraft Conversion Company into a major source for spare parts and spare engines. Originally, Dahlstrum and Darneille expected spare parts sales revenues to reach approximately $1 million. However, the high demand for spare parts during the Korean war brought the company over $30 million from these sales.[13]

The Aircraft Conversion Company proved much more profitable than expected. The high demand for aluminum and other metals extracted from the planes and sold to industry provided the salvage firm with a ready market for its goods. At its height, the entire operation had four thousand employees. The airplane salvage job was Texas Railway & Equipment's largest and most profitable project. It was proof that entrepreneurs willing to think creatively could find veins of gold in the mountains of war surplus materials being rapidly disposed of by the government after World War II.

In the Big Inch and Little Big Inch pipelines, the Browns found the mother lode. These lines were the single most valuable properties in the government's entire war surplus inventory. The attractiveness of the Inch Lines as investments, as well as their obvious importance to the Texas economy, drew the Browns into a group of entrepreneurs who put together a winning bid for the lines, converted them from oil transportation to the transmission of natural gas, and used them as the basis for a dynamic new company, Texas Eastern. This venture quickly grew into a major new investment that came to rival Brown & Root in importance in the Browns' holdings. Like the other major cross-country gas transmission companies, Texas Eastern would eventually be based in Houston, which had emerged as the corporate center of natural gas production and distribution for the nation.

Prior to 1946, the Browns' most significant exposure to the natural gas industry involved Brown & Root's work on two gas lines during the late 1930s through the mid-1940s. The firm built parts of the pipeline system for the United Gas system and the Tennessee Gas & Transmission Company. United Gas, based in Shreveport, Louisiana, was the largest distributor of natural gas in the southwest. Tennessee Gas (later Tenneco), was financed jointly by the Chicago Corporation, an investment trust based in Chicago, and the Reconstruction Finance Corporation, the "development bank" created by the federal government to hasten economic recovery during the depression and managed throughout the 1930s by Houston banker Jesse Jones. Tennessee Gas operated the first gas pipeline to extend from the Texas Gulf Coast to Appalachia.

The Brown brothers followed the progress of Tennessee Gas from its earliest history when Curtis Dall, son-in-law of Franklin D. Roosevelt, led the venture. At one point, the Brown brothers attempted to acquire a financial interest in the line. "The Brown-Root Callahan Group," reported Tennessee Gas president Harry Tower to Dall, "seems to be very anxious to put up $1,000,000 of working capital when as [sic] and if the Federal Power Commission issues the Certificate . . . and we have evidence of major financing from the Federal Government."[14] Later, a Tennessee Gas promoter, Clyde Alexander, heard that in early September, 1942, Brown & Root representatives met with Jesse Jones in Washington seeking a financial interest in the proposed line, but the Browns did not succeed in acquiring an equity interest in the gas company.[15]

After much debate about the nation's wartime energy needs, Tennessee Gas received federal approval to build and operate a gas pipeline from Texas to Appalachia. It hired several construction firms to build it. The primary contractor, Bechtel-Dempsey-Price, shared work with Williams Brothers Corporation and Brown & Root, which built the section of Tennessee Gas's line located in East Texas. Tennessee Gas broke ground for the pipeline on December 4, 1943, at the Cumberland River in Tennessee. Upon completion on October 31, 1944,[16] the line began carrying natural gas to vital defense plants.

The Browns' work for Tennessee Gas alerted them to the grand opportunities in the natural gas transportation business, which was poised for a postwar boom. All interested parties understood that fortunes would be made in transporting and selling gas produced in the prolific fields of the Southwest to the giant urban markets of the Northeast. The early advantage would go to the companies that established the first gas pipelines from Texas to the East.

As the war wound down, considerable interest came to be focused on the so-called "Inch Lines," two pipelines originally constructed during the war to transport crude oil and refined petroleum products from the major oil producing and refining centers in Texas and Louisiana to cities on the eastern seaboard for shipment to the European front. The U.S. government decided to finance the Inch Lines in the spring of 1942, after German U-boats began sinking American ships in the Atlantic. Oil tankers traveling from the Gulf Coast to the New York area were especially hard hit. On February 27, 1942, the same day Brown Shipbuilding launched its first PC, German submarines sank one oil tanker off the coast of Florida and one off the coast of New Jersey; three other tankers had been sunk earlier in the week. Alarmed by the rising tide of tanker losses, the U.S. government began planning an overland pipeline network to transport oil safe from German submarine attack.[17]

Consequently, the Reconstruction Finance Corporation (RFC) paid eleven oil companies $146 million to build the two longest and largest-diameter petroleum pipelines yet constructed. The twenty-four-inch diameter Big Inch transported crude oil over a 1,254-mile-long route from the East Texas oil fields at Longview, Texas, to Phoenixville, Pennsylvania, with extensions to points near New York City and Philadelphia. The twenty-inch diameter Little Big Inch carried refined products from the nation's largest refining center on the Texas Gulf Coast to Linden, New Jersey. A heroic push by thousands of workers completed these lines in time for them to serve a vital role in fueling the American victory over Germany. Almost immediately after the war ended, the RFC announced it would sell the lines. At about the same time, Brown & Root had formed its new Petroleum and Chemical Division, whose 1,800 employees were available to work on the design and engineering problems associated with gas compressors and pipelines. One of this division's first jobs was studying the feasibility of converting the Inch Lines from petroleum to natural gas transportation.[18]

The Inch Lines aroused considerable controversy. Oil, coal, and natural gas interests hotly debated the appropriate postwar use for these valuable properties.[19] The RFC hired the engineering firm of Ford, Bacon, & Davis to study the Inch Lines and recommend their most suitable postwar use. In addition, Congress held hearings during November, 1945, on the pipelines.[20] At these hearings, and in other venues, natural gas interests encouraged the government to sell the lines to a gas company; coal and railroad groups, fearful of competition from natural gas, suggested scrapping the lines or using them for oil transportation only; major oil companies stated that any oil company purchasing the lines would unfairly achieve near monopoly power in the oil transportation business; independent oil firms believed the Inch Lines could provide them access to the large markets for crude oil and petroleum products traditionally controlled by the major oil companies.[21] Given the competing interests, the struggle to acquire these valuable pipelines generated intense lobbying.

One man who proposed that the Inch Lines transport natural gas was E. Holley Poe, a utility and natural gas consultant. Since the mid-1930s he had worked at the American Gas Association, and during the war he helped organize and direct the Natural Gas and Natural Gasoline Division of the Petroleum Administration for War (PAW), the federal agency in charge of wartime petroleum policy. At the PAW, he worked closely with its assistant deputy Everette DeGolyer, an internationally respected petroleum and natural gas geologist based in Dallas. Poe and DeGolyer worked on postwar studies of the Inch Lines and of natural gas use in the United States, and they discussed the idea of purchasing the lines.[22]

Their interest in the Inch Lines took a more concrete form in February, 1946, when Poe undertook an assignment for a natural gas industry trade group involved in the Federal Power Commission's (FPC) extensive investigation of the United States' natural gas industry. The investigation examined the necessity of modifying the Natural Gas Act of 1938, which set forth the federal regulations governing the industry. The general counsel for the trade group was Charles I. Francis, a friend of the Brown brothers and a partner with the Houston law firm of Vinson, Elkins, Weems & Francis. Francis had been special assistant to the U.S. attorney general (1933–34 and 1938–45) and later became special consultant for the Department of the Interior (1944). Together, Poe, Francis, and DeGolyer discussed the possibility of forming a group to attempt to purchase the Inch Lines. They consulted with N. C. McGowen, president of United Gas, who took an interest in the project and recommended R. H. Hargrove, the vice president of his firm, to assist the Poe group.[23]

In mid-June, 1946, Francis, Poe, and Hargrove met at the New Yorker Hotel in New York City to discuss a possible bid for the Inch Lines. Frank Andrews, the president of the hotel, participated in the meeting as well. Related by marriage to Hargrove, Andrews offered the group especially important connections to financial and political resources. The Manufacturers Trust Company owned the hotel, and Andrews's close ties to the company provided a possible source of financing if the group decided to make a bid on the Inch Lines. Andrews invited a friend, George E. Allen, to the meeting. Allen, a former hotel man himself, was at that time a confidant of President Harry S. Truman and a prominent democratic party official. Most important, Allen was a director of the RFC, which then held title to the lines. Allen had joined the RFC the previous March when a new agency, the WAA, took over from the RFC and the Department of Commerce both the policy making concerning war surplus property and its actual disposal. According to Francis, Allen was actively involved as well with the new WAA.[24]

The group sought additional financial muscle and engineering expertise. Francis suggested that the group talk with two of his clients, Herman and George Brown. First, the group investigated a rumor that the Browns already had joined Houston banker and former New Deal official Jesse Jones in a group that would bid on the lines. On June 14 Francis wrote Judge Elkins, senior partner of his law firm, to "confidentially ascertain whether or not Jesse Jones is in any way interested in placing a bid for these facilities. Rumor has reached us that along with George Butler and possibly the Brown Brothers and Gus Wortham, he is contemplating making a bid for these same facilities."[25] Francis must have been satisfied by Elkins's reply because he subsequently invited the Browns to join forces with the "Poe group."[26]

The Brown brothers agreed to finance the group's effort in return for a promise that, if the bid succeeded, the engineering and construction contracts on the Inch Lines would be awarded to Brown & Root. With the engineering expertise inherited from Brown Shipbuilding, Brown & Root was prepared to enter the natural gas business in a big way. Although the Browns were not well acquainted with all members of the Poe group, they had close personal ties to Francis and his law firm, and they had an obvious tie to the pipeline industry through Brown & Root. Moreover, the Browns recognized an outstanding investment opportunity. They quickly took charge of the group's efforts, making judgments about the capabilities of all the different men involved in the venture. As their friend Posh Oltorf later recalled, "Anybody that they invited to come in with them was somebody who could bring something to the party."[27]

After considering the various reports, studies, and government hearings related to the Inch Lines, in June of 1946 the War Assets Administration announced an auction for these pipelines. The WAA advertised the Inch Lines for sale or lease nationally in thirty-eight newspapers and five oil trade journals; bids were due on July 30. Unfortunately for those groups interested in using the Inch Lines to transport natural gas, the WAA, bound by the disposal policy dictated by the Surplus Property Administration, stipulated that "first preference will be given to continuing the Big and Little Big Inch in petroleum service, thereby assuring availability of the lines in the event of a national emergency."[28] This criterion did not stop natural gas promoters from going forward with their own proposals, but it did suggest that the line would end up with an oil company.

The WAA received sixteen valid bids by the closing date of July 30.[29] Of these bids, seven proposed oil transportation, three were for gas, five proposed some combination of oil and gas use, and one did not state a preference. The WAA had great difficulty in determining the winning entry, however, because the agency had imposed no standards on the bid format. The bidders had offered several different combinations of cash, debentures, payments based upon gas sold, or other arrangements to finance the remaining cost, and the WAA had no way to compare the value of the different proposals.

The WAA decided to consider only the cash offered in each bid as the total bid price. After defining its pricing criteria, the WAA listed the highest bids as those proposing to use the Inch Lines for natural gas transmission. Trans-continental Gas Pipe Line and the Natural Gas Transmission Company both bid $85 million in cash. The Poe group had the second highest cash bid for natural gas use at $80 million. Charles H. Smith, representing Big Inch Oil, Inc., submitted a bid to use the Inch Lines to transport oil. This bid most closely met the WAA's surplus property disposal policy, but his cash price of $66 million was substantially lower than those for natural gas.

W. H. Leslie, Brown & Root's Washington liaison officer, kept close tabs on the bid evaluation procedures. He wrote George Brown in mid-August and relayed a conversation with Ross Gamble, a Washington-based attorney also representing the Poe group. Gamble reported that "it is extremely likely that the Big Inch Oil, Inc. bid will be recommended for approval unless our group takes early and aggressive steps to prevent this happening."[30] Leslie advised that if the WAA would not sell the Inch

Lines for natural gas transportation, "he [Gamble] thinks that steps should be taken to encourage members of congress to request or instruct that congress be given a chance to investigate the matter more fully to determine what usage the lines should be sold for. This might result in a delay which might give our group time to fight the opposition more successfully by bringing public and congressional opinion around to our way of thinking."[31] Along with the memo were profiles of the current WAA board members and their positions on the issues raised by the bid. George Brown passed the memo on to Congressman Albert Thomas with the paragraph mentioned above marked in red. Brown wrote, "I am enclosing a memorandum from Bill Leslie part of which I have marked in red. When you have time, read it and be thinking what if anything we can do about it."[32]

Reflecting the intensely political nature of the bid, nearly all the bidding companies included high-profile former, and sometimes even current, political figures. Many powerful men had aligned themselves with various groups of investors looking to acquire the lines. The Big Inch Gas Transmission Company included a former Ohio senator, an ex-justice of the Supreme Court, and an ex-chairman of the Maritime Commission. Attorneys for this company included former Roosevelt aide Thomas Corcoran, a former general counsel for the FPC, and a former FPC commissioner. Another bidder, American Public Utilities, included former trustbuster Thurman Arnold and Abe Fortas, former undersecretary of the interior for Harold Ickes and future Supreme Court justice.[33]

The highly politicized nature of the bidding attracted immediate media attention, much of which was directed at the Poe group. Harold Ickes, former secretary of the interior, wrote one column specifically about the Poe group. For his syndicated column, "Man to Man," Ickes wrote an article titled "'Uncle Jesse's' Bid for Oil Pipelines Seen Loaded Two Ways for Monopoly." Ickes suggested that former RFC chairman Jesse Jones was behind the "so-called E. Holley Poe bid" for the pipelines. Ickes warned that "if the story is true, the War Assets Administration had better do some keen sniffing because Uncle Jesse has no peer among horse traders that I have known." Ickes was more certain that George Butler, a Houston lawyer, "husband of Jesse's only heir and custodian of many of Jesse's enterprises, both business and political, is one of the E. Holley Poe crowd."[34] Ickes noted, however, that he liked Poe's bid.

Ickes's article brought quick denials from Poe, Francis, and Butler. Poe wrote to Ickes and met with him two days later, assuring him that Butler and Jones were in no way connected with the bid. Francis did the same, and Butler issued a statement denying that he was part of any group interested in the Inch Lines. Ickes responded to the denials and wrote: "I cannot see that the column did your group any harm. On the contrary, it served notice on Jesse Jones that he was more or less suspect and he might be disposed to keep his hands off. That is really what I had in mind in writing the column."[35]

Although Ickes apparently satisfied himself that he had at least warned Jones and Butler to stay clear of the Inch Lines, his article encouraged more scrutiny of the group. Newspaperman Marshall McNeil sent a telegram to George Brown: "In view of recent printed reports would you please wire me collect answer to question whether Jesse H. Jones is interested directly or indirectly in E. Holley Poe bid, with which you are associated, to buy from government Big Inch and Little Big Inch pipe lines. Thanks." At the bottom of the telegram was written: "Mr. Frensley said to hold—not answer."[36]

Next came an inquiry from the government. The day after Ickes's article appeared, the WAA requested a list of the "identity and business connections" of all individuals or firms associated with the particular bids.[37] Poe responded to L. Gray Marshall of the utilities branch of the WAA "that my proposal and supplementary data discloses the names and official connection of all persons associated with me."[38] This controversy soon quieted down as critics became satisfied that Jesse Jones was not a member of the Poe group. Unfortunately, the apparent resolution of this particular issue did not bring forth a quick sale of the pipelines.

Faced with intense political and business pressures, the WAA delayed any announcement of which group had won the bid. By late October and early November, several articles in trade journals and daily newspapers cited rumors that the WAA was ready to award the Inch Lines to a bidder for oil transportation. Repeating a rumor heard earlier by Brown, Elliot Taylor, editorial director of the journal *Gas*, claimed that the WAA was on the verge of awarding the Inch Lines to Charles H. Smith's Big Inch Oil, Inc., presumably because this bid corresponded most closely with the SPA's surplus property disposal policy.[39]

As the WAA wallowed in confusion, supporters of a sale of the Inch Lines to a gas bidder received an unexpected boost. John L. Lewis assisted their cause when he threatened to call another nationwide coal strike. Lewis's threats sparked an immediate public backlash, encouraging political support of a sale of the Inch Lines for natural gas transmission; natural gas could replace coal in northeastern markets. Both Harold Ickes in his syndicated column and Marshall McNeil flailed Lewis for threatening to strike. Ickes suggested that a good dose of natural gas competition via the Inch Lines would serve Lewis right. And Ickes chastised the WAA for simply not selling the Inch Lines to the highest bidder. Anything else, he wrote, "is just plain politics."[40]

Congress finally called for a new investigation of the Inch Line question, something gas bidders had wanted all along. On November 19, 1946, the Select Committee of the House of Representatives to Investigate the Disposition of Surplus Property met to determine the best way to dispose of the Inch Lines. On the opening day of the hearings, the chairman of the WAA surprised everyone by rejecting all previous bids for the Inch Lines. Support for this decision came from an interagency committee studying the disposal policy: "it had become evident that the interest of national defense could be met regardless of whether the pipe lines be used for natural gas, petroleum and its products, or a combination thereof . . . the bids had been invited on a restricted basis, which precluded the government from securing the maximum net cash return."[41] The WAA also estimated that the fair value of the Inch Lines was $113.7 million as of September 30, 1946, a figure significantly greater than the highest cash value bids previously received.[42]

John L. Lewis's coal strike combined with an increasingly severe energy shortage in the Northeast prompted the WAA to find a way to use the Inch Lines to provide much-needed energy. Tennessee Gas made a formal proposal on November 29 to lease the Inch Lines on a temporary basis and deliver gas to Appalachia—but not farther eastward.[43] Tennessee Gas also agreed to spend $250,000 for material and installation involved in using the lines for natural gas and connecting them with its own system. On December 2, the WAA accepted Tennessee's proposal. Tennessee's lease began at 12:01 A.M. on December 3, 1946, and extended to midnight on April 30, 1947.[44]

In the meantime, Charles Francis and the Poe group lobbied hard for the Inch Lines to be sold for natural gas transmission. Early in August, Francis met with General John J. O'Brien, a WAA official, to discuss his group's bid. Francis subsequently urged O'Brien to consider the benefits of allowing the lines to be used for natural gas.[45] The Poe group, he said, would maintain the lines in good condition, ready to be converted back to oil during a national emergency. Francis's main argument was that the Inch Lines could market the tremendous amount of natural gas currently being wasted during oil production in Texas. Francis also met with Ernest O. Thompson, head of the Texas Railroad Commission, to discuss the bid. A few days later, Thompson wrote Robert Littlejohn, war assets administrator, and verified Francis's statements about the tremendous volume of natural gas being wasted in Texas.

Francis pressed the issue on every front. He discussed the bid with H. S. Smith, a special investigator for the House Surplus Property Investigating Committee. Smith met with representatives of each of the sixteen bidders during his investigation. Francis "urged him to immediately commence an investigation of the proceedings that had been followed in offering these lines for sale [and] . . . pointed out that there were no standards for bidding, no good faith deposits required, and that the whole matter had been handled in a very unbusiness like manner, detrimental not only to the best interest of substantial bidders, but also detrimental to the best interest of the Government."[46]

Francis's bold lobbying tactics continued. On December 16, he wrote Littlejohn and made three suggestions concerning the guidelines for a new round of bidding. Francis believed that earnest money should accompany each bid, that the successful bidder should be given time to acquire an FPC-mandated "certificate of public convenience and necessity" before purchasing the lines.[47] All of these recommendations ultimately became government policy.

The Poe group carefully followed Littlejohn's movements during the next few weeks to gain insight into his thinking about the next round of bidding. On December 24, J. Ross Gamble reported that Littlejohn and L. Gray Marshall, a WAA executive, were the two primary officials involved in preparing the new bid guidelines. Gamble noted that Littlejohn had left the previous night for his home in South Carolina without releasing any information. "For what it may be worth," Gamble wrote, "we are assured that General Littlejohn intends to follow the program which he announced at his last appearance before the Slaughter Committee; that he intends to make it possible for oil and gas bidders to have an equal opportunity; that he expects, subject to action of Congress, to sell the lines to the highest bidder."[48]

For his part, Littlejohn moved quickly on the new bid. On December 27, he announced the format and procedures for the submittal of new bids. George Allen, who had previously worked with the Poe group, informed President Truman on the same day that he was resigning from the RFC effective January 16, 1947.[49] All bids were due by noon on February 8, 1947, at the WAA box located in the Washington, D.C., post office. Unlike the first auction, this time all bidders would submit their offer on a standardized form. There was also a strict timetable, broken down into four payments, set forth for the purchase of the lines. First, each bidder had to submit a $100,000 check with the bid as a deposit. Next the winning bidder would pay $100,000 after the WAA issued a letter of intent to sell the lines. A third payment of $4 million would be made on or before the day the winning bidder began transporting fuel through the pipelines. The final payment was due within nine months from the date the WAA issued its letter of intent to sell the lines.

With the Inch Lines approved for natural gas service, the WAA announced a drastic change in disposal policy. Regardless of the bidder's proposed use of the lines, the highest bid would win.[50] Bids were due by noon, Saturday, February 8, 1947, and each bid had to be accompanied by a $100,000 deposit. The E. Holley Poe group reorganized specifically to bid on the Inch Lines. They renamed their company Texas Eastern Transmission Corporation, with George Brown as chairman and Poe as president. From this point forward, George Brown had a major role in running Texas Eastern. At Brown & Root, Herman Brown always had the final say; at Texas Eastern, his voice at times carried substantial weight, but George was now the chairman of the board of his own energy company.

Texas Eastern formally contracted the services of the New York investment bank of Dillon Read, which assigned to the case August Belmont IV, a vice president of the firm and the great-grandson of August Belmont, who had served as the American correspondent banker for the European Rothschild family and had financed part of the New York City subway system.[51] Belmont had left his troubled family investment bank, August Belmont & Company, to join his Harvard classmate Douglas Dillon at the prestigious house of Dillon Read in 1945, upon his discharge from the navy. Because Belmont had previous experience with utility underwritings and restructurings, he drew the assignment of helping organize the financing for the Texas Eastern group's bid for the Inch Lines. According to the history of Dillon Read, this transaction became "one of the most important deals in the firm's history." Indeed, the history concludes that "the pipeline deal was one of the largest, most complicated, and delicate any Wall Street firm had managed in a generation."[52]

To raise the $100,000 deposit necessary, the company's directors issued 150,000 shares priced at one dollar per share. A total of twenty-eight persons purchased these shares, including a number of friends, family, and associates of the Brown brothers and the other directors. Herman and George Brown each purchased 14.25 percent of the stock and also advanced $67,500 for stock purchases by other directors and friends with the understanding that if the bid was successful, they could repay the Browns within two years. Until repayment, George Brown retained the voting rights to the stock; he therefore had effective control over the company during its crucial formative years.

The other large stockholders included Reginald Hargrove, who owned about 8 percent of the stock. Hargrove acquired 3,000 shares or a 2 percent interest for each of his three sons. His cousin by marriage, Frank Andrews (president of the Hotel New Yorker), subscribed to 9 percent of the original Texas Eastern stock. The other original stockholders included Poe, DeGolyer, Francis, Judge Elkins (of Vinson, Elkins, Weems, and Francis), J. Ross Gamble, Justin and Randolph Querbes (business associates of Hargrove's), Gus Wortham (a Houston businessman originally rumored to be part of Jesse Jones's group to bid on the lines), and eleven partners of Dillon Read & Company, including August Belmont.

Under Belmont's directions the firm's financial resources were quickly enlarged for the bidding process. The board announced that some of the original stockholders, along with Dillon Read, had contributed a $1.35 million loan to Texas Eastern to assist in financing the bid. This capital was earmarked for the $1 million second payment that would be required if the WAA issued its letter of intent to sell to Texas Eastern. The directors also marshaled letters and reports from United Gas Pipe Line Company; E. DeGolyer of DeGolyer & MacNaughton; and Dillon Read to be included as supplementary documentation regarding availability of natural gas reserves and prospects for financing in support of the company's bid. By this time Manufacturers Trust Company agreed to make a most important $4 million loan to Texas Eastern. This money was earmarked for the prescribed third payment to the WAA to be made at the beginning of the winner's interim lease of the Inch Lines. The loan and subsequent financial assistance provided by Manufacturers Trust may have reflected the influence of Frank Andrews and perhaps Judge James Elkins, whose Houston bank maintained correspondent banking ties with Manufacturers Trust, led by Harvey Gibson. Later, Manufacturers Trust would serve as a trustee and as registrar and agent for the pipelines' bonds, and it also provided a revolving construction loan.

On the night before the bid, Brown, Poe, Francis, Hargrove, and August Belmont met in Washington, D.C., to determine their bid. Belmont came to the meeting after agreeing with Douglas Dillon, chairman of Dillon Read, on top bid figures. Texas Eastern's offer had been modified several times since the original $80 million cash proposal. In mid-December, the group raised the cash value of its bid to $120.5 million and expected to invest a total of $160 million in the pipelines. In mid-January, George Brown received financial information from an engineer at United Gas Corporation. The engineer noted that "the dominant factor determining what can be bid is therefore the average price which can be realized from the sale of the gas."[53] The engineer calculated that if the lines operated at 95 percent capacity and the gas sold for $.26 per thousand cubic feet (mcf)—the best possible scenario—the maximum bid price would be $133 million.

What occurred during the meeting on the night of February 7 remains uncertain. Several different versions circulated concerning the events leading to the determination of the exact bid price. In one, George talked to Herman and told him the group was prepared to bid $133 million. Herman told George to add another $10 million. In another version, the five men wrote on a piece of paper their estimates of the proper bid, and these were then averaged. In another, August Belmont had convinced his superior, Douglas Dillon, that "the project could stand a price of $140 million" and asked Dillon to give him permission to suggest a $135 million bid, with authority to go as high as $140 million. What is certain is that before the night ended, the group decided to bid $143,127,000, and Belmont received Dillon's approval for the higher amount the next day.[54]

On Saturday, February 8, all bids were due at the WAA's post office box. August Belmont and Ted Wadsworth of Dillon Read, J. Ross Gamble, and George Pidot of the Wall Street law firm of Shearman & Sterling delivered the bid to the War Assets Administration office located at the U.S. Post Office.[55] To prove they had deposited their bid at the post office and on time, they had a photographer take their picture on the post office steps. Gamble later told Belmont that he had received permission from the postmaster general to sit in the post office gallery and watch Box 2707, where the bids were being deposited, in case anyone attempted to open the box and view or remove any of the bids.[56]

On February 10, WAA representatives opened each of the bids in a public hearing; the agency accepted five bids and rejected five. The three highest offers proposed natural gas use. Texas Eastern's winning bid of $143,127,000 was only $2.5 million less than the Inch Lines' original construction cost; and it was more than $12 million above the second-highest bid, which had been submitted by a group representing the Transcontinental Gas Pipe Line Company. Tennessee Gas and Transmission's bid was third highest at $123,700,000.[57]

The bid's outcome incensed Gardiner Symonds, the president of Tennessee Gas and a bitter rival of Texas Eastern in pursuit of the Inch Lines. At the time, Brown Booth, a nephew of Herman and George Brown, worked in the public relations department at Tennessee Gas. Booth was in Washington with the company's president, Gardiner Symonds, when Texas Eastern's victory was announced: "it was consternation that broke out. And they were trying to figure out everything they could to throw somebody out, some bidder out. It was pretty hectic. I finally just absented myself from the scene and went to the National Press Club and joined the poker game."[58] Later, Booth left Tennessee Gas and became Brown & Root's public relations director.

Years later, Ronnie Dugger, a Texas journalist and biographer of Lyndon Johnson, speculated on the possible involvement of Johnson in the sale. He asked attorney Charles Francis in 1967 if Johnson had been involved. Francis said no, although Donald Cook, assistant to Attorney General Tom Clark, did research for the group. For his own part, Francis said that "Lyndon took no part in helping out with the original sale. I talked to Lyndon. He said, 'Charley, I can't do that. Tommy Corcoran is on the other side.'" Although Francis accepted this denial, he reminded Dugger that Brown and Johnson "were awfully close," and Francis might not have known of conversations between them regarding the Inch Lines.[59]

It is known that later in the afternoon of February 10, after the bids for the Inch Lines were read, Bill Leslie sent a telegram to George Brown: "Francis contracted top man who agreed assist therefore we feel better not to contact Doug unless you advise otherwise."[60] It is not clear exactly what this memo means, although certain inferences are evident. "Doug" is apparently Douglas Dillon of Dillon Read, and "Francis" certainly refers to Charles Francis, about whom Belmont later said, "he embarrassed George Brown sometimes."[61] Years later, Belmont speculated that "'Top Man' may well have been George Allen, but I do have a feeling that it might have been a reference to Clark Clifford as I have a memory of being aware of efforts by Charlie in that direction."[62]

As the group's investment banker, Belmont knew that numerous political challenges had to be negotiated before any group could actually operate the lines. Railroad and coal companies and their labor unions opposed putting the Inch Lines in natural gas service, and a lack of an eminent domain law for gas pipelines presented a potential operating problem as well, particularly in Pennsylvania. Belmont recalled that "the big thing we didn't want was to have anybody hidden, some political guy, being offered some stock in this thing to get us through." Discovery of such a hidden political figure might have resulted in the rejection of the group's bid.

According to stock records, one political figure did end up with a sizeable share of founder's stock, although it is not known what services, if any, he may have provided to the group. A founding stock holder, Frank Andrews, transferred approximately 25 percent of his original stock subscription to George Allen, who was prominently but quietly involved in the group's effort. Since Allen was a recent director of the RFC, which owned title to the Inch Lines, public knowledge of his association with the group's plans might have resulted in the elimination of its bid. A significant amount of Andrews's stock was also transferred to accounts controlled by Manufacturer's Trust.

Allen was a prominent corporate lobbyist in Washington during the 1940s and 1950s. In 1950, he wrote a book titled *Presidents Who Have Known Me,* in which he discussed his role as a business lobbyist: "There was much speculation while I was in the White House but not on the government payroll about what I was doing there for my business associates. . . . I knew I was doing nothing for them." He also stated that "I am able to tell my associates whom to do business with, how to make their way through the maze of red tape, and how best to conduct themselves."[63] During his own Senate confirmation hearings for the RFC post, the *New York Times* reported that Allen joked "he was sure there was a dictaphone under every chair. At least he acted on that assumption, he said with a broad grin."[64] Years later, Charlie Francis acknowledged Allen's lobbying skills when he noted that George Allen was "the most *important* and *influential* lobbyist in the U.S."[65]

The bidding for the Inch Lines was a high-stakes episode in business-government relations during postwar demobilization. According to an article in the *Harvard Business Review,* the sale of the Inch Lines was "the best war-surplus deal [the federal government] ever made."[66] That Herman and George Brown along with other members of their group successfully negotiated the financial, political, legal, and technical hurdles confronting their efforts showed that these men had learned how to do business in the nation's capital and on a national scale.

After the successful bid, George Brown set in motion the process of organizing a gas pipeline company. Before Texas Eastern could actually transport natural gas across state lines, the new gas firm had to receive a "certificate of public convenience and necessity" from the Federal Power Commission. The FPC issued such a certificate only after the conclusion of lengthy public hearings during which the applicant had to prove that it had the engineering resources to maintain and operate the pipeline, that a market existed for the pipeline's capacity, that the company had access to at least a twenty-year supply of natural gas for that market, and that the company's financing plan was feasible.

This was a delicate time, as some of the unsuccessful bidders for the Inch Lines continued to seek ways to have Texas Eastern's bid disallowed. For one, the Texas Eastern group had to acquire eminent domain powers for the Inch Lines since these lines were not legally qualified to transport natural gas across all state lines through which they extended. Attorney Herbert Brownell assisted the group in finding a solution to the problem.[67] There were other potential problems as well. Bill Leslie wrote George Brown: "Ross [Gamble] has an unconfirmed rumor that the WAA Real Property Review Board may consult the state of Texas about its attitude on the pipe line deal. He wondered how close you were to [Texas Governor] Beauford Jester in case it became necessary to discuss the matter with him."[68] The Browns and their investment group had emerged victorious in a high-profile bidding process in competition with numerous other powerful, well-connected interests. They were not prepared to have their victory reversed by disgruntled parties who had lost in the bidding process.

With the pipelines secured, the group now faced the challenge of turning them into a profitable enterprise. Reginald Hargrove and George Brown took full control of Texas Eastern's operations. As the chairman and the owner of a controlling interest in the company, Brown took a leading role in planning at Texas Eastern. But once the enterprise was launched, he trusted the company's day-to-day management to Hargrove, a seasoned natural gas company executive on loan from United Gas Corporation. Hargrove agreed before the bid to become the company's president, replacing E. Holley Poe in that role. Poe's vision had brought the investment group together, and his perseverance and expert knowledge of the industry helped push the company to the edge of its successful bid. But Poe lacked the practical managerial experience of Hargrove. Although the Browns and others now viewed Poe as an unnecessary outsider, DeGolyer, with the power of his reputation behind him, stood up for his business associate and friend and managed to keep Poe on as a director of the company.[69]

Hargrove and George Brown made a good team. Brown was the aggressive entrepreneur who had grown up in the rough-and-tumble construction business. Hargrove was the professional manager, conservative in his approach to business decision making. Yet for all these differences, the Browns understood Hargrove's critical importance to their new venture. He would be the steady hand at the wheel, the day-to-day manager with solid experience in the gas industry. In return, Hargrove and his sons received a significant share of Texas Eastern's original stock issues. George and Herman Brown allowed Hargrove to establish the new company's headquarters in Shreveport, where Hargrove had spent his adult life and lived with his family. Although the Brown brothers sorely wanted their new company in Houston, they understood that Hargrove was more important to the success of Texas Eastern than a central office in Houston—at least for the time being.

In Hargrove, the fledgling corporation had a seasoned manager still only forty-nine years old and vigorous enough to stand the pressures accompanying this new venture. N. C. McGowen, president of United Gas, did not protest as Hargrove assembled his executive operating staff almost entirely from United Gas. McGowen knew that Texas Eastern would be a major customer for United's natural gas supplies. George and Herman Brown tapped their connections in Houston to help staff upper levels of the financial and accounting departments. Herbert J. Frensley, associated with the Brown brothers since the early years of Brown Shipbuilding, became secretary and treasurer. Orville "Dick" S. Carpenter, previously associated with both the Browns and Herbert Frensley, became the comptroller.[70] Most of the field employees came to Texas Eastern from the original War Emergency Pipeline staff.

Sitting atop this impressive collection of executive talent were George and Herman Brown, Texas Eastern's two largest stockholders. As chairman, George remained active in overseeing the company's operations. He enlisted the aid of Herbert Frensley, a trusted lieutenant who had daily conversations with the president of Texas Eastern and then passed the gist of these conversations to George.[71] This allowed George to keep abreast of key issues without staying deeply immersed in the details of company affairs, and he had the option to step into the picture to shape crucial decisions. Herman played a somewhat more muted role at Texas Eastern. He regularly attended board meetings, and his opinions no doubt registered solidly because of his large stock holding, his position as the big brother of the chairman, and his forceful personality. But, all in all, Herman remained content to focus his energies on Brown & Root and let George play a dominant role in looking after the brothers' interests at Texas Eastern.[72]

Texas Eastern's newly recruited workforce pushed hard to take over operation of the Inch Lines on May 1, 1947, at the termination of Tennessee Gas's lease. As Hargrove, Goodrich, and their fellow workers began converting the Inch Lines into natural gas pipelines, Texas Eastern weathered the FPC's hearings for the certificate of public convenience and necessity. The hearings began on July 7, 1947, and lasted through September 26, 1947.[73] During the hearings, Herman and George Brown faced several attacks on their involvement in the project. Under direct questioning, Baxter Goodrich, assistant chief engineer at United Gas who joined Texas Eastern as chief engineer, fielded a large number of questions designed to ascertain Brown & Root's precise role in the engineering, construction, and design of Texas Eastern's compressor stations and expansion plans. These questions sought to determine if a conflict of interest existed regarding Herman and George Brown's roles as both directors of Texas Eastern and sole partners of the construction firm building the new gas firm's compressor stations.

The commission spent a great deal of time questioning Goodrich on Brown & Root's role in building the stations. Goodrich mentioned that since 1936, Brown & Root had completed similar pipeline and compressor station work for United Gas. At

Texas Eastern founders: E. Holley Poe, George R. Brown, Charles Francis, (standing) Herman Brown, Everette DeGolyer. *Courtesy Brown & Root.*

one point, the FPC attorney asked Goodrich, "Was there any effort made to obtain competitive bids on the work to be done on the design, construction, and engineering necessary for the compressor stations covered by this contract with Brown & Root?"

Goodrich answered, "No, sir." He went on to say that with Brown & Root's top executives on the Texas Eastern board, management believed that, as an interested party, Brown & Root would do the best job possible.

The FPC attorney asked, "In truth you were not consulted very much in that regard but merely participated from an engineering standpoint. Is that right?"

Goodrich, Texas Eastern's chief engineer, responded, "That is correct, sir."[74]

After several months of intense and thorough examination of every aspect of Texas Eastern's organization, from financing to engineering matters, the FPC satisfied itself that Texas Eastern should be certified to operate in interstate commerce. On October 10, 1947, the FPC issued a certificate to Texas Eastern.

With its permanent certificate in hand, the company finalized its purchase. August Belmont—who had been named as a director of Texas Eastern—led a large staff of lawyers, accountants, engineers, and financial experts through several weeks of long days, grinding out the paperwork needed to complete the financing. Texas Eastern decided to sell $120 million in bonds and approximately $32 million in stock, and Belmont negotiated with insurance companies for their purchase of Texas Eastern bonds. He first approached Metropolitan Life Insurance Company, which had conditionally agreed to purchase $36 million in Texas Eastern's First Mortgage Pipe Line Bonds. By October 23, a total of twelve major insurance companies committed to purchase all $120 million of Texas Eastern's 3.5 percent First Mortgage Bonds at 101 percent face value. The insurance companies conditioned their commitments on a Texas Eastern commitment to raise $30 million from stock sales.[75] On November 10, Dillon Read instituted a public offering of Texas Eastern stock priced at $9.50 per share.[76] Dillon Read characterized this as the "largest single corporate deal since the market crash of 1929."[77]

The next day several newspapers featured reports about the transaction. Although the Browns may have wished to avoid publicity, the Texas Eastern story received close scrutiny from reporters. One newspaper headline read, "Texas Eastern Transmission Stockholders Stand to Make $9,825,000 for $150,000."[78] The story behind all the articles was that twenty-eight persons invested $150,000 to purchase a $143,127,000 pipeline. The controversy intensified when one stockholder, J. Ross Gamble, sold original shares that had cost him $2,500 for $166,250.

The high profits resulted from the rapid inflation of the value of the original dollar-per-share stock issue. Prior to the stock offering, the original stockholders received an additional six shares, at no extra charge, for each one purchased. This was essentially a stock "split" of seven times, as each share became seven shares purchased for the original dollar per share—or a new price of approximately 14 cents per share. Thus, when Texas Eastern offered to the public its first stock issue at $9.50 per share, the company's stock value was sixty-six times the effective price paid by the original investors. The original $150,000 investment turned into a $9,975,000 windfall (including the original $150,000), at the time of the public offering.

The furor over the stock values of Texas Eastern thrust the Browns into the national spotlight, exposing them to national attention that they had never before experienced. August Belmont and Douglas Dillon held a series of press conferences to answer reporters' questions about the transaction. Belmont spent the rest of the week discussing the financing with reporters and editors from *Time, Newsweek,* the *New York Times,* and the newspaper *PM.* The next series of articles contained a more balanced story. Texas Eastern's founders and the initial stockholders made a large profit, but the government received almost its full cost for building the Inch Lines.[79] This highly public debate over high finance was later repeated with more subtlety in two opposing articles in the prestigious *Harvard Business Review.*[80]

The public controversy over intrigue, political entrepreneurship, and dramatic profits gradually subsided, leaving Texas Eastern positioned to close its purchase of the Inch Lines on November 14, 1947. This proved to be a complex undertaking, with 150 representatives of the various parties involved in the transaction. On one side of the table sat representatives of the WAA, RFC, and the Federal Reserve. On the other side, the officers of Texas Eastern, representatives from twelve insurance companies, three banks and investment bankers, the trustee for the bonds, the transfer agent for the stock, and all their respective counsel. The meeting was at the New York Federal Reserve Building, and there Harvey Gibson, president of Manufacturers Trust Company, George Brown, and Reginald Hargrove gave Robert Littlejohn, WAA administrator, a check payable from a Manufacturers Hanover account for $143,127,000. The WAA then returned Texas Eastern's initial payments totaling $5,100,000. Others who gave the company loans were paid at this time. As one of Littlejohn's aides observed: "When we set that deadline, we never thought you or anyone else would be able to meet it."[81]

After substantial investments in preparing the lines for service, Texas Eastern began delivering natural gas to the Appalachian area, Philadelphia, then parts of New York, and finally New England. By the early 1950s, the new company was in heated competition for new markets with Tennessee Gas and Transcontinental Gas, both of which had built new pipelines from southwestern gas reserves to northeastern markets for gas. Such was the demand for natural gas that all three companies quickly moved into the ranks of the largest natural gas pipeline firms in the nation.[82]

Texas Eastern proved to be a bonanza, and it quickly took a prominent place beside Brown & Root in the brothers' business empire. As one of the largest cross-country gas transmission companies, Texas Eastern placed the Brown brothers at the heart of one of the fastest growing industries in postwar America. As the supply of natural gas and the demand for this product boomed, Texas Eastern grew rapidly. Through Texas Eastern, the Browns became major actors in a prosperous industry whose predictable, regulated earnings served as a counterweight to cyclical swings in the more volatile construction industry. In addition to more balance in their diversified holdings, Texas Eastern gave the Browns a convenient corporate vehicle with which to enter the domestic and international oil industry, the local real estate market, and the gas pipeline construction industry, both onshore and offshore.[83]

As a bonus, Brown & Root became the primary contractor for Texas Eastern's sustained expansion program in the 1950s and 1960s. As one of the three major U.S. pipelines transporting southwestern natural gas to northeastern markets, Texas Eastern underwent almost continuous expansion in the decades after its founding. New pipelines and expensive new compressor stations to push the gas through the lines were almost constantly under construction. Brown & Root received much of this business. Between 1947 and 1977, Brown & Root completed approximately $1.3 billion in natural gas system work for Texas Eastern. (See table 5.1.) The other major gas pipeline client of Brown & Root in this period, United Gas Pipe Line Company, had strong ties with Texas Eastern.

**TABLE 5.1. BROWN & ROOT CONSTRUCTION WORK FOR TEXAS EASTERN,
ON NATURAL GAS PIPELINES AND COMPRESSORS, 1950–62**

Date of Projects	Cost
1950–54	$138,225,000
1956–57	134,440,000
1955–56	70,412,000
1957–58	45,000,000
1947–51	44,425,000
1961–62	44,392,000
1956–58	35,000,000
1958–59	33,187,000
1954–55	29,510,000
1961–62	22,231,000
1960	21,173,000
1960	15,535,000
1958	12,024,000
1960–61	11,774,000
1952–53	8,678,000
1956–57	8,508,000
1958–59	5,514,000

Source: Adapted from Brown & Root Statistical Summary of projects valued at more than $5 million
between 1941 and 1975.

Despite the fact that George Brown was chairman of Texas Eastern's board and
executive vice president of Brown & Root while Herman Brown was a director of
Texas Eastern and Brown & Root's president, no federal regulatory agency ever seriously objected to the close business connection between the two firms. Brown &
Root's work for Texas Eastern not only brought the construction firm good profits, it
enabled the firm to gain invaluable experience in pipeline work that it could later
utilize for other clients around the world.

Not surprisingly, Herman Brown had a clear opinion of the government's questioning of ties between Brown & Root and Texas Eastern. When the Securities and
Exchange Commission sent inquiries concerning the interrelationship between the
two companies in response to Texas Eastern's initial stock offering, a coworker noted
his opinion: "Herman's reaction was, 'It's none of their goddamned business!'"[84] Texas
Eastern's formal response to such inquiries was a bit more restrained. It avoided the
appearance of conflict of interest by following a simple procedure. Whenever the
Texas Eastern board considered a construction or engineering contract for which Brown
& Root was a bidder, George and Herman, if present, excused themselves from the
evaluation of bids. Certainly, their mere absence from the board at these occasions

did not necessarily remove any pressure they may have exerted. Brown & Root was Texas Eastern's primary pipeline construction and engineering contractor through the 1960s.

The early Texas Eastern work made good use of Brown & Root's newly formed engineering department headquartered in Houston. Pat Anderson ran the engineering department with the assistance of second-in-command Gerry Dobelman. Formed as Brown Shipbuilding Company was being phased out, this department had a number of highly trained engineers, and a top priority immediately after the war was to find enough design work to keep these trained men busy. One of the department's first tasks was to conduct feasibility studies for converting the Inch Lines from petroleum to natural gas. Work for Texas Eastern—and other gas and oil firms—remained an important area of expansion for Brown & Root in the postwar years.

Texas Eastern's gas transmission business was highly regulated and highly profitable, a combination that led management to seek opportunities to diversify so that some profits could be invested in potentially profitable ventures not subject to regulatory controls. In its early years, Texas Eastern explored numerous diversification opportunities, but it actually pursued only a handful of ventures outside of natural gas transportation. These included the development of a petroleum products pipeline, the purchase of a liquefied petroleum gas company, and the establishment in 1951 of a production subsidiary, Texas Eastern Production Company. The subsidiary embodied the Browns' longing to get a small piece of the action in the oil industry. Although none of these endeavors had much immediate impact on the company's bottom line, the impulse to diversify finally made a decided difference at Texas Eastern when the company became involved in North Sea oil and gas production.

As with many successful business decisions, this one resulted from a series of less-than-successful ventures and a healthy dose of good luck. During the late 1950s, Texas Eastern actively pursued a chance to participate in the planning for a proposed submarine pipeline to connect the large Algerian natural gas fields with either France, Spain, or Italy. George Brown made several trips to Europe to investigate possibilities for Texas Eastern and perhaps even for Brown & Root in this project. The French-Algerian war, as well as technical considerations, greatly slowed progress on these plans. It became clear to all involved that the sub-Mediterranean pipeline proposal could not be acted upon immediately. Texas Eastern next focused its attention on northern Europe. The discovery of the Groningen Field in Holland in 1959 put a damper on immediate plans to transport Algerian natural gas because the new discovery was very large and very close to European markets.

During its formative years, Texas Eastern was a speculative venture, burdened with heavy economic, technological, and political risks. But what began as a long-shot gamble of considerable time and money gradually matured into an extraordinarily profitable enterprise. Herman Brown admitted his deep misgivings in discussions with August Belmont, who recalled that "Herman thought . . . this was a wild-assed scheme of George's, and he went on kind of grumbling . . . And when it finally came true, he told me . . . 'George is just lucky.'"[85] Luck—combined with entrepreneurial talent, the selection of good workers and managers, and the phenomenal market for natural gas—gave the Browns the business opportunity of a lifetime.

The Browns tried with little success to make Texas Eastern a domestic oil and gas producer. They had somewhat more success in the oil industry as the owners of a small producing company organized outside of both Texas Eastern and Brown & Root. Herman and George typically joked about their status as "light producers of heavy crude." But even a small producing company offered the Browns the romance and excitement of a place in the oil business. The brothers purchased Highland Oil company of San Antonio in January, 1946, and made plans to move its offices to Houston. Included in the purchase were twenty-five hundred acres in Frio County, six hundred of which were proven oil bearing. Also included was an interest in fifty other areas in southwest Texas and twenty-five hundred acres of undeveloped acreage in southwest Texas. Although eighteen producing wells were located on these properties, it was unclear whether or not the wells were included in the transaction.[86] In addition to the Highland Oil Company properties, Herman and George Brown formed a separate partnership that also engaged in the oil and gas business. The Brown brothers' gas-producing properties became fairly substantial, but here again they did not broadcast their involvement in oil. Texas Eastern began purchasing natural gas from some of Herman and George Brown's own properties.[87]

Herman Brown ran much of the oil business, and he was boss to a young employee named Ralph O'Connor, George Brown's son-in-law. O'Connor was a student at Johns Hopkins University when he married Maconda Brown, George and Alice's middle daughter. In an opinion somewhat reminiscent of the stance taken years earlier by Alice Pratt's parents, George strongly opposed the marriage on the basis that O'Connor should establish himself in a career before marrying. After the couple married in 1950, O'Connor accepted a job with Highland Oil, moved to Texas, and began working for Highland as a roughneck and roustabout. He moved frequently in his early years with the company, and, in the process, he steadily advanced within the company. Finally, the Browns choose him to run the company.

The Brown's personal oil business was never large, and employees at various times attempted to take advantage of the small firm. One of George Brown's employees at Highland who was interested in deep wells, while George Brown was only interested in the shallower wells, entered negotiations with Union Carbide about a deep well deal in West Texas. Ironically, George Brown met with a Union Carbide vice president about various matters, and the Union Carbide official mentioned the West Texas deal. George Brown became very interested in the deal and asked who the contact was. It turned out to be his own employee—who was fired the next day.[88]

During the mid-1950s, Highland drilled a discovery well in Northeast Falls, Texas, during a time when O'Connor's superiors were attending a lengthy meeting in New York. The crew drilled another well, which proved dry, and made plans to drill still another well. When the report came to Herman Brown, who was overseeing the oil and gas operations, he telephoned O'Connor and ordered him to stop drilling there. O'Connor told Brown, "Well, we got a good well. We know it's got to be someplace. Just because it wasn't on the north, we figure it's got to be in the south. So, we're trying to find it." Brown replied that O'Connor was being "ridiculous" and to shut down the well.

"Well, Mr. Herman, you're the boss," O'Connor said. "I'll do it, but can I show you why we ought to keep going?" Brown replied that he would stop by the office on his way home but in the meantime to shut down the operations. After Brown came by, O'Connor convinced him that while there was only one development well and one dry hole in the eighty acres they had explored, there was surely more oil in the entire eight-hundred-acre tract. O'Connor convinced Brown to go forward if he could "farm" out the well and let somebody else drill it for a half-interest.

Later, after drilling a core sample "dripping with oil," O'Connor wrapped it in newspaper and drove to Herman Brown's house to show him. O'Connor recalled that:

> The butler answered the door, and I'm sitting there with this greasy sack of crap in my hand and obviously they were having a dinner party and I didn't realize it and they said, "Come in." . . . I came in and he [Herman Brown] said, "What do you have?". . . I said, "Well, I just wanted to show you the core.". . . I don't know who his guests were but they were very clean, tablecloths all set up, wine glasses and everything. . . . He took it [the core] and put it on his plate and looked at it and smelled it and fiddled around with it, got [oil] on his hands and he passed it around to everybody! He said, "Good." He shook my hand and said, "I'll see you tomorrow."[89]

Highland continued to grow, and at one point, George Brown and Orville S. Carpenter considered merging Highland into Texas Eastern.[90] The merger never took place; instead, the Browns gradually expanded the company, finally transforming it into Highland Resources, Inc., and managing it as part of Brownco, a firm organized to coordinate the activities of a variety of Brown family ventures. Through it all, the Browns' involvement, however small, in the oil production business allowed them to enjoy the status and romance—and absorb the risks—of participation in the independent oil industry.

In the postwar years, the Browns created a number of interesting and profitable enterprises separate from Brown & Root. Profits from Brown & Root and Brown Shipbuilding provided them the capital for significant investments outside the construction industry. Some of these, such as Fort Clark, they pursued for socializing and fun. Others, like Texas Railway Equipment Company, engaged their ability to see profit in what might be thought of as industrial-style garage sales. Some of these minor investments appear almost as hobbies, diversions from the high-pressure work at Brown & Root. Even their longtime involvement in Highland Oil remained in part a diversion in the playground of oil exploration and production.

Such was not the case with Texas Eastern. In response to a spectacular—if speculative—opportunity, they organized a determined and resourceful group of investors and won the prize despite great odds and well-connected competitors. Texas Eastern served as a counterpoint to Brown & Root in the Browns' investment portfolio and to Herman's dominance at Brown & Root in their shared workload. The predictability of the regulated gas industry fit nicely with the ups and downs of the construction industry. The new company gave George Brown a more prominent management role than he played at Brown & Root. It also exposed the Browns to new opportunities for diversification within the gas and oil industries as well as to a new level of high-powered government and business connections. Finally, the gas pipeline company gave Brown & Root the inside track to a giant new customer in a rapidly growing industry. This was only one of many fast-growing markets for Brown & Root in the postwar years, and all of these markets helped Brown & Root remain the centerpiece of the Browns' increasingly diverse business empire.

CHAPTER SIX

Brown & Root's Postwar Diversification

Although the brothers pursued a number of ventures outside of Brown & Root, the construction company remained the centerpiece of their growing economic empire. The postwar boom presented almost unlimited prospects in construction, especially in the Houston area, and the Browns were in the right place at the right time to build Brown & Root into one of the world's leading construction companies.

The key to the firm's expansion after 1945 was diversification. Before 1945 Brown & Root focused on public works, primarily roads, dams, and ships. In the postwar era, the company expanded and diversified its public works construction with a series of large, innovative, and often much-publicized projects. But the Browns also led their construction company into an array of ambitious and profitable projects in the private sector, including extensive construction in the oil, gas, petrochemical, electric, and paper industries. As it diversified, Brown & Root grew steadily, and at times, spectacularly, becoming one of a handful of large American construction firms capable of completing all sorts of major projects around the world. Its assets, total revenues, and net profits increased impressively (see table 6.1), reflecting the transformation of the company from a regional concern into an international giant.

As Brown & Root expanded, the Browns' roles within the company changed. No longer could Herman keep a close personal watch on all aspects of all major projects; no longer could George remain deeply involved in engineering. The brothers came to rely more heavily on trusted subordinates, who took greater responsibility in an increasingly decentralized management system. Herman and George continued to commit most of their working hours to providing top-level guidance and pursuing promising projects. The years from 1945 to 1962 were the heyday of the Browns' years together at Brown & Root, and the company's success raised them to new levels of wealth and influence.

TABLE 6.1. BROWN & ROOT FINANCIAL DATA, 1946–62
(IN THOUSANDS OF DOLLARS)

Year	Assets	Revenue	Net Profit
1946	$ 5,971	$ 12,459	$411
1947	11,808	23,377	2,020
1948	15,601	59,215	1,980
1949	19,285	78,117	2,057
1950	22,969	97,019	2,134
1951	26,653	115,921	2,211
1952	30,336	134,823	2,288
1953	28,477	153,724	2,364
1954	28,677	117,453	694
1955	28,744	111,069	1,792
1956	32,609	96,815	1,819
1957	40,147	261,854	1,656
1958	37,482	142,957	2,439
1959	44,105	112,070	4,076
1960	44,885	156,000	794
1961	47,619	122,228	2,791
1962	60,895	137,079	3,371

Source: Compiled from "Brown & Root Group Financial and Employment Data, 1946–1979," based on annual audited financial statements.

A thrust toward diversification was almost inevitable at Brown & Root, given the opportunities available to the company during the 1940s and 1950s. As a postwar boom transformed Houston into one of the nation's largest metropolitan areas, hundreds of major companies migrated to the region; massive construction of factories, pipelines, and civil works facilities were under way. Recognizing the central importance of Houston in the future of Brown & Root, in 1946 Herman Brown moved from Austin to Houston, buying a home next to that of his brother in River Oaks. As Herman managed Brown & Root's day-to-day operations from its Houston headquarters, George came into his own as a salesman with a string of triumphs in obtaining major projects. As a large, well-known Houston-based construction company, Brown & Root was in a strong position to compete for business in its booming region, throughout the nation, and around the world.

The Browns' experiences in the 1920s and 1930s taught them to embrace these opportunities to diversify. Herman Brown once told reporters that "I learned during the depression that we had better broaden our base. One construction field gets crowded before you know it. You had better be a little broader than your competition or you will be sunk with them when this happens." Brown also noted that it was important to tackle new projects: "Look for harder jobs if you want to keep ahead."[1]

According to a *Business Week* article written in 1957, the Great Depression had encouraged the Browns to diversify. Remembering Brown & Root's struggles in the 1930s, "the brothers deliberately began to branch out into oil, real estate, office buildings, hotels, pipelines, mines, even a dude ranch." Such ventures generally reflected a desire to learn to use existing resources and personnel in areas at least indirectly related to the company's core business, construction. George Brown recalled that diversification "in the early days not only spread the risk for us, it also permitted us to acquire a lot of knowledge we didn't have much of."[2]

Successful diversification required excellent engineers capable of designing projects they had never done before. As the Browns pursued larger, more complex, and more diverse projects, they developed a growing corps of technical specialists at Brown & Root. At war's end, Brown Shipbuilding employed about one hundred engineers, giving Brown & Root the nucleus around which to build a large, permanent engineering design department.[3] The Browns recognized that this was a giant step forward, one that required aggressive expansion to generate the ongoing stream of creative work needed to guarantee full-time employment for these specialists. The high overhead required to support a permanent engineering design team carried real risks; but the Browns focused instead on the potential rewards of becoming a "full-service" construction firm able to offer would-be clients a wide range of design and construction services. In addition to their desire for profits, the brothers also felt a sense of responsibility to provide jobs for loyal employees who had worked with them before and during the war.[4]

A tradition of decentralized management facilitated Brown & Root's diversification. The Browns always encouraged their project managers to be independent, since construction often required quick decisions in out-of-the-way locations. They sought strong-willed men capable of taking charge; in the words of George Brown, they wanted to "find good people and turn them loose."[5] The Browns wanted above all to retain flexibility of operations, believing that the employee should define the position rather than the position defining the employee. Like Henry Ford in an earlier era, Herman Brown held formal organizational charts in disdain. He once remarked that "I've seen more people not do anything on account of those organizational charts. . . . Everybody here does what they need to do and what I want them to do."[6] Herman wanted his managers to develop autonomy to the point of developing their own clients. Holding the sprawling operations together were monthly meetings of project managers and central auditing and accounting combined with Herman Brown's oversight.[7]

The most obvious result of this extremely decentralized approach to management was a corporate culture at Brown & Root that stressed individual initiative and fostered excellent morale. Thomas J. Feehan, who rose through the ranks to become the president of Brown & Root, recalled that "my most vivid impression of Brown & Root was the freedom granted the employees in carrying out their assignments."[8] Technical employees as well as managers were given considerable leeway to innovate. Longtime engineer W. R. "Rick" Rochelle recalled: "Many of our most imaginative and difficult projects were started by one of the Brown brothers' most trusted thinkers . . . walking out in the bull pen and asking one or more young engineers and/or field men to consider how he might go about achieving some seemingly impossible project, and to record his ideas, no matter how unconventional. Quite often these ideas . . . would become the foundation for a successful project."[9] The Browns encouraged all employees—from managers to engineers to construction workers—to think of themselves as part of a large extended family in which all members had a responsibility for success. In the heady days of postwar expansion, the company could offer even its most ambitious employees ample opportunities for advancement.

Brown & Root's extraordinary growth in this era reflected the efforts of a cadre of executives who grew up inside the company and remained loyal to the Browns. W. A. Woolsey, one of Herman Brown's original field superintendents from the late 1920s, remained Herman's right-hand man until his death in a horse-riding accident in the early 1950s. L. T. Bolin, a civil engineering graduate from Texas A&M who had worked with the Browns since the road building days, eventually rose to the position of executive vice president in charge of worldwide Brown & Root operations after 1955. Herbert Frensley, the company's longtime financial officer, first joined the Brown's during World War II at the Brown Shipbuilding Company; he subsequently served as the company's president from 1963 to 1974. Other executives who played prominent roles in the company's postwar expansion included Louis H. "Preacher" Durst, vice president for heavy construction; Jimmy Duke, the company treasurer; Carl Burkhart, senior vice president in charge of administration; and M. Pat Anderson, vice president and chief engineer. These—and many others—provided the management skills that allowed Brown & Root to move rapidly into new markets all over the world.

Leadership came, of course, from the Brown brothers, who brought complementary talents to the task of building Brown & Root into an international force. Herman remained the company's president and its undisputed boss; George remained second-in-command while providing vital talents in marketing the firm's services. Herman was the "storekeeper" and the "inside man" who generally could be found at the company's Houston headquarters looking after day-to-day operations; George was the "outside man" and the "mover and shaker" who traveled extensively representing the company to the outside world. Herman Brown delighted in focusing on the details of a project while George Brown quickly grew impatient with details and would say, "Never mind that, where do we go from here?"[10] Simply put, Herman managed the firm and George sold its services. The combination of the brothers' personalities and talents gave Brown & Root a flexible "two-headed" management team at the top.

In the rough-and-tumble world of heavy construction, Herman Brown took the lead in watching costs and encouraging efficiency. In a rare public speech in 1952, he summarized his philosophy of cost-cutting as follows: "We have never forgotten the lessons we learned in the depression. Dollars were scarce in those days and we learned to part with them with a great reluctance. This aversion to wasting money has stuck with us—whether it is our money or our client's money."[11]

An often-noted symbol of Herman's obsession with preventing waste were his memos; he would simply write out a short note on the bottom of a page from a legal-size yellow pad, initial it "H. B.," then use a ruler to tear off the note so that he could write as many as four or five memos from a single page of paper.[12] To ensure such efficiency in project planning, Herman Brown "was extremely interested in the detail of how they [Brown & Root] were going to accomplish a project. What the equipment was going to be, who they were going to put in the top jobs of management on these large projects . . . and he was a master at it. . . . He just had a way of talking to people and asking the right questions so that he had a real strong sense of whether they were telling him the truth and whether they knew their subject or not."[13] Another former employee, Howard Counts, recalled how Herman Brown figured the value of bids on construction projects. "His mind was such," recalled Counts, "that he could size up a job four hundred miles away or a thousand miles away." Typically, in the preparation of a bid, Brown inquired about the dollar amounts allocated to labor, equipment, and materials and then asked the manager for his estimate of the total bid. In response, Brown suggested adding or subtracting a certain amount in regard to labor, materials, or the territory where the project was located.[14]

Herman was a demanding taskmaster who believed in directly confronting potential problems. He was more apt to point out deficiencies than to acknowledge good work. One man who worked under the Browns recalled that "about the only way you had occasion to think that you were doing all right as far as the Browns were concerned was because they weren't on your ass. If it was fairly calm, you figured you were doing okay."[15] The tensions generated by close contacts with the hard-driving, blunt-speaking brothers led some of the younger employees to joke about nervous stomachs and ulcers as "Browns' disease."[16] Ralph O'Connor made a distinction between the approaches to "employee relations" of the two brothers: "[Herman Brown] would yell at you sometimes, but he'd pat you on the back . . . you felt he either loved you or hated you and he didn't mind kicking some people in the butt." George Brown was "more sophisticated and deliberate in his feeling. I think he never really wanted to get his arms around anybody."[17]

Herman Brown's employees respected him, but they also feared his response to disappointing results. He had "a bad temper, a quick temper," recalled one employee, "and the one thing he couldn't stand [was] people vacillating and skipping all around a subject and he knew what the problem was and he wanted them to come out and say it and he would keep you there, or keep individuals there, for a long time until they would fess up to it. . . . There were men working for that company in high positions that would rather take a whipping than tell the truth when they'd made a big mistake. But he would get it out of them."[18] Although both Herman and George could dress down any employee, they earned a reputation as loyal employers who were generous when an employee or an employee's family needed help.

They also had great difficulty firing anybody, even those caught stealing from the company. Thus, "their reputation was [as] absolute holy terrors for chewing people out, but for also, perhaps, [being] constitutionally incapable of firing somebody. They'd get somebody else to do it."[19] At times this caused problems for their companies. With decentralized authority, a manager left in place despite failures sometimes became a barrier for other managers to maneuver around to accomplish their goals. Often this problem was resolved by easing the person in question over into another job; seldom was the person simply let go. This practice reflected Herman's deeply held belief in the mutual loyalty owed to employees and expected from employees.

George Brown differed from his older brother in temperament and in his approach to management. George's academic training in engineering made him more interested in the technical details of projects. He was proud of his work as an engineer, and took particular pride in professional awards, such as the Engineer of the Year award given to him by the Texas Society of Professional Engineers in 1956. But he was a manager and a salesman as well. George delegated authority more readily than did Herman, who "was more inclined to step in if he saw somebody getting out of step or out of line . . . and say, 'You're not going to do it this way.'" George "was more inclined to let them run their course of what they were trying to accomplish and if they didn't make a success, then sitting down with them and talking to them about it."[20] In this sense, Herman demanded success from his employees; George seemed more interested in teaching employees to learn from their mistakes.

Given their differences, it was not surprising that tension at times developed between the brothers on business issues. One arena for conflict was the monthly Brown & Root management meetings attended by various company officers. Regular attenders included the firm's board of directors—all of whom were "inside" men in the privately held firm—and members of the operating committee. At these monthly meetings, which typically lasted one and one-half days per month, the two brothers sometimes engaged in extremely tense and loud arguments. These arguments usually seemed to clear the air, allowing the brothers to resolve disagreements and maintain their cordial working and personal relationship. One employee recalled these battles:

"He [George] and his brother used to have some real good ones . . . you'd think it was going to get into a physical damned fight. . . . Yes, I'd prop my chair against the wall, it was a spectator's sport! You know, you don't get involved! But, when they'd agree on something, they'd finally reach an agreement and both of them would work as hard as they could to see it come about the way they'd agree on."[21] Brotherly management was not always easy, but it proved effective in guiding Brown & Root toward successful diversification.

The brothers had plenty to argue about in the years after 1945, as Brown & Root undertook hundreds of new projects. The postwar boom presented a staggering array of opportunities for growth, and the Browns grabbed with both hands. Herman Brown's strategy of diversification was simple: keep the company's personnel and resources occupied by taking on any interesting work anywhere it might be found. He summed up this approach as follows: "We try to impress upon our men that Brown & Root can do anything and we pretty much have done just that."[22] Brown & Root expanded its traditional work on defense facilities, dams, and roads while undertaking new types of projects for a variety of private businesses. The company moved quickly into international markets, in the process adapting to different political demands and labor conditions. It entered a succession of very large projects around the world, generally as a partner with several other large construction companies in joint ventures. It became adept at negotiating new types of financing, from defense-related projects in a Cold War environment to public works in developing nations paid for by the World Bank. In short, the company grew in almost every direction, with the notable exception of high-rise buildings, a market avoided by Brown & Root in part because of the high concentration of unions in this area. Its sustained expansion in the twenty-five years after World War II earned Brown & Root the ranking of the largest engineering and construction firm in the United States in 1969.[23]

Brown & Root's first major postwar projects symbolized its new diversification into giant international ventures. Military construction was this first big step into international markets. In 1946, Brown & Root joined a consortium that reconstructed Guam, an American territory in the South Pacific that suffered great damage in World War II. In the Cold War, Guam took on added significance as a bulwark of U.S. defense in the Pacific, and the military decided to construct a military outpost and surrounding city there. This job flowed naturally from the company's previous work on the Corpus Christi Naval Air Station and its wartime work on two other military facilities, a naval munitions storage station in Oklahoma and a Texas-based ordinance depot. But the Guam project was twice as large as those three projects combined. Brown & Root remained active on Guam until completing the work in 1957, and its share of the more than $300 million in overall construction amounted to almost $82 million. This work introduced the company to the problems, as well as the opportunities, presented by work in distant areas and foreign cultures. It also led directly to the company's inclusion in various joint ventures to build military facilities around the world.

Table 6.2 shows that as the Guam project went forward in the 1950s, Brown & Root joined two other projects to build huge military installations in Spain and France. The combined costs of military construction involving Brown & Root in Guam, France, and Spain in these years totaled almost $800 million, with the company's share in the joint ventures amounting to almost $240 million. Brown & Root had joined the ranks of major construction firms tapping into an important new market for public works, the Cold War defense buildup in which the U.S. government constructed military facilities around the world.

The company's involvement in Spain illustrates the combination of diplomacy, politics, and labor management required in such projects. The U.S. government sought naval and air force facilities in Spain to give U.S. and North Atlantic Treaty Organization (NATO) forces quick access to both the Mediterranean and the North Sea. After a long period of negotiation, Spain and the United States signed a treaty on

TABLE 6.2. MILITARY PROJECTS CONSTRUCTED BY
BROWN & ROOT, 1940–62

Client	Project	Date	Cost*
U.S. Navy	Corpus Christi Naval Air Station	1941–44	$78,799,000 29,549,000
U.S. Army	Red River Ordnance Depot, TX	1941–44	26,238,000 13,119,000
U.S. Navy	Naval Ammunition Storage, OK	1942–45	58,000,000 29,000,000
U.S. Navy	Guam Air Naval Base	1946–57	306,427,000 81,714,000
U.S. Air Force	French Air Bases (9)	1953–54	164,500,000 54,833,000
U.S. Army	Tank Repair	1953–58	23,601,000
U.S. Navy	Spanish Air Naval Facilities	1955–61	308,000,000 102,667,000

* The first amount is the cost of the total project; the second amount is the amount paid to Brown & Root.

September 26, 1953, allowing the United States to provide military weapons and material to the Spanish army and to train its personnel in the care and use of weapons. In return, General Francisco Franco's government allowed the United States to build air and naval bases on Spanish territory for the joint use of Spanish and American forces.

About 250 American construction companies organized joint ventures to bid on these Spanish bases. The cost-plus-fixed-fee project was one of the largest heavy construction jobs of the postwar period. The project called for the greatest possible use of Spanish labor as well as the coordination of the activities of a variety of military and civil agencies in both the United States and Spain. This was to be a most complex undertaking. The U.S. Navy, the contracting agency for the project, narrowed the applicant pool to nine before awarding the contract to a joint venture of Brown & Root, Raymond International of New York City (and later of Houston), and Walsh Construction Company of Washington. One justification for the choice of Brown & Root was its recent participation as project manager in France for the construction and maintenance of nine military bases for the U.S. Air Force as part of the United State's contribution to NATO. These bases cost $200 million, and Brown & Root's successful work there in 1953 and 1954 gave the company considerable experience in building European military bases.

The project in Spain led to frequent culture clashes between American and Spanish workers, forcing Brown & Root's managers to adapt to demanding conditions. Ward Dennis, a young American who worked briefly for Brown-Walsh-Raymond (BWR), described these labor conditions. "The relationship between the American personnel and the Spanish personnel," he recalled, "was cruddy, to put it mildly."[24] The Americans were used to "snap-to" type workers, and Spaniards were not accustomed to being treated like workhorses. In defense of the workers, Dennis recalled that "the Spaniard is a very, very good worker, as the present economy of Spain demonstrates, if you give them the respect."[25] Being fluent in Spanish and familiar with both Spanish and American culture, Dennis at times acted as a mediator in disputes between the Spanish workers and American personnel. His abilities prompted his promotion to foreman.

H. N. Hockensmith, whose prior experience included work for the U.S. Army Corps of Engineers, directed the project. He faced a variety of problems, including recruiting and training the Spanish workforce, obtaining materials, and maintaining good relations with the U.S. Navy. The Spanish workers had to learn how to operate modern equipment with which they were unfamiliar. Some road crews there still constructed roads literally by hand with "pick, shovel and bags, and wicker baskets that you'd [pack] up on the back of a donkey."[26] The Spanish workers had a strong incentive to adapt, since they received close to three times the local wage.

Work on the five Spanish air bases began in September, 1954, and lasted until early 1960. The final cost was $300 million. The original contract called for a four-year term, but it was extended to six years, and the consortium built a wide assortment of related projects. These included five air bases, three naval bases, a network of air control and warning stations, a 425-mile cross-country underground petroleum pipeline system (POL), harbor facilities, underground fuel storage with a capacity of 5.5 million barrels of fuel and lubricants, a large ammunition storage farm, and many military buildings. The largest base, built for the Strategic Air Command at Torreon near Madrid, cost $70 million. The main naval base at Rota on the Atlantic Coast near Cadiz cost $120 million. The entire project employed some eighteen thousand Spanish workers. The major military installations in Guam, France, and Spain were big projects by any measure, and they gave Brown & Root a strong international presence that the company had not previously enjoyed in a defense market growing rapidly during the early years of the Cold War.

The construction of large dams also pulled the company into foreign lands in the postwar years. The Marshall Ford Dam represented a breakthrough into giant projects for Brown & Root in the 1930s, and in subsequent decades the company applied its knowledge of dam construction to numerous other large projects (see table 6.3). In the five years immediately after World War II, the company worked on five major dams in the United States at a total cost of over $100 million. This experience gave Brown & Root the technical know-how to respond to similar opportunities outside of the United States.

Yet the company at times encountered unexpected problems applying this technical expertise in new settings. One of its most exotic dam projects in these years was the Peligre Dam in Haiti. As the company completed this work in the early 1950s, it encountered a variety of labor-related issues peculiar to major construction work in a developing nation. Most pressing was the need to train some thirty-five hundred local workers in the use of modern construction equipment. A Brown & Root employee later described the process of change: "As a result of an intense training program, Haitian workmen are now wheeling big earthmovers, bulldozers, and other heavy equipment around like experts. . . . The project will probably mean the gradual end of such standard Haitian construction methods as breaking up rock for roadbeds with small hammers in the hands of an entire family."[27] Even more difficult was the challenge presented by the poverty of many of the workers: "They were so malnourished that accidents could be fatal a lot of times that you'd never think that they would be. It was just one of those things, they wouldn't survive."[28] The company addressed this problem by providing special diets to the workers to increase their stamina. There was, however, no ready solution to a final labor-related problem: the workers refused to work on traditional holidays and festivals, adding to the already

TABLE 6.3. BROWN & ROOT DAM CONSTRUCTION PROJECTS, 1936–62

Dam	Location	Awarded	Cost
Marshall Ford	Texas, USA	1936	$20,000,000
Nimrod	Arkansas, USA	1941	2,000,000
Granby	Colorado, USA	1946	12,300,000
Fena River	Guam, US Territory	1948	11,000,000
Bull Shoals	Arkansas, USA	1947	39,188,000
Granite Shoals	Texas, USA	1949	(cost below)
Marble Falls	Texas, USA	1949	8,100,000
Morelos	Mexico	1949	6,350,000
Canyon Ferry	Montana, USA	1949	12,700,000
Artibonite River Valley	Haiti	1953	28,000,000
Bhumiphol Dam & Power Plant	Thailand	1958	20,500,000
Flaming Gorge & Power Plant*	n/a	n/a	29,602,497
Tantangara & Tunnels	Australia	n/a	17,125,000
Big Bend Powerhouse & Substructure	South Dakota, USA	n/a	17,461,918
Troneras	Columbia	n/a	4,252,000

* This project was constructed by Mid-Valley, Inc., a wholly owned subsidiary of Brown & Root that, unlike Brown & Root, was a unionized company.

Source: *Brown & Root: Engineers and Constructors* (Houston), various years, Herman and George R. Brown Archives, Houston, Texas.

lengthy delays from bad weather. The management of Haitian laborers presented challenges to Brown & Root that made the NLRB appear tame by comparison.

The project's funding also represented a departure for Brown & Root. The United States agreed to provide $30 million through the American Export-Import Bank to the Republic of Haiti for the dam and associated irrigation equipment. The dam was intended to assist local rice farmers in the eighty-five thousand acres located in the lower Artibonite River valley, near the border between Haiti and Santo Domingo. The dam itself was to be the highest buttress-type dam in the Western Hemisphere, stretching 282 feet high and 1,200 feet long. The combined dam and irrigation network together required more than 20 million cubic yards of excavation and 350,000 cubic yards of concrete. The hydroelectric portion provided electricity to the capital city of Port-au-Prince.[29] This was obviously an ambitious project for which a small nation such as Haiti required outside financing. The Export-Import Bank was one of several international lending agencies sponsored by the United States to help developing nations build much-needed infrastructure; greater economic prosperity was seen as a bulwark against third world radicalism. Brown & Root found itself as the sole contractor of a substantial project in a new, U.S. government–sponsored market for heavy construction in the developing nations.[30]

The company also found itself in the midst of political turmoil during 1956 as it completed its work in Haiti. Merritt Warner, the project accountant, was the last Brown & Root employee to leave the Artibonite project. The dam was finished just as Haiti fell deeply into political chaos, and the president, Paul Eugene Magloire, fled the country to Jamaica during the final days of the dam construction. Warner was also trying to leave the country, but he had to wait for some of his company checks to clear. He managed to enter the firm's bank with the assistance of a bank employee, a former American marine who had remained in Haiti since the U.S. occupation of 1917. He allowed Warner to confirm that the checks had cleared. However, as the last Brown & Root employee on the island, he was unable to write a company check to pay for the shipment of the project records back to Houston since the checks required two signatures. So Warner used his own money to pay the shipping costs. He then raced to the airport and fled to the United States.[31]

A second giant dam constructed by Brown & Root at about the same time proved more routine. The Bhumiphol Dam located in Thailand was, when completed, the largest dam in Southeast Asia and the seventh largest of its type in the world. This dam cost $53 million and was part of the larger Yanahee hydroelectric project, costing about $175 million. The completed concrete dam on the Ping River rose 505 feet and was 1,592 feet long. Its structure contained 1.3 million cubic yards of concrete. The Yanahee project had a significant effect on Thailand's economy. Its reservoir contained a 128-mile-long, 5-mile-wide body of water equalling 9.8 million acre feet. The electricity produced by the dam provided enough energy for half of Thailand's population of thirty million. The irrigation potential of the dam was 600,000 acres, providing the water necessary for the nation's massive rice production. In May, 1964, King Bhumiphol Aduldet threw the switch that initiated the dam's electricity generation.[32] As had been the case in the United States, the building of dams and associated hydroelectric plants in other parts of the world were the big projects that brought much publicity to Brown & Root while transforming the living conditions of millions of people.

Of course, the traditional strength of Brown & Root was in the construction of roads, another significant type of infrastructure that greatly affects people's lives. Herman Brown remained especially fond of reminding those around him that road building was "the field in which Brown & Root started and we feel we practically wrote the book on road construction in Texas."[33] After the war, the company continued to add new chapters to this book, as new construction evolved to include modern highways and ever larger bridges. The firm continued to work on roads in Texas and around the world. The "Highway, Tunnels, and Bridges" division of Brown & Root remained a leader in its field and a significant contributor to the company's growth. One much-heralded road project in the Houston area was the construction of the Gulf Freeway, a fifty-mile-long stretch of four-lane highway from Houston to Galveston. Built at a cost of $28 million, this freeway officially opened in August, 1952. It represented the culmination of almost fifteen years of planning. Built along an abandoned right-of-way once used by an interurban rail line, the Gulf Freeway was hailed by the local press as "the longest toll-free superhighway built in the nation since the end of World War II."[34] This was one of the first phases of a massive highway construction program that made Houston one of the first "freeway cities." Ironically, the freeway ran through existing offices of Brown & Root, forcing the company to relocate as it completed the project. The new highway greatly reduced the driving time from Houston to points south, opening up the entire area between Houston and Galveston to accelerated development. In this sense, the new Gulf Freeway helped prepare the way for the future Manned Spacecraft Center a decade later.

Perhaps the company's most well-known postwar civil transportation project was the Lake Pontchartrain Bridge extending from New Orleans north across the lake. During late 1954, the Louisiana parishes of Jefferson and St. Tamany jointly offered for bid the planned Greater New Orleans Expressway, which included a two-lane bridge approximately twenty-eight miles long, as well as the south and north approaches. This was one of the world's longest bridges. The Louisiana Bridge Company, a joint venture led by Brown & Root and including T. L. James & Co. of Houston, won the contract with the low bid of $30.7 million.[35] (The final price of the bridge rose significantly to nearly $50 million.[36])

The bridge attracted great popular and technical interest, and the local press followed its progress. For Brown & Root, the project's most forgettable day may well have been the first. On that first day at the north side of the lake near Mandeville, a large group of reporters assembled to observe the first pile driving. The piling was ninety feet long and sixty inches in diameter, and it fell and broke into two pieces as the crowd looked on.[37] The bridge consisted of "three precast concrete elements: a pair of prestressed hollow cylindrical piles capped with a reinforced concrete beam forming the piers that support the prestressed monolithic deck spans."[38] Brown & Root built a $6 million shore-based plant to mass-produce the prestressed concrete members of the 2,246-span structure. The company designed mechanically operated steel forms to cast fifty-six-foot-long, 185-ton railway sections. Brown & Root finished the bridge four months ahead of schedule, and it opened for traffic amid much fanfare on August 30, 1956. The wide publicity surrounding the construction of the Lake Pontchartrain Bridge reinforced Brown & Root's regional reputation as an efficient and technically innovative builder.[39]

Despite the company's continued success in public works, its greatest departure in the postwar era was its aggressive expansion into construction for private businesses. Responding to a boom in demand for its services, particularly in the rapidly growing Houston area, the company quickly established a strong position in numerous markets for industrial construction, including natural gas pipelines, petroleum refineries, chemical plants, hydroelectric plants, paper mills, gasoline plants, steel mills, marine work (including docks, offshore pipelines, and drilling rigs), and mining facilities.[40] Industries of all sorts migrated to the Texas Gulf Coast in the 1940s and 1950s, and Brown & Root met them with a broad range of construction services.

In 1952, as an invited speaker to an audience of McGraw-Hill editors who were meeting in Houston, Herman Brown surveyed his company's status. His remarks reflected his pride in Brown & Root's presence throughout the world, but he left no doubt where he considered home: "Our first love, however, is the Southwest where I started this business in 1914. You can't do business all over an area for 38 years without feeling you have a certain proprietary interest in the locality." Herman Brown offered a straightforward explanation of the company's success in its home region: "Every rooster is king of his own dung-hill, and in the Southwest Brown & Root can simply do a job cheaper than most of its competitors."[41] According to one longtime employee, he delivered this same message in a more pointed way to one potential competitor, Henry Kaiser. When Kaiser seemed to be interested in bidding for jobs in the Houston area, Herman apparently took direct action. "He knew old man Kaiser and he just told him cold turkey, 'you come over here and I'm going to whip your butt.'"[42]

For a time in the boom years of the late 1940s and early 1950s, Herman Brown even employed an assistant to undertake "contact work with outside companies seeking to locate plants in the Southwest." To help such companies in their planning, this assistant "had a wealth of information in his files on nearly any given location in the Southwest."[43] Such services no doubt paid dividends when it came time for these migrating companies to choose a contractor to build their new facilities.

The closely related oil, gas, and petrochemical industries were obvious markets for construction. Refinery construction proved a profitable business for Brown & Root. Herman Brown explained the simple logic of diversification into this business: "Building a refinery is just a good-sized pipe fitting job. If he could follow a blueprint, a plumber could build a refinery." As early as the late 1930s, Brown & Root began to undertake construction work on refineries for Humble Oil & Refining Company, and it greatly expanded its involvement in this booming regional market in the postwar years.

From refinery construction to the design and construction of petrochemical plants was a small step the company gladly took. The production of petrochemicals was one of the nation's fastest growing economic activities from the late 1930s through the

1960s, and the Gulf Coast was a favored location for petrochemical manufacturing. An oil industry publication (1951) summarized the region's attraction for this fast-growth industry: "The Gulf Coast . . . had what it took to foster petrochemistry. There was a bountiful supply of natural gas and refinery gases for raw material and economic fuel. There was a reservoir of trained manpower. . . . There was access to deep salt water for shipping and certain base chemicals; there was also plenty of fresh water for industrial purposes."[44] This list might have been appropriately extended to include the availability of the specialized construction knowledge and experience needed to build petrochemical plants.

Brown & Root was a prominent local construction company, and it quickly entered this promising field. Its first major project came in 1946, when it won the engineering contract for a chemical plant on the Houston Ship Channel near Pasadena, Texas, for Diamond Alkali. Other contracts followed, climaxed by a $100 million project for a polyethylene plant for Union Carbide at Seadrift, Texas. On the Gulf Coast and in other manufacturing regions, chemical and petrochemical plant construction remained an important activity for Brown & Root (see table 6.4). The organization of a "Petroleum and Chemical" division showed that plant construction had arrived as an important market for the company by the early 1950s. The company also won contracts to design and build other types of industrial plants, including numerous paper mills, power plants, and parts of Armco's large Houston-area steel mill.[45]

Equally significant were other petroleum-related activities housed in the "Pipeline, Oilfield, and Marine Work" division. The company found new work in various aspects of petroleum-related construction.[46] Perhaps the most important was the construction of facilities for offshore oil and gas production. As oil exploration firms took their first tentative steps into the unexplored Gulf of Mexico, Brown & Root was there to assist them.

The company participated in the creation of the modern offshore oil industry. It first entered the Gulf of Mexico in 1937 and 1938 with the construction of a pier out into the ocean for Humble Oil's efforts to produce oil at McFadden Beach, near High Island, Texas. At the same time it built a giant wooden platform for use by Pure Oil and Superior Oil in developing the Creole field a mile off the Louisiana coast near Cameron, Louisiana. Then in 1947 it designed and built the platform for the world's first offshore producing well out of sight of land. This marked a significant step in the industry's evolution, since offshore oil and gas production grew steadily as an important source of domestic oil. Brown & Root's client for this historic project was Kerr-McGee, an Oklahoma-based oil company. A team led by George Brown designed the first platform, completed in November, 1947, in the gulf about twelve miles south of Morgan City, Louisiana; it stood in less than twenty feet of water.[47]

TABLE 6.4. PETROCHEMICAL PLANTS CONSTRUCTED BY BROWN & ROOT, 1950–62

Company	Year	Plant	Location	Amount
Creole Petroleum	1955–58	Pressure Maint.	Venezuela	$94,260,000
Canadian Chemical	1951–54	Chemical Plant	Canada	76,700,000
Azienda Nazionale	1955–59	Chemical Plant	Italy	40,000,000
Union Carbide	1957–59	Polyethylene	Puerto Rico	28,400,000
Union Carbide	1952–55	Chemical Plant	Texas	64,009,000
Union Carbide	1959–62	Chem. expansion	Texas	55,380,000
Celanese	1951–53	Chemical Plant	Texas	18,596,000
Diamond Alkali	1956–59	Chlorinated Prod.	Texas	16,110,000
Geigy Chemical	1962–64	Diazinon Unit	Alabama	12,797,000
Union Carbide	1956–59	Chemical Plant	Brazil	11,196,000
Diamond Alkali	1951–53	Chlorine Plant	Texas	10,609,000
Celanese	1961–62	Chemical Plant	Texas	10,100,000

Note: Only projects valued above $10 million are included.

Source: Adapted from "Brown & Root, Inc., and Subsidiaries Summary of Projects—$5 Million and Above," Dec. 31, 1977.

Decades later, in November, 1982, Brown received the American Petroleum Institute's Gold Medal for Distinguished Achievement for his contribution in designing and building this historic platform. The API's award gave Brown personal credit: "It was your vision that led to the design and construction of the first producing offshore drilling platform thirty-five years ago. This effort pioneered what has since become extensive exploration, drilling, development and production of petroleum in waters all over the world—today a vital element in the critical search for petroleum and energy."[48]

In the post–World War II years, this search fed a boom in offshore construction, pushing the industry farther and farther out in the gulf. To help in the building and servicing of offshore platforms, Brown & Root bought several surplus ships from the U.S. military and converted them into derrick barges. Brown & Root became an acknowledged leader in offshore platform construction and laying underwater pipelines. The Greens Bayou facility, located on part of the former Brown Shipbuilding grounds, fabricated some of these products. Brown & Root later used lessons learned in this pioneering work to remain a leader in the construction of specialized equipment needed to produce and transport deepwater oil and gas around the world.[49]

Brown & Root recorded many firsts in its offshore work in the Gulf of Mexico. It installed the first platform in 100 feet of water off the Louisiana coast. The firm set a new record by erecting a drilling platform in 285 feet of water in the gulf, and the depths continued to increase. The firm set another record when it laid in the gulf nearly three and one-half miles of concrete-coated oil pipeline in one day. During one remarkable year, Brown & Root constructed twenty-five permanent drilling platforms.[50] By the mid-1950s Brown & Root, along with New Orleans–based rival J. Ray McDermott, had gained preeminence in the construction of offshore platforms and pipelines.

A hurricane in the mid-1950s that knocked out a large number of Gulf of Mexico rigs became a good advertisement for Brown & Root work. George Brown recalled:

> *At first many oil companies built their own drilling platforms. All had different ideas on how to save money. The hurricane in the '50's knocked out about half of the platforms in the Gulf, but the ones we had built survived the pounding, and after that it was easy sailing for us. The oil industry saw what our platforms could do. . . . We had studied tides, currents, waves, and winds, and we knew that it wasn't the winds that hurt, even hurricane-force winds; it was the waves. The waves can batter hell out of a platform. So we built our platforms high off the water so the waves couldn't get to them. Our platforms were higher than the others, and we rode out the storm.*[51]

The natural gas industry—both onshore and offshore—was a major market for Brown & Root, which became deeply involved in this business after Herman and George Brown acquired a substantial interest in the Texas Eastern Transmission Company. Indeed, while the Browns studied the feasibility of purchasing the Inch Lines from the government and converting them to natural gas, they called on technical expertise from their newly created petrochemical and engineering division at Greens Bayou. One of the first projects undertaken by the engineers who had transferred to Brown & Root from Brown Shipbuilding involved the Inch Lines. Once the Browns completed their acquisition of the lines and created Texas Eastern Transmission Company (TETCO) in 1947, Brown & Root took a large and ongoing role in building and maintaining the pipelines and compressor stations required by the new company.

Launching a jacket in the Gulf of Mexico. *Courtesy Brown & Root.*

This proved to be a substantial market for Brown & Root, since TETCO pursued a vigorous policy of expansion in its formative years. As one of the three major gas pipelines carrying southwestern natural gas to the Northeast, Texas Eastern expanded rapidly to keep up with the surging demand for its product. Both companies, and other Brown & Root clients, followed the supply of natural gas offshore in the Gulf of Mexico in the 1950s and 1960s, and Brown & Root played the leading role in building the massive pipeline complex that effectively tied these offshore gas fields into the existing market for natural gas. As it expanded both onshore and offshore, Texas Eastern enjoyed the luxury of having a major construction company that would pay close attention to its special needs. Brown & Root had the inside track on a growing and lucrative market for its services with one of the nation's fastest-growing gas transmission companies.

In the 1950s this relationship brought Brown & Root excellent opportunities to build gas pipelines and related facilities throughout much of the United States. By the late 1950s, it followed Texas Eastern into several significant international ventures involving the construction of gas pipelines, and George Brown made numerous trips to Europe in these years to investigate the prospects for both companies in building and running substantial pipeline projects.

This was simply one part of the broader movement of Brown & Root throughout the world in the postwar years. The company's search for both public sector and private sector construction jobs took it all over the globe. In addition to the giant military bases built in Guam, France, and Spain, the company expanded into Mexico in 1950 to construct a pipeline and roads, Canada in 1951 to build a petrochemical and synthetic fiber plant, and Venezuela in 1951 to build a series of gas injection plants.[52] Such projects carried the name of this Texas company into nearby nations, a trend that took Brown & Root all over the world in subsequent decades.

By any measure, diversification into a variety of markets after the war brought an era of unprecedented expansion, propelling Brown & Root into the ranks of the nation's major construction firms. All of this expansion and diversification altered the Browns' roles within the company. No longer could they stay closely involved with all aspects of its operations. Decentralization of management responsibilities allowed Herman to monitor the results, if not the daily activities, of the various project managers. George became less involved in engineering, but he had more than enough to do pursuing new projects around the nation and the world. Both enjoyed the bustle of activity that was Brown & Root in the postwar years, and both took great pride in the far-flung activities of what had once been a regional "road building" company.

In the early 1960s, Brown & Root had more than ten thousand employees active all over the world. It was one of the largest engineering and construction companies in America, and it had carried the Browns to exotic places throughout the world while making them wealthy men. Yet despite the growing size and prominence of Brown & Root, the Brown brothers worked hard to keep a low profile for themselves and for their company. A *Business Week* cover story about the company in 1956 noted that "headquarters for one of the largest and most versatile construction companies in the world are in Houston. Yet, if you call its switchboard there, the operator will say simply and anonymously: 'Fairfax 3-7121.' That's the way brothers Herman and George R. Brown who head and personally control Brown & Root, Inc., want it."[53]

Yet this low-profile approach to management did not apply in one key area—labor relations—that kept the Browns and their company in the spotlight of public scrutiny in the decade after World War II. As Brown & Root's managers faced the new challenges posed by diversification, they also faced the old challenges posed by labor-management relations in a new context. Indeed, in the postwar years the struggle with newly aggressive outside unions for control over the Brown & Root workforce became a central concern of Herman Brown.

Despite the growing acceptance of unions in the American political economy, Brown was reluctant to modify the approach to labor that he had learned on the highways of Central Texas in the formative years of Brown & Root. His strongly-held antiunion views were, in his own words, "close to my heart."[54] For twenty years he had lived and worked with his men, and he felt that he understood their interests more clearly than did labor unions.

He also felt a strong personal responsibility to ensure full-time employment for his loyal workers, and he feared that unions would hamper his efforts. In particular, he feared that if Brown & Root went union, longtime Brown & Root workers might suffer from a loss of seniority and job choice. He also dreaded the creation of a time-consuming grievance process. One of his managers summed up the company's attitude in a letter to the editor during a strike in Austin: "Nor do we need a 'grievance committee.' The Brown & Root men seldom have a 'grievance, but when they do, they have free access all the way up from the foreman to the president of the company."[55] Critics scoffed at such claims, but they misunderstood Herman Brown if they considered him a hypocrite who talked of loyalty as a smoke screen to allow him to exploit of his workers. He deeply believed that he would treat his workers more fairly than would a union, and he deeply resented the idea that a union representative wanted to stand between him and the individual worker.

Economic realities in the labor-intensive construction industry reinforced Brown's philosophical belief in the open shop. He noted that "on a union job if there is no work for carpenters on a certain day, they are laid off. On our jobs we can put carpenters to work doing something else and they stay on the payroll. A union would not permit crafts to cross lines." Brown voiced disdain for union practices that limited workers to strictly defined crafts: "None of these practices add to the excellence of the finished product and they certainly add to the cost. We think this is a needless and shameful waste of a customer's money."[56] Even after Brown decided to match the basic wages won by unions in regions where he had projects in order to recruit workers and dampen their incentive to organize, he still believed that nonunion labor gave Brown & Root lower labor costs and greater flexibility than its rivals who used union labor.

Changes in Brown & Root's work force and in the rules governing labor-management relations meant that Herman Brown had to defend his traditional views of labor in an environment decidedly different from that of the early years of road building in Central Texas. The most obvious change within Brown & Root was the scale of the company's operations. From the Marshall Ford Dam project through the ship-building years and on into the era of postwar diversification, the company hired tens of thousands of new workers for jobs around the world. The close personal ties between owner and worker that had grown in a small company doing business in a single area were no longer possible. Like Andrew Carnegie and others who had built the first giant, modern factories fifty years before, Herman clung to the notion of a "family" of workers and managers even as Brown & Root's growth into one of the first giant, worldwide construction companies created the need for the mobilization of large numbers of temporary crews who could be at best stepchildren within Brown & Root.[57]

As the company grew rapidly from the mid-1930s forward, the laws regulating labor relations also changed rapidly, giving new federal government agencies expanded powers to mediate labor disputes. Herman Brown often remarked that his workers could always come to him with a grievance, since "there was no disinterested third party between us." But after the creation of the NLRB in 1935, a federal regulatory agency stood between every employer and every employee. This mediation board established a new structure and new rules for labor-management relations in the United States, creating a government-sanctioned process for workers seeking to create independent unions. The NLRB served a useful societal purpose by taking labor-management disputes off of the streets and into a hearing room. In the process, it also tilted the balance of power in such disputes sharply toward organized labor, at least in the years before the passage of the Taft-Hartley amendments of 1947. While creating the NLRB in 1935, the Wagner Act enumerated a long list of illegal "unfair labor practices," thereby taking from the hands of management numerous tools traditionally used to combat independent unions. It also established NLRB-managed procedures for holding binding elections in which workers could choose to create independent unions. Thus, after 1935 a "disinterested party" did stand between Herman Brown and his workers, although Herman and many other employers regularly complained that the newly created NLRB favored unions and was far from neutral in its interpretation of the law.[58]

Herman Brown experienced the impact of such changes firsthand in the late 1930s and the war years. During the construction of the Marshall Ford Dam, disagreements over the use of union workers and the proper wage scales to be paid to various classifications of workers caused recurring tensions between the contractors and several federal agencies that supervised this government-financed project.[59] At Brown Shipbuilding, Herman regularly skirmished with labor organizers and the NLRB over the union certification process. Although the unions won a series of NLRB victories over Brown Shipbuilding, the company's management at times invoked the overriding need to build ships for the war effort in order to stall the implementation of the results of these elections. First and foremost a pragmatic businessman, Herman Brown proved willing to bend toward the acceptance and use of union workers during the war. After the war Brown & Root even took the practical step of organizing a unionized subsidiary, Mid-Valley, to use on jobs that could not be undertaken without unions. But Herman Brown never abandoned his philosophical commitment to his vision of a family of workers and managers—with him as the strict but fair father.

This vision was threatened by the spread of unions across the industrial landscape with the end of World War II. Brown & Root and other large employers of industrial labor faced fundamental choices regarding labor relations. Independent unions had grown rapidly in the decade from 1935 to 1945, and war's end brought a surge of new organizing activities. On Brown & Root's home turf in Texas, strong new CIO and AFL unions spread quickly among the refineries and petrochemical plants. The heavily industrialized area from the Houston Ship Channel to the Beaumont–Port Arthur region some one hundred miles to the east, quickly became a stronghold for organized labor. Near the center of this region was the Browns' Greens Bayou site, where Brown & Root's primary fabrication yard quickly replaced Brown Shipbuilding facilities after the war. From this vantage point and through discussions with other Houston area business leaders, the Browns had a front-row seat from which to observe the ongoing war between owners and unions. As employer after employer along the ship channel negotiated truces with unions or gave up the fight when confronted with orders from the NLRB, the Browns decided to stand and fight.

They did not take the threat to their own company lightly, since they strongly believed that unions would hurt the regional economy as well as the owners and workers at Brown & Root. Manning the barricades in this war against the spread of unions with all the economic, political, and legal weapons at their disposal, they emerged as symbols of successful antiunionism in their region and in the national construction industry.

Herman Brown brought an intensely personal passion to the struggle. In the years immediately following World War II, he seized the initiative in fighting the spread of unions in Texas. The first battlefield was in Washington, where he joined many other business leaders led by the National Association of Manufacturers (NAM) in lobbying for changes in the Wagner Act. The result was the 1947 passage of the Taft-Hartley Act, which swung the power of the federal government back toward the side of management.[60] Lyndon Johnson, who by this time had developed close ties to the Browns, confounded some observers by voting for the act in the House. This vote undoubtedly helped the Browns convince like-minded businessmen in Texas that Johnson was worthy of their support. This vote was a litmus test of sorts for Johnson, for Herman Brown cared too deeply about this issue to continue to support politicians who took the other side.

The Taft-Hartley Act reserved the states' rights to pass open shop laws—that is, laws that forbid the requirement of union membership as a condition of employment. Herman Brown believed fervently in the "right to work," and immediately after Taft-Hartley passed, he joined the campaign for a similar right-to-work law for Texas. In this crusade, Brown certainly was not alone. One survey in Texas (1946) found that 71 percent of Texans surveyed favored the abolition of the closed shop, and sentiment in the state legislature strongly favored this law.[61] Indeed, Herman Brown was not prominently mentioned in the newspaper accounts of the vigorous campaign to pass a wave of new laws restricting union power to organize and operate in Texas. One labor leader in 1947 summed up the situation in Austin: "You might turn any legislator upside down and shake him and an anti-labor bill would fall out of his pocket."[62]

When the new legislature convened in 1947, a well-organized lobbying campaign by business groups in Texas led by the Texas Association of Manufacturers wasted no time in pushing to the forefront a cluster of antiunion bills. In a frenzy of activity, no fewer than nine new antiunion laws passed rapidly through the legislature, all approved by Governor Beauford Jester, a friend and political ally of Herman Brown. The centerpiece was the Texas Right-to-Work Act. It stated that "the right to work is the right to live" and proclaimed the "inherent right of a person to work and bargain freely" before stipulating that "no person shall be denied employment on account of membership or nonmembership in a labor union."[63] Other new laws forbade the "check-offs" of union dues without the written consent of the employee, blocked picketing by more than two people within fifty feet of a plant entrance, and outlawed secondary boycotts in support of strikers. The crusade to make unions subject to the state's antitrust law also succeeded.

Taken as a whole, these laws dramatically tilted the legal balance of power between owner and worker in Texas. The National Labor Relations Act of 1935 had created new rules for collective bargaining that had given labor unions in the United States new legal and regulatory weapons to use in organizing workers. The postwar wave of restrictions on labor's activities swung the balance of power in labor relations back decidedly toward business. So total was the antiunion victory in the Texas legislature that one bitter prolabor legislator introduced an amendment to one bill to "abolish unions, confiscate union members' property, line all union members up against a wall and have them shot, and send their families to concentration camps."[64]

Herman Brown no doubt joined his colleagues in a good laugh over this amendment. They had plenty of reason to smile in the aftermath of their rout of unions before the Texas legislature. But the time to celebrate their political triumphs was short, since outside of Austin the struggle over union organization continued. This was particularly true for Brown & Root, which faced a series of pitched battles in Arkansas and Texas with the building trades unions in the years from 1947 through the early 1950s. Herman asked and gave no quarter in his defense of the principle of the open shop at Brown & Root.

This became clear in the opening skirmish between the company and organized labor in this era, the prolonged and bitterly fought battle of Bull Shoals in 1947 and 1948. Brown & Root was a member of an eight-company consortium, the Ozark Dam Constructors, which contracted with the Little Rock District of the Corps of Engineers to build an ambitious combination flood control and power generation project in the Ozark Mountains near the Arkansas-Missouri border. The same group of eight companies formed Flippin Materials Company in November, 1947, to supply crushed stone aggregate mined from a nearby quarry for use in construction of the dam. One general superintendent managed both projects, and Brown & Root

Bull Shoals Powerhouse under construction, 1951. *Courtesy Brown & Root.*

was "empowered as attorney-in-fact to direct all the operations of both joint ventures, and to exercise control of their personnel and labor relations."[65]

Brown & Root was the only member of the consortium that remained in 1947 as an open shop company with a strong commitment to fight against the closed shop arrangements preferred by the local building trades. As Herman Brown was quick to point out, Arkansas had passed a right-to-work law that made the closed shop illegal in the state, but the unions replied that this did not prevent the creation of de facto closed shop in which contractors agreed informally to such arrangements while avoiding overtly violating the right-to-work laws in the formal contract. Unlike his partners in the consortium, Brown was not willing to play this game, and he quickly emerged as the point man in the effort to combat the demands of the union. The partners agreed that the exclusive use of unions would significantly raise the cost of the projects, and they deferred to Brown & Root's desire to resist the demands of the local building trades councils.[66]

These AFL-affiliated unions felt that the Bull Shoals project represented an excellent prospect for expanding their presence in construction in Arkansas and throughout the South, while also holding out the possibility of a highly publicized victory over one of the region's most vocal and visible proponents of antiunionism, Brown & Root. Organized labor was on the ascent throughout the nation, and significant gains had already been made with other members of Ozark Dam Constructors. Even the political environment in the state might provide some support for their efforts, since these were Arkansas local unions challenging large construction companies based outside of the state.

After the failure of their initial attempts to gain a voluntary agreement by the consortium to use only union labor, the building trades turned to the NLRB certification process. In March, 1948, Ozark Dam Constructors received notice of a petition by the AFL Joint Council in the region seeking an NLRB election. Accompanying the notice was another appeal for a conference to seek a cooperative solution. Alvin Wirtz, whose firm of Powell, Wirtz, Rauhut & Gideon, with offices in the Brown Building in Austin, took charge of the legal work generated by this dispute, quickly replied that the NLRB had no jurisdiction in this matter, since construction of the dam did not involve interstate commerce. He then wrote to Herman Brown informing him of the exchange of letters and reporting his view that "the General Counsel of the National Labor Relations Board has determined to make a 'guinea pig' of the Ozark Dam Constructors in order to establish jurisdiction over general construction" on the grounds that the project required the movement of large quantities of materials in interstate commerce. He urged Herman to seek the input on strategy of others in the consortium and the industry, "since this matter will no doubt result in a test case of vital importance to the construction industry."[67]

As events unfolded at Bull Shoals, Herman Brown received regular advice from Wirtz in Austin and from Brown & Root managers at the site. In June, 1948, Wirtz counseled Brown to try to win the certification election instead of filing suit against the NLRB on the jurisdictional issue since "there is a possibility that we will win the election" and Bull Shoals does not afford "an ideal case for a test" of the NLRB's jurisdiction over original construction.[68]

After Brown accepted this advice, he received a strongly worded handwritten letter from longtime Brown & Root employee Tommy Tompkins in early July, several weeks before the NLRB certification election for machinists. Tompkins had examined the books of another construction company active in the region around Bull Shoals, and he told Brown that "after looking at the rates—I felt it was a wonder we didn't have more trouble." Brown already had authorized pay raises for the project, but Tompkins noted his belief that "the present rates are much more out of line with the rest of the country, than the differentials authorized by your wire." He argued that "the only way to fight fire is with fire—and the project Mgr. should have whatever fire is necessary to bring wage rates in certain trades in balance with others, as well as to feed small raises, when necessary to offset Union propaganda." He closed by noting that "I don't believe any one can appreciate more than I the tremendous cost of operating a union shop that has nothing to do with the rates of wages." The answer was that "the union threat can best be dealt with by the man on the job having the ammunition." Tompkins' advice summarized a basic antiunion policy that gradually emerged at Brown & Root in these years—the most effective way to prevent the organization of unions was to pay prevailing wages, especially in heavily unionized areas or when faced with a serious threat of organizing. But the raises suggested in his letter proved too little and too late for Brown & Root at Bull Shoals, and the AFL won the right to represent the building trades on the project in the elections mandated by the NLRB.[69]

The day after the machinists election, Brown & Root embarked on a new strategy based on delaying serious negotiations with the AFL's representatives while moving forward as quickly as possible with work on the dam. One of the company's supervisors at Bull Shoals wrote Herman Brown: "Now that we have the Union to deal with it appears to me that I should not be in a position of bargaining with the Union Agents." Herman agreed, advising that "your answer to them should be that you are not discussing the bargaining until you have been advised by your attorney, Senator Wirtz." Brown continued that it would probably take Wirtz a "month or more before we will get to the point where we will have to go through the motions of talking to the agents." As the unions demanded "good faith bargaining," Brown & Root used its excellent lawyers to delay negotiations.[70]

In late August, Wirtz informed lawyers for the Joint Counsel of the AFL that bargaining could begin "with the stipulation that by so doing the Employer does not thereby admit that the National Labor Relations Board had jurisdiction," an issue that was now being placed before the courts. Wirtz then advised that Ben H. Powell Jr., who would negotiate for Ozark Dam Constructors, would not be available to begin until after his vacation ended in the middle of September. Once negotiations did begin, the two parties found numerous areas of fundamental disagreement, including the inclusion of an open shop article in the proposed contract.[71]

In October the NLRB added a new element to the conflict by charging the consortium with illegally firing about three hundred known union sympathizers on the eve of the certification election in July, 1948, refusing to bargain collectively with the unions after their victory in the election, and threatening employees attempting to exercise their legal rights to organize. This charge set in motion a round of legal challenges and responses that went forward throughout the remainder of the Bull Shoal controversy and was not fully resolved until the early 1960s.[72]

In response to the slow pace and the lack of movement in the negotiations, union representatives called for a strike in early December, 1948. Pickets went up throughout the job site, and the conflict entered a more bitter phase. Over the next nine months, strikers and workers frequently clashed as work proceeded despite the picket lines. At the height of the picketing, the NLRB estimated that about half of the approximately one thousand workers on the payrolls of Ozark Dam Constructors and Flippin Materials ceased working. The outcome of the strike remained in the balance for several months as a test of wills ensued. Recognizing the seriousness of the situation, Herman Brown personally spent two weeks of the strike at the site helping to boost the morale of workers who were crossing the picket lines. According to one longtime Brown & Root employee, "it was a dangerous situation. Herman's presence showed his loyalty to the workers."[73]

More than three months into the strike, other partners in the consortium began to exert strong pressures to settle. J. S. Bonny, the head of the Idaho-based Morrison-Knudsen Company, sent a letter to Herman Brown and all of the other partners on March 16, 1949, calling for a meeting of the consortium to discuss ending the strike. After noting the "confusing labor picture" at Bull Shoals, Bonny said that he had received "a number of calls from Union representatives in different locations" threatening to picket other jobs if "satisfactory negotiations were not concluded" at Bull Shoals. He also reported on a telephone call in which the president of the Building Trades Council of the United States, Richard Gray, informed him that Senator Robert Taft felt that every member of the consortium was liable for legal action for violating the Taft-Hartley Law.[74] Herman Brown responded to Bonny's plea for a settlement with a forceful statement that there would be no retreat, no surrender. He reminded Bonny that they had agreed before the project began that it would be operated on an open-shop basis and that to change this now might have serious financial ramifications. Brown characterized the statements of the president of the Building Trades Council as "pure bluff. . . He knows that they have lost the battle at Bull Shoals and that, as of today, he could not win an election in any craft." Gray's assertation about violations of the Taft-Hartley Act was "a plain lie." Brown summarized his view of the matter by saying that "I consider Gray's squack [Brown's word] to you as one last effort to put pressure on us . . . since he knows that he has lost out . . . He or some of his group has gone to every personal friend that I have in both the Senate and the House in Washington to try to get them to put pressure on me to let up." Brown said that he would be glad to meet with his partners to discuss "the progress of the job" and he concluded that "I do not want to see any of the joint venturers hurt by this operation and I do not think you will be if you insist on your rights to invest in any joint venture you see fit without being harassed by them [the unions]."[75] In the face of Brown's obvious determination, his partners backed away from demands for a settlement and let the strike take its course, despite the threat of union retaliation at other sites.

The conflict moved to other venues as the picketing continued. Decisions in a series of court cases established the NLRB's jurisdiction over construction. In a separate court action, the consortium made effective use of a traditional weapon of business against strikers by obtaining a court injunction in May, 1949, against picketing at Bull Shoals. The consortium's negotiator, Ben Powell, maintained a hard line in his talks with representatives of the unions. The two sides remained far apart on issues of wages and conditions of labor, the subcontracting of electrical and plumbing work to contractors using union labor, and what Powell referred to as the "under-the-table closed shop agreement desired by the AFL." He later summarized the gist of his negotiating stance with a simple statement: "we weren't going to give in."[76]

By the end of the summer of 1949, support for the strike had waned. Negotiations continued, with Ozark Dam Constructors holding the upper hand. Progress on the dam moved forward as workers gradually recognized the inevitability that the strike would fail. Striking workers who sought to return to work generally were not rehired. The collapse of effective picketing meant that union negotiators bargained on Herman Brown's ground. In late September this proved to be literally true, as Brown and his attorney Ben Powell met with four labor leaders at the Brown's Suite 8-F in the Lamar Hotel in Houston. The labor representatives offered to call off the strike in return for a promise that strikers would be reinstated. Powell's uncompromising reply that some, but not all, might be reinstated assured that no agreement would be reached.[77] It had become abundantly clear that the terms of any settlement would be dictated by Herman Brown.

The week before Christmas 1949, more than a year after picketing had begun, one of the strikers, R. L. Ricketts of Gassville, Arkansas, tried a more personal approach to resolving the dispute by addressing a letter to Herman Brown setting forth the point of view of those who had manned the picket lines. Given Brown's great pride in his ability to empathize with workers, it was not surprising that he took the time to read the letter and write a long personal response. This exchange well summarizes the two points of views underlying the postwar battle between union organizers and Herman Brown, a self-made man who still saw himself as a man of the people.

Ricketts begins with a long description of the low wages and poor working conditions at Bull Shoals in comparison to similar construction jobs "under contractors who know that employing competent men, union members or not, and paying a respectable, living wage will get a maximum of work done." He notes the bleak Christmas faced by the strikers: "Personally, unless the Lord provides better than I can foresee at this time, it will be biscuits and beans in our house, unless I can get a few squirrels." Pleading the case for "a grand Christmas present" in the form of a contract calling for a union job and wage scale, Ricketts argues that "you will find that employing competent craftsmen, at decent wages will benefit the company more than the individual worker. . . . Consider labor as something you are buying. Pay for quality and you get it. At cheap prices, you get a cheap product, what is more expensive in the end."[78]

Herman Brown's answer begins by asserting that "I will be as frank with you as you were with me" before informing Ricketts that "the difference between us and the Unions is that they wanted a closed shop. This is against the law in Arkansas . . . therefore we have not come to terms on that basis." In thirty years experience in construction, Brown continues, he has had "just as good a quality workmanship out of non-union men as I have union and have worked both kinds all of these years." Warming to the task, he adds that wage scales at Bull Shoals are "the prevailing wages in that particular part of Arkansas." And "I am also sure they are higher than the wages being paid on State Highway work in Arkansas today." The conclusion moves beyond such corrections of "misinformation" in Rickett's letter to a more personal level: "If I were you and was dissatisfied and unhappy with my work, I would seek employment with which I would be happy—employment where I thought they were paying me what my skill is worthy of."[79]

Herman Brown fought and won the battle of Bull Shoals by taking such an un-yielding stance. Viewing the struggle as defense of a personal commitment to the open shop, he was not prepared to be gracious in victory. He had correctly gauged the situation at hand, taking advantage of the nature of the heavy construction industry to fashion a strategy that proved successful against high odds. He knew that in his industry—unlike in manufacturing industries where industrial and craft unions were making rapid progress—legal delays could buy enough time to complete a project and move on to a new location, leaving the unions to protest before the NLRB and in the courts after it was too late to have a permanent impact on a project. Such an approach required the resolve to keep the work progressing in spite of pickets, and he had demonstrated such resolve in the face of determined opposition from the unions as well as the skepticism of partners in the consortium who had long since made their accommodations with organized labor. A note from one of his managers from the site summed up the sentiment among Brown loyalists within Brown & Root: "[You] sure gave these people good for sure. Wish more people had the same sort of guts. Trust you find some solution to your Beaumont troubles."[80]

This note reminded Brown that there was no time to savor the victory, since the struggle with the AFL quickly had moved to another battlefield, this time in the union stronghold of Beaumont, Texas, amid the industrial plants of the upper Texas Gulf Coast. Angered by Brown's attitude—and his success—at Bull Shoals, the national Building Trades Council encouraged opposition to Brown & Root on projects through-out Texas. If Herman Brown wanted a fight, the Texas AFL was happy to oblige in an all-out effort to force him to recognize unions.

The opening salvo in this new battle came in the spring of 1949 in Beaumont, where Brown & Root had contracted to build a compressor station for Texas Eastern. When the twenty-five-man crew for the project arrived at the job site, they found local trade unionists spoiling for a confrontation on the unions' home turf. Each day the tension and finally the violence escalated. By the fourth day, according to one of the Brown & Root workers, "It became pretty violent. . . . the unions just invaded the job, ran everybody off, put up picket lines with clubs and axe handles. The sheriff was on their side and he was looking the other way all the time. And they finally just ran us off the job."[81] An article by syndicated columnist Westbrook Pegler described the episode as a "riot" in which "law-abiding carpenters were attacked by a mob of 150 goons." He also reported that the police escorted Herman Brown away from the site and out of Beaumont, telling him that if he and his crews did not leave town "they 'could not guarantee to handle the trouble that might ensue.'"[82] Round one clearly went to the unions.

Faced with superior strength in a well-defended locale, Brown decided to fight round two on ground where he enjoyed an advantage: state politics. On June 1, 1949, he wrote a personal appeal to the head of the Texas Rangers requesting "protection for men attempting to work without intimidation and bodily harm under the laws of the state of Texas." Two days later, he met with his friend and political ally Texas governor Beauford Jester for forty-five minutes in Austin, repeating his appeal for "outside help" to curb the violence that had stopped work on the compressor station. A month later, his lawyers filed an appeal with Texas attorney general Price Daniel arguing that the unions in Beaumont had violated the state's antitrust laws.[83] When none of these appeals produced a decisive response capable of permitting Brown & Root to return to the job site, the company decided to quit the field, moving the site of the compressor station across the Sabine River into the state of Louisiana. Brown & Root had been routed in the short-lived battle of Beaumont.

Even before this skirmish had been concluded, a second front opened in the company's home city of Houston. As a member of the Rice University Board of Trustees, George Brown had committed Brown & Root to build a modern, seventy-thousand-seat football stadium on the Rice campus. This project was a highly visible monument to the "coming of age" of Houston as a "major league city," and it attracted much attention in the region and even around the nation. Work was to go forward on a nonprofit basis, with an incredible goal of building the stadium in ten months. Accounts of the origins of the ensuing dispute between Brown & Root and the local Building Trades Council differ, as should be expected in the case of labor-management conflict. Labor representatives later recalled that they had met with Herman Brown to offer their the services "of the best mechanics so that they could get the job done at as low a cost as possible." They saw this offer as a sort of civic gesture similar to that of Brown & Root, and the various supply companies had offered to do the job at cost. They reported that Brown seemed agreeable to their suggestion and promised to talk with others involved in the project and then get back in touch with them. When he failed to contact them, they decided to picket the project as unfair to union workers.[84] Brown & Root's version of this meeting was quite different. Longtime Brown & Root lawyer Everett L. Looney recalled that union representatives at the meeting made a single demand concerning the work on the Rice stadium and on another large, local project, the building of the Baytown tunnel for vehicular traffic: "all union men or none." They also let it be known that they would picket Brown & Root jobs around the Houston area if this demand was not met. When Herman Brown refused this challenge to the open shop, the strike began.[85]

What followed became the focus of an extended court case. Construction on the stadium continued at a breakneck pace despite pickets. Other pickets at Brown & Root's Clinton Drive headquarters and at various of the company's job sites heightened tensions. At one such site, the Stauffer Chemical Company, union leaders thought that they had struck a blow against Brown & Root by convincing workers of the Southern Pacific Railroad to refuse to deliver materials across a picket line. According to these union representatives, they had the job stopped "until Herman Brown had come out to the job . . . and said that he would have the cars in there, and immediately thereafter the officers [of the Southern Pacific] had come to the plant and put them in."[86] Such direct action was the stuff from which Herman Brown's fame and infamy as an antiunion crusader was made. As will be discussed in chapter 8, despite the pickets, Brown & Root met the deadline for completion of the stadium in late September, 1950.

At this time, Brown & Root adopted a new strategy to combat the growing union challenge to its open-shop policies. It moved the dispute into the courts by filing a blanket suit against numerous unions in the state—including the Building Trades Councils in Beaumont, Houston, and Austin; the Texas State Federation of Labor (made up of AFL locals throughout the state); two carpenters councils; and the Houston Labor and Trade Councils. The suit sought relief from what Brown & Root called the illegal activities of the unions aimed at putting the company out of business. Herman Brown explained that the company was "suing them under Texas Statutes— one, that prohibits the closed shop; two, prohibits secondary boycotts and three, violation of the antitrust act." In his view "they [the trade unions] openly made the statement that they were going to make us go closed shop or put us out of business and they started in to do just that without much success. We let them harass and picket us for about ten months. By that time, we had accumulated enough evidence to file suit." He added that "we expect a rather long drawn out affair but we intend to carry it through."[87]

The progress of the case was much discussed in the media. Local newspapers reported the testimony in detail; lawyers from each side argued the merits of their cases in highly reported debates before an annual meeting of the Texas Bar Association.

Both sides brought in their legal heavyweights, with Brown & Root mobilizing the efforts of the firms of both Alvin Wirtz and Edwin Clark. When the smoke had cleared and the appeals had been heard, both sides claimed victory. Brown & Root took comfort in the fact that the courts had ruled for the open shop and against the unions by granting an injunction against illegal picketing that required unions to gain court approval before picketing Brown & Root projects. The unions countered that the courts had upheld their freedom of speech by ruling that legal picketing of Brown & Root could proceed.[88] All in all, this case demonstrated the value to Brown & Root of the antiunion laws Herman Brown had helped put on the books in Texas in the late 1940s, while also proving again the verity that good lawyers are worth their weight in gold in the American legal system.

At considerable cost in legal fees and personal time and energy, Herman Brown had preserved the nonunion shop at Brown & Root in the tense decade after World War II. In the process, he helped to blunt the spread of unions in the construction industry in Texas. Brown was not alone in his strong commitment to the open shop or even in his success at stopping the growth of unions within his company and his home region. Humble Oil (later absorbed by Exxon) and DuPont succeeded in blocking industrial unions in their large Gulf Coast plants in the same era. Brown was also not alone in his ideological commitment to the open shop; numerous executives wrote letters of encouragement to him as he led the fight against unions. Nor were his antiunion attitudes particularly unusual in the business community of postwar Houston.

What was noteworthy was the passion and intense personal commitment of a man who went to Bull Shoals to shore up the troops, was led out of Jefferson County by local police, and stopped the unions from blocking the movement of Southern Pacific trains into his job site. As the unions fought tooth and nail against him, even their members granted a grudging respect to the determination of this salty old man who said what he meant and then backed it up. Among union members, his name became synonymous with virulent antiunionism in postwar Texas; among businessmen he was seen as a throwback to the values of a past era when owners had almost complete control over their properties and their workforces. Within Brown & Root, he was seen simply as "Mr. Herman," the man who had built Brown & Root and who had placed the values of independence and antiunionism at the heart of the company's corporate culture.

In a sense, the successful antiunion crusade of the postwar years merely reenforced Brown & Root's can-do image. Just as the company could go anywhere in the world and complete any engineering task, it also could stand—at times alone—against the wave of unionization that swept over America in this era. In withstanding this wave, he admittedly had more than his personal zeal in his favor. The state of Texas as a whole remained a bulwark of antiunionism despite significant union successes on the upper Texas Gulf Coast. The nature of the construction industry meant that Brown & Root was a moving target, since it completed its work on a specific project and then moved on to new locations. But wherever Brown & Root operated in this era, it left a legacy; the company and its founder remained deeply loved or deeply hated and long remembered by those who participated in or observed the company's no-holds-barred battles with organized labor.

In the years from the end of World War II until his death in 1962, Herman Brown symbolized Brown & Root's self-image as a company of rugged individualists who could get things done. His leadership helped catapult Brown & Root to the top of two lists: the list of the largest diversified engineering and construction companies in the nation and the list of the most visible antiunion companies in the nation. He was proud of both, which he saw as vindication of the lessons he, his brother, and their loyal employees had learned on the roads of Central Texas and applied to giant construction projects throughout the world.

CHAPTER SEVEN

Building Circles of Influence

The hard-earned success of Brown & Root and Texas Eastern gave the Brown brothers a substantial economic base. Their personal wealth, which approached the $100 million mark for each brother in the late 1950s, placed them among the nation's wealthiest businessmen.[1] In Brown & Root's formative years, the construction business—with its heavy involvement in public works—drew the Browns deeply into lobbying for government contracts. With the diversification of their economic empire, the Browns' interest in civic and political affairs also broadened. As Lyndon Johnson ascended from the House to the Senate to the vice presidency and finally to the presidency, the Browns received frequent public mention as Johnson's behind-the-scenes supporters.[2] Equally significant was their ongoing involvement in civic and political affairs in Houston and Austin, the Texas state capital.

The Brown brothers shared a strong lifelong interest in government. As young adults they enjoyed debating questions of political economy. Were the New Deal programs good for the country? How could the United States avoid the worldwide trends toward fascism or communism? What was the proper role of business in American democracy? This last question was far from theoretical to the Browns; they provided a practical answer in the way they lived their lives. They helped elect political leaders at all levels, fostered their vision of a "healthy business climate" at home, and championed free enterprise abroad. They felt that successful members of society had an obligation to use their wealth for social good. They understood that what was good for Houston was good for the Browns, and they used their influence to push for policies that they thought would make their city a better place in which to live and work. As a consequence, they actively supported the arts, higher education, medical facilities, the building of a modern airport, and the effort to bring NASA's Manned Spacecraft Center to the Houston area.

To the Browns, "influence" was not a dirty word but a practical reality of doing business in a democratic system. Government regulation and procurement were especially significant in two industries important to the Browns, construction and natural gas, and they approached politics with the same pragmatism and passion with which they built Brown & Root and Texas Eastern. George and Herman Brown were aggressive, mid-twentieth-century Texas practitioners of business lobbying, an accepted, if often controversial aspect of interest group politics. The height of their success came during an era of wide-open business influence in American politics. Before the Watergate scandal brought fundamental reforms in campaign contribution laws in the 1970s, the Browns learned to operate effectively in a world with few clear, strictly enforced rules governing a businessman's political activities.[3]

The Browns' home base in the postwar years was Houston, but they maintained outposts of influence in Austin and Washington. In each city they established a distinctive presence, complete with an influential circle of friends and a convenient location for entertaining. Suite 8-F in downtown Houston's Lamar Hotel became their most publicized meeting place. In Austin, they frequented the Driskill Hotel, a prominent downtown establishment they owned for a time, and there was Fort Clark, a West Texas ranch to which they retreated from both Austin and Houston. For Washington-area gatherings, they purchased Huntland Farms, a country estate within driving distance of Washington, D.C., near Middleburg, Virginia. The Browns moved regularly and easily among their different home bases and among the often overlapping circles of influential businessmen and politicians.

Suite 8-F in the Lamar Hotel, their home base in Houston, garnered a measure of popular and historical fame. "The 8-F crowd" became a widely used designation for Houston's power elite in the postwar years. The phrase referred to a number of business, civic, and political leaders who used the Brown's Lamar Hotel suite as a convenient place to relax, play cards, and discuss the day's issues. They also came together to guide Houston's development. They played a leading role in many aspects of Houston's growth because they shared a general vision of the city's future and they had the resources, connections, and energy required to act on that vision.

"Membership" in this group was by no means fixed; individuals moved in and out of this circle of influence as their careers and interests changed. Most, but not all, of those who regularly visited the suite were not actually part of the core 8-F group. Perhaps the best way to provide a snapshot of the group is to make a distinction between the core group that met regularly over several decades and a broader group that came together on specific issues. The core is easily identified: Herman Brown, George R. Brown, Judge James A. Elkins, Gus Wortham, Jim Abercrombie, and William A. "Bill" Smith. The broader group included at different times Governor William P. Hobby, Oveta Culp Hobby, Jesse Jones, Walter Mischer, Leopold Meyer, Robert Henderson, Wesley West, George Butler, Naurice Cummings, Charles Francis, R. E. "Bob" Smith, Leon Jaworski, Howard Keck, and Colonel Ernest O. Thompson.[4] These influential individuals had ties to most areas of the Houston economy, and they often mobilized broader support from friends and coworkers throughout the city and the state on issues of importance to the 8-F crowd.

Notable by their absence from the list of the 8-F crowd are the names of numerous prominent Houston business leaders of this era. Those missing include Hugh Roy Cullen, the independent oilman who was deeply involved in the growth of the University of Houston; Glenn McCarthy, another independent oilman who built the Shamrock Hotel; and Judge Roy Hofheinz, who helped attract major league sports to Houston with the construction of the Astrodome. Also not generally mentioned in discussions of 8-F are Morgan Davis, Carl Reistle, and other leaders of Humble Oil and Refining and other national oil firms who were active in Houston's civic affairs yet remained somewhat different from the "loyal" leadership in outlook and involvement. Their absence serves as a useful reminder that Houston was a maturing urban region in the postwar years with numerous centers of power and influence. At times these groups spoke with something akin to a unified voice on important issues such as the need to create strong educational and cultural institutions. But at other times, they fought bitterly over issues ranging from political campaigns to building a new multipurpose stadium for the city's various sports events. 8-F was not the only suite in town, although it was certainly the most highly publicized.

Within 8-F, there was by no means consensus on every political candidate or every significant issue. But this group shared a broad vision of Houston and a predisposition to use their collective economic, political, and civic influence to shape the city's future. They sought to make Houston a "major league city." They watched their city grow dramatically in the prewar years, and they sought to augment Houston's expanding economic base with the amenities found in the nation's larger, more mature cities. They felt that this required above all else the continued growth of business, and they moved aggressively to promote a "healthy business climate" characterized by a minimum of government regulations, a weak labor movement, a tax system favorable to business investment, the use of government subsidies and supports where needed to spur development, and a conservative approach to the expansion of government social services.[5] They also recognized that a major league city needed well-developed civic institutions—art museums, symphonies, ballets, operas, universities, medical facilities—and their time and money were readily available to foster such endeavors. They exerted a strong influence over several decades in shaping Houston's evolution.

Critics protested that their concentrated power contradicted the basic premise of democracy. "There was talk in Texas in the 1940s and 1950s," wrote Texas historian George Norris Green, "that state affairs were handled by card-playing multimillionaires who convened in Herman Brown's suite," but he acknowledges that the 8-F crowd "did not leave much of a trail for the historian to follow."[6] Journalists and historians following the 8-F trail have been quite critical. Writing in 1976, journalist Harry Hurt asserted that "their rule was a virtually unchallenged and—they would emphasize—very 'civic-minded' gerontocracy."[7] James Conway argued that "during the 1940s and 1950s they exercised a concerted influence in Texas that was unparalleled."[8] In *Free Enterprise City*, a historical account of the role of elites in Houston's development, sociologist Joe Feagin argued that the 8-F group's narrow, business-related definition of what was good for the city led Houston down a path characterized by underdeveloped public services, a mediocre educational system, and a harsh climate toward labor. Feagin concluded that 8-F "appears to have been the most powerful elite in the city's history."[9] Almost all of these critical accounts suggest a sort of soft conspiracy of a few powerful men to shape Houston in their own image.

But Louie Welch, who served as mayor of Houston during the 8-F crowd's heyday, gives a quite different view: "You'll hear, I'm sure, all about 8-F, like that's some mysterious, sinister meeting place where people got together and figured out what to let the common people do. It wasn't anything of the sort." To Welch, 8-F "was a sort of brainstorming group" that focused on things they felt Houston needed, such as a medical center. According to Welch, "they created the initiative and the locomotive to pull the train . . . They were movers and shakers, but they were not self-serving in anything that I ever saw them do."[10]

No doubt, the 8-F crowd wielded considerable power over important aspects of Houston's development in the 1940s and 1950s. But both their cohesiveness and their power have been exaggerated. Indeed, in historical perspective, the Browns and their circle of friends represented a broadening, not a narrowing, of the Houston elite. The city traditionally had been ruled by a combination of real estate developers, lawyers, and bankers. In the late nineteenth century several influential corporate law firms asserted strong civic leadership, in part because of their strong ties to the railroads and other northeastern-based companies migrating to Texas. Working with and through their affiliated local banks, these Houston law firms became deeply involved in all areas of economic, political, and civic affairs. Captain James Baker, a longtime partner in the leading law firm in the region, Baker & Botts, emerged as the symbol of the Houston elite in these formative years. His influence touched most of the significant endeavors in the growing city from the 1870s until his death in 1941.[11]

In the early twentieth century, the relatively small circle of lawyers and bankers expanded to include other businessmen drawn to the city by cotton- and oil-led development. Chief among these was Jesse Jones, who migrated to Houston in 1898 as a young man and made his fortune in real estate and banking. He also owned one of the city's major newspapers, the *Houston Chronicle*. Jones joined Baker and others in 1914 to help complete a deepwater ship channel from Houston to the Gulf of Mexico, an event that symbolized business power in Houston's development. In the 1920s and

1930s, Jones arose as the city's very visible spokesman, first by bringing to it the 1928 Democratic nominating convention and later as a prominent member of the Herbert Hoover and Franklin Roosevelt administrations. "Uncle Jesse" became "Mr. Houston" for most Americans in those years, but while he spent long periods of time in Washington, a younger generation of business leaders, including the Browns, emerged in Houston.[12] They represented the new city's wealth as well as the new industries that were reshaping its economy.

Like their counterparts in the Baker and Jones generations of leaders, most of the 8-F members migrated to Houston from small Texas towns. Opportunities to pursue their ambitions on the larger stage drew them to Houston, and they tended to embrace their adopted city's future as their own. Collectively, they demonstrated that Houston remained an open city that judged newcomers more by their energy and their pocketbooks than their family trees. The Browns and Bill Smith represented the area's booming construction industry. The Browns also had ties to one of the fastest growing regional industries after World War II, natural gas. Oil and oil tools and services found a spokesman in Jim Abercrombie of Cameron Iron Works. Gus Wortham built American General into one of the largest insurance companies in the Southwest. As managing partner of Vinson & Elkins and chairman of the board of First City National Bank, Judge Elkins represented the legal/banking interests that traditionally guided Houston's evolution, but Elkins's law firm and bank were both aggressive, entrepreneurial concerns that broke rank with the city's more established law firms and banks in tone and client list. Beyond this core list of 8-F regulars was a broader list of influential business leaders that included numerous oil men and representatives from the Houston offices of many national corporations that had migrated there. Suite 8-F hosted Houston's new elite; the "crowd" that met there represented the new industries feeding the boom. The 8-F leaders recognized that their self-interests were tightly bound with the continued expansion and maturity of their city.

Very strong personal and business ties linked these men. Jesse Jones served as a kind of godfather to the group, lending the benefits of his long experience in city building and broad contacts with business and political leaders around the nation. He owned the sixteen-story Lamar Hotel, and he and his wife took up residence on the hotel's top floor upon its completion in 1927. At about the same time, Herman Brown began leasing Suite 8-F so that he would have a convenient base of operations in downtown Houston (at the corner of Main and Lamar) on his frequent trips from Austin to consult with George on Brown & Root business. With the expansion of Brown Shipbuilding during World War II, Herman made more frequent use of the suite, which had two bedrooms, a large living room, and a small dining area and kitchen.[13] It provided a comfortable home away from home for Herman, and it was available for other guests when Herman was not in town. Its central location made it a logical daytime meeting place for those who had downtown Houston offices. When Herman and Margarett Brown finally moved to Houston after World War II, the Brown brothers retained Suite 8-F for such meetings. Their well-known upstairs neighbor seldom came down to 8-F to talk with the Browns. Instead, "Mr. Houston" called the younger men up to his apartment above them.

Gus Wortham, a protégé of Jones and a friend and business associate of the Browns, became one of the fixtures in Suite 8-F. For a time, Wortham rented Suite 7-F, directly below the Browns' suite at the Lamar. From its earliest years in Houston, Brown & Root was a major client of Gus Wortham's American General Insurance Company, which supplied workman's insurance for the Marshall Ford Dam project that Wortham organized in 1926 with capital supplied by Jesse Jones, Judge James A. Elkins, and others. As his company grew into one of the largest of its kind in the Southwest, Wortham led the Houston Chamber of Commerce in the 1930s; he also devoted much time and money as a patron of educational and cultural institutions. In the 1940s and 1950s, George Brown and Gus Wortham spent long hours together on the Rice University board of trustees. Among his friends at 8-F, Wortham was considered the quiet man who could be counted on to do his share behind the scenes. He also set the tone for the group as a whole with an oft-repeated distinction between "business dollars," which had to be watched closely, and "entertainment dollars," which were meant to be spent lavishly.[14]

Wortham's ties to another prominent member of 8-F's inner circle, Judge James Elkins, dated all the way back to Wortham's childhood in Huntsville, Texas. In the 1910s, both left Huntsville for Houston, where they quickly established themselves as members of a business and civic elite that remained quite open to outsiders. Elkins subsequently joined Jesse Jones in providing financial backing for the original organization of American General. "The Judge" played a central role in the dynamics of decision making at 8-F; all members of the group valued his counsel. This reflected his well-earned reputation as a decisive man capable of persuading others to agree with his strongly held opinions on political, legal, and economic issues. It also reflected his broad connections into all areas of the Houston political economy through his bank and his law firm.[15]

The Browns came to play an important role in Elkins's First City National Bank through their association with another bank, First National. As prominent stockholders in First National, Houston's oldest bank, the Browns led the search in the early 1950s for a merger partner to strengthen the bank's declining competitive position. They first approached Jesse Jones, the primary owner of the National Bank of Commerce. George Brown took the lead in these negotiations, and he later recalled the outcome: "We negotiated and negotiated, and I would go back and tell Herman, 'Herman this thing is taking too much of my time. We ought to talk with someone

Herman Brown and Judge James Elkins. *Courtesy Brown Family Archives.*

else.'" But Herman convinced George that "we still owed it to him [Jones] to merge the First National Bank with the National Bank of Commerce to create the dominant bank in Houston." George finally thought he had reached an agreement with Jones, and he went home and told Alice that he had completed "the most trying business negotiation I have ever had." But when George arrived at work the next day, he received a phone call from Jones, who said, "George, I've been thinking more about that deal we struck last night and I'd like to talk to you more about it." George had reached his limit: "That did it. . . . I got in my car and went down to City National Bank and talked to Judge Elkins. In maybe two hours, we had reached a deal to merge First National . . . with the City National Bank, creating the largest bank in Houston."[16]

This major bank merger in 1956 cemented the Browns' existing business ties to First City and Vinson & Elkins. The Judge's law firm was particularly important to Texas Eastern, where Vinson Elkins lawyers such as Charles Francis played significant legal and managerial roles.[17] The lives of the Browns, Judge Elkins, and Gus Wortham intertwined on many levels, and the strong bonds connecting these men held 8-F together.

Jim Abercrombie was a fifth important member of the group who gave his 8-F friends insight into the oil business. In this sense, he represented the heart of the "new" Houston, the complex of industries that grew to produce oil and to build the tools and provide the services needed by oil companies. He was one of the key figures in the growth of Cameron Iron Works, a major Houston-based oil tool company. Abercrombie also became a prominent independent oil producer. His ties to the Browns were more personal than business-related. Naurice Cummings, a mutual friend and regular 8-F visitor, had introduced Abercrombie to the Browns. "Mr. Jim" quickly became a prominent regular himself.[18] The men all shared a passion for hunting, and Abercrombie helped cement the ties that bound them together by sponsoring an annual hunt, usually at the Blackwell ranch near Cuero in West Texas or the Palomas Ranch near Falfurrias, where the Browns, Abercrombie, and Bill Smith shared a forty-thousand-acre lease for many years. The treks to the Palomas Ranch provided the hunters a chance for relaxing, drinking, joking, and shooting. The ranch had an old, rundown "camp house" where the men stayed. Most slept in a large central room aptly called the "bull pen," but the loud snoring of Herman Brown and Jim Abercrombie often forced the two of them into smaller bedrooms so that those in the bull pen could sleep. Recalled one visitor: "It was a very masculine atmosphere down there. The cook was a man. There were no women around. They would stay up all night and play cards and tell stories and have a wonderful time. The shooting was important to them, but I think the camaraderie, the friendship was just as important."[19] Each hunt provided raw material for still more outrageous stories on the next trip, and photographs of the hunters dressed in their khakis remained prize possessions decades after the trips.

Abercrombie also helped organize another popular 8-F event, the group's annual trek to the Kentucky Derby. The Browns invested in racehorses with Abercrombie, daughter Josephine, and brother Bob. The friendships developed through such activities went far deeper than business ties alone.

One prominent Houston couple remained near 8-F's inner circle while seldom venturing to the Lamar Hotel. The Hobbys—Governor William P. Hobby and his wife, Oveta Culp Hobby—were good friends of Herman and Margarett Brown and George and Alice Brown. William P. Hobby served as Texas governor from 1917 to 1921, when the Hobbys moved to Houston from Austin and purchased from Jesse Jones the *Houston Post,* which competed with Jones's *Houston Chronicle.* Oveta Culp Hobby had first encountered the Browns in Central Texas, and she renewed ties with them when she moved to Houston. During World War II she went to Washington to head the Women's Army Corps, and she later served in President Eisenhower's cabinet as secretary of Health, Education, and Welfare. She was the only woman generally included in the inner workings of 8-F, although she seldom visited the suite itself, which remained a male domain. As experienced public officials, prominent Housto-

George Brown and Jim Abercrombie at Fort Clark.
Courtesy Brown Family Archives.

nians, and publishers of the *Post,* the Hobbys commanded respect on issues concerning the development of the city.[20]

Like economic elites in other cities, the 8-F crowd shared a variety of bonds. Their economic interests often overlapped, a situation symbolized by their memberships on various "interlocking directorates." In practice, this meant that many of them met regularly as directors of banks and other businesses based in Houston. In general, they belonged to the same downtown businessmen's clubs, where they often ate lunch together. The Browns regularly hosted lunches at a table reserved for them at the Ramada Club. After the meal, those present could, if time permitted, walk back to 8-F for a midday visit. Many belonged to the same country clubs. They often supported each others' favorite charities and civic causes, just as they closed ranks in support of political candidates.

Beyond that, they became good friends who enjoyed spending time together. Many afternoons of cards, regular hunting expeditions to various retreats, and an annual trek to the Kentucky Derby gave them time to relax away from the pressures of demanding jobs. This point was brought home to Ralph O'Connor on one of his trips to Fort Clark. After seeing one of the other men "very drunk," O'Connor was called to Herman Brown's room for a brief discussion. Herman told him, "It's very easy for people to, without knowing it, to say things, to say so-and-so was drunk, sloppy drunk. That's why these people come down here. He's a very important man. He's got to let off steam. I don't want you to say anything."[21]

Within this group, strong-willed men who spent most of their working lives making hard decisions under intense pressure could relax. In a society made up almost exclusively of "bosses," these prominent men could banter and cut up like kids. Strong bonds of camaraderie and friendship developed within the group, cementing ties based on economic and political self-interest. A closeness and mutual respect gained from long and often fun-filled hours spent together helped smooth over disagreements on the issues of the moment, fostering a cohesion that was crucial to the sustained influence exerted by this group. The 8-F crowd worked hard and played hard, building a sense of unity somewhat similar to that of a college fraternity.

Part of the "dues" expected from its members was money collected for shared interests. Each might be asked to ante up a contribution to the campaign funds of political candidates at the local, state, and national levels who had passed muster with the group. Similar contributions could be expected when one of the group became highly committed to a particular charity or civic cause.[22] Individuals could pass on a particular candidate or on a request for a charitable contribution, but all recognized that a part of their mystique as a group was their often demonstrated capacity to close ranks behind favored causes.

A short note from Albert Thomas to George Brown in 1948, when Thomas was beginning to solidify his hold on his seat in the U.S. House, suggests how support for an effective politician spread throughout 8-F. Thomas begins by telling George of a letter to Judge Elkins "asking him to put in the ballot box about two bushels of votes for us. He has a host of friends, and if he does a little talking it will be of inestimable value." Thomas then asks if "you and Herman can find time to ask him for a little contribution. . . . My thought is simply this: he, being very practical, might back it up with a little good talking on my behalf." Included is the statement that "if this would be embarrassing to either of you, don't do it." Coming from a sitting congressman who had known George Brown since they were both teenagers at Rice and presumably knew something about the inner workings of 8-F, this suggests that consensus of political thought and action could not be assumed.[23]

Politics was one of the favorite sports at 8-F. Although most who visited there were conservative Democrats in the traditional one-party system of Texas, a Republican or two was tolerated. Politicians who sought support underwent what amounted to job interviews at 8-F, and those who passed could expect vigorous support. The 8-F crowd's unofficial endorsement was a potent weapon for annointed "moderate" candidates who sought local and even state offices. The campaign funds and the newspaper support that generally followed made such candidates difficult, though not impossible, to defeat.[24]

The 8-F crowd sought to keep a low profile in politics. As Gus Wortham was fond of saying, "You don't go hunting with a big brass band."[25] But on several occasions, controversy over political disputes spilled out into public. In the early 1950s, 8-F supported Roy Hofheinz for two terms as mayor of Houston. But his efforts to raise property assessments on downtown buildings cost him their continued support. In the 1955 mayoral race, the group threw its weight behind Oscar Holcombe, who previously had held the office for ten terms. Hofheinz responded by going public with the contents of a meeting in 1952 with the Browns, James Elkins, and Gus Wortham to both the *Houston Chronicle* and the *Houston Post.* Hofheinz reported that Herman Brown told Holcombe he was through as mayor and then informed Hofheinz that "we will support you for mayor and all you have to do is call them down the middle."[26] When Hofheinz later disagreed on what constituted "the middle" on issues such as taxes on downtown buildings, he lost 8-F's support.[27]

This group clearly had the power to direct strong financial support to a favored candidate, helping to swing elections. Many prominent local, state, and national politicians found financial support there. Lyndon Johnson, Albert Thomas, and, later, Lloyd Bentsen maintained ties to many of those who frequented 8-F and on occasion visited the suite.[28] The Browns' close connections to the increasingly powerful Lyndon Johnson afforded the brothers and their 8-F friends a measure of political access on the national level.[29]

The 8-F group often used their wealth and influence for broad social purposes as they aggressively shaped Houston's development. The group supported such initiatives as Jim Abercrombie's impulse to build and support Texas Children's Hospital and George Brown's commitment to Rice University. They sought to bring major league baseball to Houston as early as the 1950s, when they enlisted the aid of New York investment banker August Belmont in an unsuccessful attempt to buy the Saint Louis Browns and move them to Houston.[30] They supported the efforts of fellow Houstonian and sometime collaborator Hugh Roy Cullen in contributing to the University of Houston. In short, 8-F was interested in almost any project that supported their vision of Houston as a major league city.

One of the (broadly defined) 8-F group's most highly publicized initiatives in "city building" involved the creation of a new jet airport for Houston. The Browns had long been involved in lobbying at several levels for better air transport for Houston. By the mid-1950s, the Houston Municipal Airport (later renamed Hobby Airport in tribute to Governor William P. Hobby) had been renovated with a new terminal building. Yet those who frequently flew, including several members of the 8-F group, discussed among themselves the need for a larger airport farther from downtown

Shooters, 1972: John McCoy, Doc Neuhaus, Governor John Connally,
George R. Brown, Senator Lloyd Bentson, L. F. McCollum.
Courtesy Brown Family Archives.

Houston. A group including the Brown brothers, Jim Abercrombie, R. E. "Bob" Smith, William A. Smith (all generally considered 8-Fers), plus Hugh Roy Cullen, Ralph Johnston, and a dozen others commissioned studies of potential airport sites. The studies noted that fog was more intense and lasted longer south of Houston where the existing airport was located than north of the city. A survey conducted during the early 1950s indicated that a site north of Houston was the most appropriate to locate a new jet airport, although no plans developed out of the survey.[31]

The group decided to acquire the site and then attempt to convince the city that a new airport should be built there. In late 1956 and early 1957, members of the group called on Archer Romero, who was president of Archer Grain Company and an expert on real estate, and asked him to begin purchasing as a trustee land north of Houston. Romero negotiated with approximately twenty heirs of an estate that owned the majority of the land.[32] The group put the land into a corporation they formed, originally intended to be called Jet Era Ranch Company. But the typist made a typing error on the application for the charter, replacing "a" with an "o" to form the word "Jetero." The corporation thus became known as the Jetero Ranch Company (of which Ralph Johnston was president), and the main road leading into the airport became Jetero Boulevard. Romero sought to keep the deals as quiet as possible to prevent landowners from asking more for their property.[33] Romero ultimately purchased about three thousand acres for less than $2 million without any land requiring condemnation. The average price per acre was approximately $619.[34]

After acquiring the land, George Brown and several others involved in the purchase met privately in early 1957 with Houston mayor Oscar Holcombe and other city officials, including the city attorney, the public works director, and the aviation director. One city council member, Louis Welch (who later became a mayor of Houston) was also present. During the meeting, those who had purchased the land described the deal they had completed. The city officials were somewhat skeptical. They noted that Houston's existing airport had just received a new terminal building. The airport group insisted that Houston needed a new jet airport or the city would not grow. They purchased the land with borrowed money and promised to hold it until the city was ready to buy it. The group offered the city of Houston a one-year option that extended until February 1, 1958, to buy the land.

Although Mayor Holcombe subsequently lost his reelection bid, he indicated that he wanted the city to go forward with the purchase of the land before he left office. He engaged a member of a New York–based private consulting firm, Wainwright & Ramsey, to investigate the possibility of funding the airport construction through bonds. In November, 1957, the city council endorsed an agreement between the Jetero Ranch Company and Mayor Holcombe for the purchase of the three thousand acres of land for about $1.9 million. The letter of agreement set forth a straightforward arrangement in which the city agreed to complete the acquisition by February 2, 1960, and to begin building an airport within seven years.

Before the city could complete the transaction, however, Holcombe left office. Louis Cutrer succeeded Holcombe as Houston's mayor in an election that also replaced all of the previous city council except for Welch and Lee McLemore.[35] Welch recalls that when he approached Cutrer on the airport land deal, Cutrer responded: "You're not going to tell me that that bunch of high rollers isn't in this for money. They're looking for a profit on this."[36] Welch countered that they were interested not in making money but in helping the city build a new jet airport. Cutrer, however, remained skeptical of the deal. As time passed, the members of the Jetero Ranch Company had to carry the considerable interest on the loan used to help acquire the land, and tensions began to grow between the group and Mayor Cutrer.

In March, 1958, the Jetero Ranch Company held a special meeting to consider what to do in light of the city's inability or unwillingness to close the deal. The city's failure to make the first payment of accrued interest meant that the agreement was now technically void. Mayor Cutrer maintained that "the City was still interested in acquiring the subject properties, but the matter was contingent upon the City Council's approval and the availability of the necessary funds."[37] This forced the airport group to come up with a new strategy while continuing to meet interest payments of up to $100,000 per year.

According to the minutes of the meeting, Herman and George Brown voiced the general sentiments of the group when they said that "the undertaking to acquire the subject properties and the sale to the City of Houston was one in the nature of a civic service."[38] Their opinion was that the sale should go forward if at all possible financially.

But it would not be easy to stay the course. The city had difficulties in solidifying both financial and political support. The city swung between resolutions commending the airport group in September, 1958, to a request for an extension for completing the deal in December, 1959. Tensions increased after Mayor Cutrer remarked to newspaper reporters that he assumed an extension would be granted, since "the City had always had such excellent cooperation from the group" and he "believed the parties would continue to cooperate with the City." In response to Cutrer's stated concern that "some of the parties misunderstood my remarks and felt that I had taken entirely too much for granted," he wrote to George Brown and others in the group offering a personal apology and explaining that he "was simply trying to be complimentary rather than forward and dictatorial."[39]

Such apologies were irrelevant if the city failed to pass measures assuring funding of the purchase. After much delay, the purchase was finally completed, thirty days after the original deadline had passed. The airport group had exhibited the patience required to guide the land deal through the minefield of local politics. By the early 1960s, much of the political conflict had been forgotten as construction began on Houston Intercontinental Airport, which became Houston's major airport when it opened for operations in 1969.[40]

Even strident critics of the 8-F crowd at times shook their heads in wonder at the capacity of these men to get things done. On many critical issues, they provided the leadership—and often the funds—to push Houston up the ranks of the nation's largest cities. Others might have preferred a different path toward the same general goal, but the Browns and their friends collectively held the power to craft the direction of Houston's growth.

On the issue of power, the recollections of an outsider provide an interesting perspective. August Belmont IV, the fourth generation of his family to take a prominent role in New York banking, first encountered the Brown brothers when he came to Houston in 1947 to organize financing for Texas Eastern while serving as a vice president of the Wall Street investment bank of Dillon Read. "They'd all go up to there to lunch," Belmont stated, "and get other people up there [8-F] . . . and making all kinds of deals and talking girls and every other damned thing, and . . . Herman was very mixed up in the politics of the area. They really . . . controlled the politics in Houston."[41] Belmont remained fascinated by this powerful group of Houstonians, who reminded him of the men who had peopled the world of his grandfather, a leading "power broker" in late nineteenth-century New York. In a city experiencing tumultuous change brought on by rapid development, a close-knit business elite could impose its sense of the proper order and direction.

Belmont's memories no doubt reflect the excitement of a young man's introduction to what was to him the exotic world of Texas oil and gas men. His recollections, however, serve as a powerful reminder that leadership in Houston during these years was for white men only. At that time the voices of the city's growing population of minorities went largely unheard in civic and political affairs. The only blacks in 8-F were the drivers of the white men present. White women were heard, if at all, through their husbands—with the clear exception of Oveta Culp Hobby. There was as yet little organized, effective voice for labor and other groups who held sharply different views than that of 8-F on what constituted a healthy business environment. The broadening of Houston's civic leadership would come later, after 8-F passed from the scene to be replaced by a more open, diverse approach to local civic influence.

Equally as important as their presence in Houston, the Browns' influence in Austin helped them protect their business interests. The state government had a considerable impact on the construction industry on issues ranging from awarding contracts to passing labor laws. Herman Brown took a special interest in the state legislature, particularly in the area of labor law, and he spent much time and energy during the 1940s and 1950s tending to the brothers' political interests in Austin.

In those years the Austin political scene was wide open to aggressive lobbying. Ronnie Dugger, publisher of the liberal *Texas Observer* and a frequent commentator on Texas politics in that era, described Austin's politically charged environment. "What must surely be the most wide-open lobby in America," he wrote, "moves into Austin every two years to entertain and coerce the men who write the laws. Bribery, cajolery, pheasant hunts, Mexican vacations, airplane junkets to the Kentucky horse races, bus junkets to the border brothels—you name it, the legislators have had it."[42] The *Texas Observer* built a small but dedicated following of readers by criticizing trends in Austin with often biting satire of the corrupt political process. Herman Brown and Brown & Root were frequent targets of its criticism. Indeed, observers throughout the political spectrum considered Herman Brown to be a most effective practitioner of the not-so-subtle art of political influence in Austin.

The Browns owned two important pieces of real estate in Austin, the Brown Building and the Driskill Hotel. The stately old Driskill Hotel had been a downtown Austin landmark since its construction in the late nineteenth century. The Browns purchased it in October, 1953. It remained a favored meeting place for politicians and businessmen visiting Austin. After Lady Bird Johnson purchased the Austin radio station KTBC in 1953, the Johnsons initially located their radio and television station in the Brown Building, another downtown building owned by the Browns, before later moving it to the Driskill. Lyndon Johnson as well as the state and national press corps regularly watched election returns at the Driskill during his many campaigns.[43]

The Brown Building had been built in downtown Austin in 1938 to house the headquarters of the Texas Employment Commission (then called the Texas Unemployment Commission).[44] Brown Securities Company, the Brown & Root subsidiary originally organized in the 1930s to acquire paving notes, owned the building. In the late 1950s, controversy arose over the TEC's continued rental of the Brown Building, since the agency generally owned its own properties in other locations. Eyebrows were raised over what appeared to be political maneuvering to assure that the Browns would not lose their main tenant.[45]

The Browns' West Texas property at Fort Clark became a frequent weekend retreat for the brothers and their friends in business and government. The ranch was a short plane flight away from the state capital, and Brown & Root's DC-3s regularly ferried passengers from Austin and Houston to Fort Clark. In addition to serving important business purposes after the war, Brown & Root planes provided convenient transportation for the Browns and their friends in an era before widespread and inexpensive airline service in the Southwest. Jack Robbins, the son of Brown & Root's first pilot, Reg Robbins, who was Herman Brown's personal pilot, recalled that he often flew politicians and businessmen to Herman Brown's "powwows" at Fort Clark on the weekends. "If it wasn't hunting season," Robbins recalled, "why, it was always political season. That is to say we'd go by Austin and pick up a group of congressmen or senators or something and go out and spend the weekend at Fort Clark and they'd have their big political powwows."[46]

Searcy Bracewell, a onetime state legislator who later worked on political issues in Houston for several 8-F members, later recalled some of these "powwows." "He [Herman] not only invited the senate and their wives . . . but he invited some other people." Bracewell remembered that of the thirty-one state senators in the 1950s, "usually fifteen or sixteen or seventeen of them . . . would come." Frequent guests at these retreats also included political writers for the state's major newspapers. Festive barbecues around a large, spring-fed swimming pool highlighted the activities. Looking back at these gatherings, Bracewell concluded: "Now, in this day and time, that would have been a scandalous situation that Brown & Root had the state senators out to their place. . . . In those days, it wasn't."[47] Indeed, under the general custom of the time, the reporters never wrote about the meetings.

Group hunting trips at Fort Clark might include Lyndon Johnson and other assorted national and state legislators, state officials such as Colonel Earnest O.

Thompson of the Texas Railroad Commission, businessmen from 8-F and around the state, and lobbyists for Brown & Root. Then senator Lyndon Johnson voiced the appreciation of many of the visitors to Fort Clark when he wrote to Herman and Margarett Brown in 1957: "I want to go on record that last week at Fort Clark was one of the most enjoyable weeks I have ever spent in my life."[48] Fort Clark provided an excellent retreat where lawmakers and other friends of the Browns could go for a long weekend or a vacation week filled with fun, rambling political discussions, and the building of personal ties that eased the way for effective lobbying.

In pursuing the Browns' interests in Austin, Herman Brown could call on the assistance of a talented and well-connected trio of lobbyists, Alvin Wirtz, Ed Clark, and Frank Oltorf. After his involvement in the protracted bargaining over the funding of the Marshall Ford Dam, Wirtz remained vital to Brown & Root's Austin affairs in the late 1930s and 1940s. As a trusted adviser to Herman Brown and to Lyndon Johnson, Wirtz worked for a time as Brown & Root's lawyer. Brown's "tremendous trust in Wirtz's discretion and his integrity and his mind" led him to rely heavily on Wirtz for all sorts of legal and personal advice.[49] As a partner in the prominent Austin law firm of Powell, Wirtz, Rauhut & Gideon, he was a most effective member of Brown & Root's lobbying team in Austin. He died unexpectedly in October, 1951, while watching a University of Texas football game with Herman Brown at Memorial Stadium in Austin, and his death left a void in Herman Brown's life and Brown & Root's legal and political work.

Ed Clark was another prominent Austin attorney who worked closely with Herman Brown, assisting him in drafting proposals for legislation and arranging for their consideration by the Texas legislature. Clark became Brown's close personal friend, and the two spent long hours discussing business on the telephone. They also shared a strong interest in Southwestern University in Georgetown. Clark had known Alice Brown while both were students at Southwestern, and, later, Clark and Herman Brown both spent time on the university's board of trustees. Like Wirtz, Clark also worked closely with Lyndon Johnson, who rewarded Clark for his loyalty by naming him ambassador to Australia. By the 1960s, Clark had become a symbol of the Texas establishment in Austin politics, and his earlier work for the Browns contributed to his initial reputation as a man who could get things done in Texas. Frank "Posh" Oltorf described Clark as Herman's "personal crony," a man greatly valued for his ability to make Brown laugh. Oltorf remembered that "he [Herman Brown] loved talking to Clark on the phone. They'd talk maybe an hour at a time. . . . Margarett Brown one time said, 'You know, when Herman dies, he thinks if he's good, he will go to you [Oltorf] and Ed Clark and Gus Wortham.'"[50]

Oltorf was the final member of Brown's lobbying team in Austin. Though his name was Frank, he was universally called "Posh." Oltorf grew up in the Central Texas region that also was home to the Browns and Alvin Wirtz. He came from a prominent political family in the town of Marlin, some thirty miles south of Waco on the Brazos River. While still in the army during World War II, Oltorf was elected to the Texas House of Representatives. Upon his return from military service in India, he entered the Texas legislature as a twenty-three-year-old freshman member of the house.

Oltorf first met Herman and Margarett Brown when his aunt, Mary Rudder, and Margarett met to discuss the possibility of establishing a symphony in Austin. Rudder was active in the affairs of the Houston symphony, and she came to Austin and stayed for two nights with the Browns to share her experience. Posh Oltorf was invited to join his aunt at Margarett Brown's home for dinner, and after dinner, Herman came in and joined the group. "He was fascinated with the legislature and we sat down and visited and had a couple of drinks and just liked each other immediately."[51]

As he went about his work in the legislature, for a time Oltorf did not see Brown, who opposed him on a variety of pending bills. Oltorf encountered Herman again while working against a Brown supported bill to place Texas labor unions under the state's antitrust laws. His recollections of Herman's actions suggest how he went about the task of collecting political allies. As a member of the labor committee, Oltorf voiced the opinion that the legislature should not pass the Brown-sponsored bill. Herman approached him about it.

"I can't understand how you could be against that bill," Herman told Oltorf.

Oltorf responded, "I just think the very nature of a union is a violation of antitrust. And if you're outlawing it, you're outlawing unions in effect."

Brown asked Oltorf, "Do you think that I would advise you and tell you to do anything in the world that wasn't right?" to which Oltorf said, "No."

Brown continued: "Well, I'm telling you this is badly needed and it's good legislation and your fears are groundless."

Oltorf stood his ground: "Well, Herman, I know you feel that way and you well might be right, but I just can't bring myself to vote for it." Oltorf did not vote for the bill, although it passed. Surprisingly, Oltorf found that Herman Brown seemed to hold him in higher regard after the vote.[52]

Soon after this encounter, Herman Brown asked Oltorf to be the campaign manager for former Texas senator Ben Ramsey in his run for lieutenant governor in 1950. Senator Ramsey, according to prolabor journalist Hart Stilwell, "was a key figure in engineering the restrictive labor bills through the senate in 1947."[53] Although Oltorf and Ramsey were, according to Oltorf, "on opposite poles politically," Herman Brown's insistence on the arrangement ultimately convinced both Oltorf and Ramsey to agree. After Ramsey won the election, Herman Brown asked Oltorf to work for Brown & Root.[54] Oltorf began his career as a Brown & Root representative in Austin and later in Washington.

Herman Brown's efforts to secure Ramsey's election as lieutenant governor provide insight into the nature of political power in Texas in this era. The strategic fulcrum of Texas state politics on many key issues was the senate, which remained a small body of thirty-one members. This body remained quite conservative, in part because the lack of reapportionment for decades left the more conservative rural regions of the state overrepresented in comparison with the growing cities. As president of the Texas senate, the lieutenant governor held the crucial power to select committees and name their chairmen. Thus a little influence with the lieutenant governor and a few critical senators could go a long way toward shaping the legislative agenda. Herman Brown proved adept at using well-targeted lobbying in the Texas senate to block the passage of unwanted bills.

Demonstrations of political power by Herman Brown in Austin drew statewide and even national attention, particularly from those who lost political battles with him over labor laws. In 1953, a *Reader's Digest* article quoted an unidentified source that "if Herman Brown is against something, there's no use for the Senate to meet."[55] Labor writer Hart Stilwell wrote a much harsher assessment in a pamphlet entitled "The Herman Brown Story," published by the *Labor News*. The pamphlet was an expanded version of a similar article originally written for the November 10, 1951, issue of the *Nation* magazine. Reviewing events since 1947, Stilwell characterized Herman Brown as a "multi-millionaire contractor, banker, utility man and oil man," with extraordinary power to shape antiunion legislation in Texas and to use the courts to limit unions' activities. He noted that Brown & Root had engaged in a running battle with the National Labor Relations Board on a variety of fronts. He questioned why "year after year huge federal contracts are handed out to Brown" while his companies were "simply ignoring the government's over-all labor goals."[56]

Historians writing about Brown's political activities in Texas during this era have supported many of the conclusions, if not the tone, of contemporary critics such as Stilwell. They cite the influence of Wirtz, Clark, Everett Looney (a partner in Clark's law firm), Lieutenant Governor Ramsey, and Herman Brown in spearheading the passage of the toughest antilabor legislation in Texas history. One political historian goes so far as to argue that the antiunion legislation passed in Texas after World War II was "no small event in the postwar decline of the union movement in the United States."[57] This conclusion exaggerates both the long-term impact of the state's antilabor laws on the growth of labor unions in Texas, where union membership grew steadily in the 1950s, and the importance of Texas in shaping national trends in the growth of unions.[58] Although Herman Brown did not reverse the tide of unionism in the state or the nation, his highly visible role in this fight cemented his reputation as a well-connected power broker in Austin.

His deep personal involvement in state politics was much more than a part of his crusade against unions. Throughout his life, Herman was fascinated by politics. In many ways, politics was a type of hobby for him. Searcy Bracewell recalled seeing Herman sitting alone in the Texas senate gallery one day in the mid-1950s. After fifteen or twenty minutes, Bracewell noticed that Brown remained in the gallery watching the proceedings. When Bracewell approached him to ask if he was interested in anything in particular, Brown replied that he was just sitting there observing, and then added, "You know, I guess I've done pretty well in my lifetime, but there's something I always would have liked to have done. I always wanted to serve in the legislature."[59] Even his harshest critics probably would have agreed that Herman Brown, the bulldozer, would have been something to behold as a freshman congressman or later as the speaker of the house.

In national politics, the Browns became closely identified with Lyndon Johnson and his rise to the presidency. Although far and away the most visible symbol of the Browns' influence in Washington, Johnson was hardly the sole reason for their deepening involvement in national politics. The growth of Brown & Root and Texas Eastern into nationally and internationally active companies gave the Browns strong incentives to establish a stronger presence in Washington. The Browns' interests had grown far beyond the Texas road building business of earlier decades, and the national government held influence over more and more issues of importance to the success of their companies.

George Brown took primary responsibility for overseeing the brothers' interests in Washington, much as Herman took primary responsibility for political affairs in Texas. For a time, George had the able assistance of Posh Oltorf, who became primarily a Washington lobbyist for the brothers' interests in the 1950s. During World War II, the Browns maintained a row house in Washington as their base of operations; after the war, they sold the house and acquired a suite at the Carleton Hotel. Oltorf later became a fixture for six years in the Hay-Adams Hotel. He specialized in establishing contacts with an array of midlevel government officials at government agencies such as the Export-Import Bank and the Pentagon. These men often proved quite useful in providing important information to Brown & Root in its never-ending search for construction contracts. Upon his departure to Washington, George Brown instructed Oltorf "under no circumstances to do any lobbying up on the Hill. . . . He had found that they would muddy the water more often than not." Oltorf's job for Brown & Root was to "have someone in every department that I had made friends with and wanted to help me."[60]

Oltorf avoided approaching the members of the influential Texas congressional delegation—such as Lyndon Johnson, Albert Thomas, and Sam Rayburn. He understood that "as far as Lyndon was concerned, the Browns were so close to him that the last thing they would want would be a go-between. That's something they could attend to with a telephone conversation themselves."[61] Critics of the Browns have argued that Johnson used his contacts to make the Browns rich, although Johnson himself later claimed that "I never recommended them [the Browns] for a contract in my life."[62] George Brown addressed this issue as follows: "In a material way, there was no way for him to help us because we had to be low bidder on everything that we got from the government. So the only thing he could do at all would be to give us information that might become available to him as to what appropriations they were thinking about."[63] Of course, as Brown acknowledged later in the same interview, numerous projects were not awarded strictly on the basis of the lowest bid but rather on the more subjective judgment of which company would do the best job at a reasonable cost.

As George Brown's stays in Washington became more frequent, he decided to acquire a more substantial place to entertain and to get away for weekends. The Browns purchased Huntland Farms to serve as their base of operations in Washington. Posh Oltorf recalled that the initial decision to look for a Washington-area estate came in the early 1950s, when George Brown told him, "Posh, you know, why don't you some time look around and see if you could find a country place in Virginia? And that's a place you could use to entertain some people that Brown & Root want to have business with. And Lyndon could use it. And Oveta Hobby could use it some. And when I get up to Washington, I could get some pleasure out of it." Brown felt that he and Herman needed a place to stay on the weekends after trips to New York or Washington so they would not have to make the ten- to twelve-hour flight required for prop planes to fly them back to Houston. George Brown gave Oltorf little guidance as to the style or size of the dwelling to be purchased, stressing only that "if it's more than an hour's trip away, you'll never use it. So, make sure its within an hour."[64]

After about a month of looking at various properties, Oltorf found what he considered an excellent estate that could be purchased at a bargain price. George Brown quickly agreed to purchase Huntland Farms, a historic 413-acre property complete with a beautiful five-bedroom home, two gatehouses, a gardener's cottage, stables, gardens, and even a dairy. In the late 1950s, the Browns purchased approximately 140 additional adjoining acres of land. After discontinuing the dairy farm operations, they exchanged the dairy cows for cattle. Over the years, they added a swimming pool and other improvements designed to make Huntland Farms a comfortable place for entertaining guests. The addition of air conditioning made the main house more comfortable during the summers, making it possible for Margarett Brown to stay at the Huntlands despite her severe asthma. The dredging of a fishing pond made it possible for guests to enjoy fishing without leaving the property for ponds on neighboring farms. This helped satisfy Speaker Sam Rayburn's desire to have a convenient place to fish on his visits to the Huntlands, and the pond became known as "Lake Rayburn."[65]

This retreat for those weary of Washington became a favorite place of George and Alice Brown and their family. Herman and Margarett Brown stayed at Huntlands less often simply because they traveled east much less frequently. At times the grounds were used for official government meetings, such as negotiations between the Dutch and the Indonesians to establish an independent Indonesia; on other occasions, government advisory committees on which George Brown sat convened at Huntlands for meetings. The retreat briefly entered the national news spotlight in 1955, when Lyndon Johnson suffered a serious heart attack there at a time when his name was emerging as a possible presidential nominee.[66] Huntland Farms served as a sort of home away from home for George Brown as he became more deeply involved in national economic and political affairs. In addition to his frequent trips to Washington, George also took on the obligations of several major corporate boards and a variety of prestigious government commissions. Huntland Farms was a convenient place to stop over and rest on weekends stranded away from home.

Directorships on nationally active corporations tied both Brown brothers into a broad circle of influential business leaders from all over the nation. Herman was a director of Judge Elkins's First City National Bank and of ARMCO Steel, which invited him to join its board after Brown & Root completed construction on a Houston-area plant for Sheffield Steel, which was later purchased by ARMCO Steel. After Herman's death in 1962, George replaced him on the ARMCO board. George also held seats on the boards of International Telephone and Telegraph (IT&T), Trans World Airlines (TWA), Southland Paper (later Champion Paper), and Texas Gulf Sulphur.

George took his seat on the IT&T board at a pivotal time in the company's history. In 1946 the longtime chairman of IT&T, Colonel Sosthenes Behn, faced a challenge from a group of dissatisfied stockholders led by Clendenin Ryan, the grandson of financier Thomas Fortune Ryan; Robert Young, who owned the Alleghany Corporation railroad holding company; and his nephew, Robert McKinney (whom George earlier introduced to Lyndon Johnson), chairman of Davis Manufacturing and a director of the Missouri Pacific Railroad. By late 1947, this group had accumulated sufficient stock to demand changes in the board. Behn announced the resignation of six current directors and the addition of six new directors, including Ryan, McKinney, Allan Kirby (a McKinney associate and Woolworth Company heir), Joseph Powell (a Boston shipbuilder), W. Randolph Burgess (from the National City Bank), and George R. Brown. In June, 1954, Brown was named to the IT&T executive committee, which had the responsibility of "defining goals and making key decisions regarding policy." In 1958, Brown was a member of the committee that began searching for a new CEO of IT&T. The group picked Harold Geneen, who would lead IT&T for the next two decades and become a close friend of George Brown.[67]

Homer Thornberry, Sam Rayburn, George R. Brown, Posh Oltorf
at Huntland Farms, mid-1950's. *Photo by Isabel Brown (Wilson).*

George Brown's next major corporate directorship was at TWA, where he again joined the board in a time of turmoil. When Howard Hughes, the controlling stockholder of TWA, encountered financial difficulties, he and board members favorable to him were forced from active involvement on the board. George Brown was the first of the new members named to the board. Former TWA chairman Charles Tillinghast recalled that he was not certain exactly why George Brown was selected but that "I suspect from things that were said from time to time that people were not unmindful of his connection with President Johnson."[68] Tillinghast remembered Brown as a fairly quiet member of a quite distinguished board that included Ernest R. Breech, Clifford F. Hood, former president of U.S. Steel; Barry T. Leithead, president of Cluett, Peabody & Co.; Hughston M. McBain, director and former CEO of Marshall Field & Co.;

and John A. McCone, former chairman of the Atomic Energy Commission (AEC).[69] George later helped arrange for Robert McKinney to join the board, and after McKinney assumed the position of chairman of the nominating committee, he invited Henry H. Fowler and George McGhee on the board as well. Fowler later became secretary of the treasury and McGhee became undersecretary of state in the Johnson administration. According to one longtime friend of both of the Brown brothers, Herman joked about George's membership on the TWA board, suggesting that his brother simply wanted to associate himself with big shots. But the friendships, contacts, and knowledge gained through service on major corporate boards were valuable assets to the Browns.

George Brown also served on numerous significant government commissions. One of his first and most prized memberships was at the state level, on the Texas Board of Engineers. George worked hard to improve the licensing of engineers in Texas, and he remained proud throughout his life of his own license—the fourth issued by the state. During August, 1937, George Brown's friend Gus Wortham nominated him for the board of engineers in a letter to Texas governor James Allred. State senator Weaver Moore also recommended Brown. Herman Brown, apparently already privy to Allred's intention to appoint George to the board, wrote Allred and told him, "I wish to assure you that I deeply appreciate the honor you have given George in this appointment and I am sure that you will have no occasion to regret it." Governor Allred, apparently letting Herman know beforehand that George would be appointed, wrote Herman Brown and told him that "as soon as I complete the appointments will announce all of them to the press. In the meantime, I would prefer that no publicity be given the matter."[70] As a dedicated engineer, George took special pride in this appointment.

Membership on several high-level federal government commissions gave George much greater national visibility. The first and most important of these was the President's Materials Policy Commission, commonly referred to as the Paley Commission, to which President Harry Truman appointed George Brown. The commission's chairman was William S. Paley, chairman of the board of Columbia Broadcasting System (CBS). The other members included Arthur H. Bunker, a partner of Lehman Brothers; Eric Hodgins, an author and editor from Connecticut; and Edward S. Mason, professor of economics at Harvard.[71] Truman constituted the commission on January 21, 1951, with the purpose of studying "the materials problem of the United States and its relation to the free and friendly nations of the world."[72] Supported by a large staff, the commission produced a highly publicized, landmark report—*Resources for Freedom*—detailing the nature of the natural resource problem including an analysis of available materials and energy sources as a survey of technologies capable of exploiting necessary materials.

During the Eisenhower administration, George Brown became a member of an eight-member civilian panel designed to study atomic energy. The Joint Congressional Committee on Atomic Energy, chaired by Senator Clinton P. Anderson, formed the committee and offered the chair to Robert McKinney. At this time McKinney owned a newspaper in Santa Fe, New Mexico, was close to many Washington insiders, and had served as assistant secretary of the interior under President Truman. Brown joined McKinney on a committee that included Ernest R. Breech; Sutherland Dows, chairman of Iowa Light & Power; Dr. John R. Dunning, physicist, dean of engineering at Columbia University, and director of the Oak Ridge Institute of Nuclear Power; Frank M. Folsom, president of the Radio Corporation of America (RCA); Dr. T. Keith Glennan, president of Case Institute of Technology and former member of the AEC; and Samuel B. Morris, manager of the Los Angeles Department of Water and Power and chairman of the Atomic Energy Commission of the American Public Power Association. Later, Eisenhower appointed labor leader Walter Reuther to the panel.[73]

Reuther, the head of the United Automobile Workers and one of the leading union leaders of the era, and George Brown, owner of the aggressively antiunion Brown & Root, came to a meeting of the minds during their work on this commission. Both enjoyed intellectual exchange on major issues of the day, and they came to enjoy each other's company despite their obvious disagreement on many issues. After one meeting at the White House, Reuther asked Brown for a ride back to his hotel, only to find that Brown also had taken a taxi to the meeting. Brown was waiting for a ride from his daughter, Isabel, who was then working in Washington. He offered Reuther a ride home in his daughter's car. When Brown's daughter pulled up to pick up her dad in a Mercedes Benz, Reuther remarked, "I sure hope nobody sees me in your car." It could not have been too good for Reuther's image to be seen riding around Washington with one of the archenemies of organized labor in a foreign car.[74]

The panel's mandate was to evaluate the peaceful use of atomic energy in the United States and, if possible, recommend legislation and policy to guide peaceful atomic energy development.[75] The commission held most of its meetings in Washington, but George Brown arranged for the last meeting to be held at the Huntland Farms. The panel's report to Congress, dated January 31, 1956, urged the development of peacetime atomic power under the guidance of a broad partnership of government, independent scientists, and industry. The panel also called for an end of the AEC's centralized control of the atomic energy industry.[76]

President Eisenhower called on George Brown's assistance on several other expert commissions. Brown had supported Eisenhower in the 1952 presidential election, and he was close personal friends with Eisenhower's secretary of health, education, and welfare, Oveta Culp Hobby, and his attorney general, Herbert Brownell, who had worked on the early organization of the Texas Eastern pipeline.[77] Eisenhower appointed

George Brown to chair a thirteen-member commission designed to study several rivers in Texas from the perspective of water and land conservation.[78] Eisenhower also included Brown among a large group of prominent U.S. business leaders invited to the White House on January 27, 1958, for a White House luncheon designed to encourage support of the Eisenhower administration's foreign aid program.[79] In 1959, Eisenhower's secretary of commerce appointed eight persons—including George Brown and Stephen D. Bechtel of the Bechtel Corporation—to join the Commerce Department's Business Advisory Council. The 150-member council of industry leaders advised the secretary of commerce of business viewpoints on issues facing his department.[80] Eisenhower appreciated Brown's service on such committees and showed his gratitude by agreeing to come to Houston in 1960 and give a major address at Rice University, where George Brown served as chairman of the board of trustees.[81]

George Brown's influence as an advisor to government officials on business matters continued to grow with Lyndon Johnson's election to the vice presidency and his subsequent ascent to the presidency. In the 1960s, Johnson appointed Brown as the only representative of business on the Space Council, which advised the government on space-related issues such as the site of NASA's manned spacecraft center. During the Johnson presidency, Brown regularly visited the White House to attend luncheons for prominent businessmen whose advice—and financial support—were sought by Johnson. After such occasions, the Browns frequently slept at the White House.[82]

Such visits to the White House must have spurred reflections by the Browns on their long relationship with Lyndon and Lady Bird Johnson. Since the 1930s, the Brown families and the Johnsons had led intertwined lives. Johnson's biographers noted this ongoing relationship while stressing the mutual benefits from the friendship and support of the contractors and the politician. Their complex relationship encompassed both the practical and the personal, but thirty years of friendship and shared experience was at its core.

One of the high points of the Browns' support of Johnson came during December of 1957, when the Browns hosted a much-publicized dinner for Senator Johnson in Houston. In a letter of thanks to Herman Brown, Johnson noted that "the most memorable occasion of my life will be the dinner in Houston the other night, and the most and fondest remembered among my friends will always be you and George."[83] This was an important time in Johnson's life, for he faced pivotal choices about the trajectory of his political career. In November of 1959, he wrote to George Brown, "This next year will be full of crucial decisions where I will very much need *your* judgment. I have reached a point where I can either go on as I am now doing or change the course of my life and career."[84] In such times, Johnson turned instinctively to the Browns for support and advice.

Johnson's reliance on the Browns was based on far more than political expediency.

For more than thirty years, the three men from Central Texas shared a passion for politics, as well as for power. They hunted together, both for game and higher political offices. They shared many longtime companions and political allies. And above all, they talked incessantly about political strategy, the uses of governmental power, and the future of Houston and of Texas. Johnson understood early on that Herman Brown was not impressed by flattery, and Johnson relished the direct, even gruff conversational style of the self-made man from Belton. Herman Brown told his own blunt truth on any issue, whether he was talking to Lyndon Johnson the Central Texas boy in the 1930s or Lyndon Johnson the vice president of the United States in the early 1960s.

Johnson was closer in age to George, with whom he enjoyed a warmer relationship. A central part of his relationship with both brothers was intellectual; all three enjoyed exchanging opinions about political and economic affairs that marked their long-term friendships. George described his relationship with Johnson as a "complete meeting of the minds" by two men who "became close due to our philosophy of life." He and Johnson found common ground on the education issue, since both strongly believed that this was one of the key issues that would determine America's future. George and Lyndon spent many hours in rambling discussions of "the burning questions of free-enterprise and socialist form of government, autocratic form of government versus the democratic form." George Brown recalled other occasions where he and his older brother and Lyndon Johnson would argue about such issues, and "most of the time I was on Johnson's side. We had a lot of disagreements, but a lot of times I'd argue on his side."[85] It is certain that much of the attraction that kept these three men close for forty years was their shared excitement in debating the day's issues.

These men cared deeply about each other. When Lyndon Johnson's mother died in September of 1958, Herman, George, and Alice Brown attended the funeral. Johnson, who had written kind words to the Browns after their mother's death in 1947, thanked them for their consideration at a time of personal anguish: "I borrowed heavily of your strength and fortitude, Alice and George, on the day of my mother's funeral." A second letter to Herman acknowledged that "you gave me strength by being at the graveside, Herman."[86] The Browns established a memorial fund for Mrs. Johnson, whom they had visited regularly in their travels through Central Texas. George remained a frequent visitor to Johnson's ranch after the former president retired from office. These three friends and their families led intertwined lives for decades. Accounts of their relationships that stress only the financial and political ties that bound them together miss the depth of their friendships. Despite his well-developed penchant for flattery, Johnson should be taken at his word when he wrote in 1957 that "every once in a while—and now is one of those times—I try to take stock of myself. I invariably find that my chief asset is that I have George Brown as a friend and I am particularly grateful for that fact."[87]

Johnson's quest for the presidency placed new strains on his friendship with the Browns, and particularly with Herman. In the late 1950s, Johnson decided to plan a strategy to achieve his ultimate goal, to be elected president of the United States. Early in their careers, Herman recognized Johnson's intense political ambition. Although he admired Johnson for his dedication, he feared that Johnson would have to become a "devious" politician pandering to special interest groups for votes. Herman felt that Johnson was more valuable to the nation as the powerful Senate majority leader, and he feared that Johnson's ambition for higher office was misplaced, if not unrealistic.

To put in motion his plans to capture the presidency, Johnson met with a group of 8-F members at the Palomas Ranch at Falfurrias, Texas. The group included Herman Brown, Bob Henderson, Bill Smith, Bob Abercrombie, Naurice Cummings, and Mike Stude. While the group was gathered around the fireplace, Johnson said, "I want to tell all of you fellows something. I'm thinking about running for president and if I do, it's going to cost all of you a lot of money. So, I want you to think about it." Herman responded first. "Well, I don't have to think about it. I just want to say that will be the biggest mistake you've ever made." Cummings recalled that Herman always spoke directly and forcefully to Johnson in a way that Johnson "didn't particularly like," although George "always seemed to straighten everything out."[88]

Johnson at one point turned to Stude and said, "Here Mike, take this pad. I'm going to go through every state and we're going to count the delegates and you have a column for Kennedy and you have a column for Johnson. Okay?" Stude took the pad and wrote down the information as Johnson reviewed each state in alphabetical order, following with a summation of each state's delegates plans for voting. "Well, of course it came out in favor of Johnson when he added them all up, when I added them all up," Stude recalled. Johnson discussed the significance of his delegate count to the men, and he asked for their support. These men were not ones to get immediately excited about Lyndon's pitch; they would think about it before acting. They also seemed to become somewhat bored by the entire presentation. As Johnson continued to talk, Herman abruptly stood up and said, "Well, let's go hunting." Johnson, somewhat offended by Herman's remark, replied, "Hell, I don't want to go hunting, I want to talk." Herman repeated, "It's time to go hunting." The other men followed Herman, as did Lyndon, who continued talking about his ideas for running for the presidency.[89]

Once Johnson's campaign for the Democratic nomination was under way, the demands of politics drove another wedge between him and Herman Brown. Johnson trailed John F. Kennedy in the polls, and in his effort to broaden his popular appeal, he decided to abandon his support of a national right-to-work law. Herman Brown was not pleased.

At one point in the campaign, Johnson challenged Kennedy to a debate and Kennedy accepted. Working with Robert McKinney, Johnson arranged a place and time for the debate. The television company approached the Johnson campaign and requested sixty thousand dollars to televise it. Johnson tried to collect half the amount from the Kennedy campaign, which refused on the grounds that Johnson wanted the debate and he should pay for it. As the scheduled date approached, Johnson became increasingly anxious about raising the money. He told McKinney, "You've got to call George Brown and tell him, that's all we can do."[90]

McKinney spoke with Brown. George wanted to know what Johnson really felt his chances were to get the nomination. Johnson, once very confident, realized now that his chances were not high. He estimated that he had about a 35 percent chance to win the Democratic nomination. George told McKinney, "All right, if Lyndon thinks he's got at least a 35 percent chance, we'll go ahead and you write Lyndon Johnson a check for sixty thousand dollars." McKinney replied that he had only about sixty thousand dollars cash available. Brown told McKinney, "Well, Lyndon's got to have the money today, so you write a check for sixty thousand dollars out of your bank account and give it to Lyndon and that's the end of that."[91]

Tension grew between Johnson and his supporters after Johnson realized that he could not win the Democratic presidential nomination and agreed to accept the position of Kennedy's running mate. Herman Brown strongly opposed Johnson's decision, and many different associates of the Browns long relished—and, no doubt, embellished—stories of how vehemently Herman expressed his disapproval. August Belmont recalled being with Herman Brown in 8-F when the radio broadcast the announcement that Johnson would be Kennedy's running mate. "Herman Brown," recalled Belmont, "who had sort of a face the color of those red flowers anyway, he jumped up from his seat and said, 'Who told him he could do that?' and ran out of the room."[92] Others recalled that Herman Brown called Johnson a "goddamn traitor."[93] Although George Brown was ambivalent on this issue, he recalled that his brother "was dead set against him [Johnson] taking it [the vice presidential nomination]."[94] Herman thought that Johnson was making a mistake, since his position as Senate majority leader seemed to offer more power than the vice presidency. Despite his strongly voiced disapproval of this particular choice, Herman continued to exchange pleasant letters with Johnson, including a letter of congratulations after the Democratic ticket's victory in November of 1960.

Most historical accounts of Johnson's career acknowledge the strong personal ties between the Browns and Johnson but stress the more practical side of their relationship. Robert Caro's best-selling biography of Lyndon Johnson stresses the brothers' role in bankrolling their friend's political career. In writing about Johnson's early political career, Caro concludes that his "power base wasn't his congressional district, it was Herman Brown's bank account." Caro portrays the Browns as moving political corruption to a level unprecedented in American history. Their actions in support of Johnson are viewed as a departure in the level of campaign financing by businessmen.[95]

Less strident than Caro's treatment of the Browns is that of Ronnie Dugger, a journalist who covered state politics for the *Texas Observer* in the 1940s and 1950s. Dugger's highly readable biography of Lyndon Johnson discusses many of the same episodes subsequently described in Caro's dramatic narrative. Like Caro, Dugger portrays the Browns as Johnson's eager supporters willing to stretch existing campaign contribution laws past the breaking point. He notes that Johnson regularly supplied valuable insider information on potential construction projects to Brown & Root and that the Browns pledged and delivered "100 percent" support to Johnson. Yet the tone of Dugger's biography and of his treatment of the Browns differs significantly from that of Caro's books. While acknowledging that Brown & Root's close ties to politics were at times disturbing, Dugger notes that such ties were a part of a broader pattern in postwar America, when "politics went in partnership with business." Although highly critical of this emerging partnership, Dugger presents it as an unfortunate fact of history, not as a long-covered-up scandal that marks a shocking departure in American political history. He concludes that the Browns "poured money into Johnson's campaigns, won fat contracts in work that paralleled his career, gave him a personal-political interest in higher military spending, and rooted his politics in their own antiunionism."[96]

Clearly, money for political campaigns was one significant ingredient in the strong glue that bound Johnson to the Browns. Money has long been characterized as the "mother's milk of politics," and the brothers took a straightforward, aggressive approach to campaign contributions. Both Browns understood the value of well-placed contributions to political campaigns as a logical and vital business expense for their varied economic enterprises, which often were greatly influenced by government policies.

The one publicized case in which the Browns stepped over the line of existing laws—if not existing practices—occurred in 1940 and 1941, as Johnson faced a hard and ultimately losing campaign for the Senate. In examining the Browns' support of Johnson, the Internal Revenue Service (IRS) found numerous cases of Brown & Root or a subsidiary paying a "bonus" to an employee, who then contributed the funds to the Johnson campaign or used it to pay directly for campaign expenses. Brown & Root caused itself further problems by charging off such bonuses as business expenses, which could then be deducted from its federal tax liability. The U.S. Internal Revenue Code forbade tax deductions or exemptions for political contributions.

After the investigation began, the Browns employed Alvin Wirtz as their attorney in this case. An IRS investigation of Brown & Root's 1940 and 1941 accounts found numerous cases of bonuses or other fees making their way indirectly to the Johnson campaign, including large payments to top Brown & Root executives. Fearing both a damaging investigation of his own campaign funding and a possible criminal indictment against the Brown brothers, on January 13 Johnson and Wirtz met privately with President Roosevelt. Although Johnson later denied discussing the Brown & Root case at this meeting, that afternoon, the assistant secretary of the treasury telephoned the Dallas IRS office and stated that Roosevelt had just called to say he wanted a complete briefing of the Brown & Root case at ten o'clock the next morning. After the briefing, on January 17 the IRS sent an agent from its Atlanta office to Dallas to close down the Brown & Root case. Soon thereafter, the IRS ascertained that Brown & Root underpaid its taxes by $1,099,944, and the government added a 50 percent penalty for fraud, amounting to $549,972, for a total of $1,649,916 owed to the IRS. After a series of additional conferences, IRS officials indicated that they lacked sufficient evidence to prosecute a fraud case. The IRS required Brown & Root to pay a reduced tax fee of $372,000 but otherwise dropped the case.[97] This was as close as the Browns came to serious legal problems over campaign contributions.

In addition to vital campaign contributions, the Browns made various "soft" campaign contributions in the form of services provided to the candidate. Johnson understood the mutual benefits of his ties to the Browns, and he seldom hesitated in asking for a variety of favors. Of special significance to a Texas politician holding national offices in Washington was the availability of free airplane transportation. Johnson and his top aides frequently took advantage of this "service," which was a legal and widely used business contribution to politicians until the IRS ruled that a company that allowed a friend, politician, or businessman free use of a plane had to either report the use of the plane as a gift or pay a tax on the plane's use. Jack Robbins, one of Brown & Root's pilots, recalled that in days prior to the rule changes he "got to know Lyndon Johnson very well. I flew him quite a bit. And Tom Clark, Lieutenant Governor Shivers, and then finally Governor Shivers [and] Governor Beauford Jester. Speaker of the House Sam Rayburn. Even the minority leader at that time."[98] Concern about his heart caused Johnson to insist on flying at nine thousand feet so that he did not subject his heart to the stress of flying at higher levels in the unpressurized DC-3. Robbins "didn't have any trouble getting nine thousand feet assigned to us. I'd put on the flight plan, 'Senator Johnson aboard' and they'd always give me nine thousand feet."[99]

The Browns sometimes attended to Johnson's friends and coworkers. During his congressional career, Johnson occasionally turned to the Browns for assistance in finding employment for friends; he seemed to consider Brown & Root a sort of employment service of last resort for well-connected relatives and acquaintances of politically significant people. With its far-flung operations around the world, Brown & Root could and did "absorb" new employees as favors to Lyndon Johnson. The correspondence between the Browns and Johnson on this issue suggests some tension and a measure of resentment by the Browns on occasion, but they generally sought to comply with their friend's requests, accepting them as part of the price of their friendship.

Johnson matched such favors with a variety of services and generally assured them access to national political influence. On several key issues, they could be confident of Johnson's vote and his powerful assistance in securing additional support. Labor issues remained a high priority for Herman Brown. In early 1953, for example, he wrote Senator Johnson looking for assistance in shaping federal labor laws: "I was wondering if you thought you would be able to get Senator Taft to not try to override State laws which was the original idea under the Taft-Hartley Act and this has been construed by the Courts. It was my understanding that the Republican platform specifically played up State rights and State police powers."[100] The Browns could also expect Johnson's close attention to legislation affecting Texas Eastern and the natural gas industry.

Johnson and his aides generally gave prompt consideration to requests or inquiries from the Browns, who were careful to limit such requests to issues they deemed of special importance. Thus George Brown received assistance in putting forward the ultimately unsuccessful case of a consortium of Texas universities for the location of an atomic energy laboratory in Texas. Inquiries from the Browns on the status of new plane routes into Houston were quickly relayed to key congressmen and regulators. Pleas from Texas Eastern's lawyers to fight for the removal of Leland Olds from the Federal Power Commission, which regulated the natural gas industry, received a sympathetic response from Johnson.[101] On these and other similar issues, the Browns did not instruct Lyndon Johnson how to vote, but they knew that their opinions registered forcefully with one of the most powerful politicians in postwar America.

The publication in 1997 of heretofore secret tapes made by Johnson as president gives added insight into the give-and-take of political influence. One tape that captured the voice of George Brown involved a personal plea by Brown's associate John Jones, the nephew of Jesse Jones, for Johnson's help in pushing for approval of a pending merger between two large Houston banks from antitrust officials. Brown, who had no interest in either bank, mediated the negotiations between Johnson and Jones. Jones finally got the help he wanted, but only after writing a letter to Johnson promising that he would have the support of the *Houston Chronicle,* Jones' newspaper, in the future.[102]

In Washington, Austin, and Houston, the Browns built circles of influence useful in pursuing their varied interests. Their methods were hardly unusual in the time and place in which they lived; nor were their political goals particularly extreme. Perhaps to spice his story with the sort of J. R. Ewing flavor that readers seem to expect of larger-than-life Texans, Caro goes out of his way to label the Brown brothers as "ultra-reactionaries both."[103] This portrayal makes for lively reading, but it is not an accurate depiction of the political views of either Herman or George Brown in the context of the political spectrum in Texas or even in the United States in the 1950s.

On the most significant political issues of their place and time, the Browns were conservatives, not ultrareactionaries. The Browns were never among the aggressive racial bigots who poisoned southern society in the transition years of desegregation. They never became obsessed with race as did segments of the white population. In fact, as will be discussed in more detail, George Brown took the lead in desegregating Rice University, which was one of the most visible and highly publicized episodes in the death of Jim Crow laws in Houston. Herman Brown, who had lived and worked with racially mixed road crews in the early years of Brown & Root, sought to develop apprenticeship programs for black workers traditionally segregated into all-black labor gangs. Although the Browns can be correctly criticized for accepting the racist language and attitudes of "mainstream" white southern society in the Jim Crow years, they cannot be reasonably categorized as the sort of committed racists who represented the true "ultrareactionary" forces in the South in the post–World War II era.

Of course, race was not the only controversial issue in postwar Texas. Support of or opposition to labor unions was another litmus test of political conservatism. As discussed earlier, Herman Brown and Brown & Root became symbols of antiunionism to those seeking to organize Texas workers. Herman Brown relished his image as a determined advocate for the "right to work" and a dedicated fighter of unions at Brown & Root. But even on this controversial issue, he stood out from his contemporary Texas businessmen not so much in his attitudes toward unions as in his effectiveness at putting his attitudes into action. The state was strongly antiunion, with few union defenders among its politicians or its business community.

One other key issue that distinguished conservatives from more fervent reactionaries in postwar Texas was McCarthyism or zealous anticommunism on the domestic front. Here again the Browns cannot accurately be singled out as "ultra" anything; according to the most thorough account of the red scare in Houston, neither George nor Herman was prominent in the anticommunist crusades that wracked Houston in the early 1950s.[104] Unlike some fellow Houston businessmen, the Browns never courted either Joe McCarthy or his local followers. Indeed, according to the recollections of one observer, Herman Brown was singularly unimpressed after his sole encounter with McCarthy, a brief introduction at a Washington cocktail party.[105] On this and other issues, the Browns simply were not aligned with those on the extreme right in Texas politics. They were not the sort of reactionary crusaders that many observers of Texas politics have taken such delight in lampooning.

This was most clearly evident in their choice of politicians. In national politics, George Brown originally supported Franklin Roosevelt before he and his brother hitched their wagon to the self-styled Roosevelt protégé, Lyndon Johnson. In state and local politics, they favored an array of probusiness, conservative Democrats who made up the mainstream of Texas politics during the one-party era before the 1970s. Herman Brown, the more conservative of the two brothers, was an "Eisenhower Democrat."

His pragmatic approach to politics was clearly expressed in a response to a fund-raising letter from the Committee for Constitutional Government, a nationally active group of conservatives that solicited funds from Herman in part to try to replace Lyndon Johnson with a more conservative senator in the 1954 election. Brown responded by observing that "I have been actively associated with what I consider the conservative element of Texas politics for many years and I think I fully understand Texas politics." He then argued that the committee's effort to replace Johnson "approaches gross negligence," since "Senator Johnson has personally and quietly sabotaged more radical legislation since he has been minority leader than any member of the Senate. He has been extremely successful at keeping the radical elements of the Democratic Party in line and fairly quiet."[106] This position reflected more than a blind support of Lyndon Johnson. The Browns approached politics pragmatically, seeking results rather than ideological consistency. If their position in Texas politics in the 1950s is to be labeled "ultrareactionary," then historians will be forced to invent two or three more categories of analysis to encompass the many prominent Texans to the right of the Browns on the political spectrum of postwar Texas.

The importance of the Johnson connection for the business enterprises run by the Browns is not subject to quantification, but easy access to powerful friends in Washington no doubt facilitated the extraordinary expansion of those aspects of the Browns' businesses that relied heavily on government contracts and that were directly affected by government regulations. The growth of the Browns' empire, however, was not solely or even primarily shaped by government policies. After 1945, much of Brown & Root's expansion was in private sector projects, such as the building of industrial plants and offshore facilities. Even in such areas as construction projects directly controlled by government agencies, Brown & Root's hard-earned reputation as a company that brought in well-constructed projects on budget and on time remained an important explanation for its success. In addition to its can-do reputation, the firm also remembered a key lesson about public works first learned in its road building days. A well-developed network of information about potential projects and expertise in preparing bids was crucial in laying the groundwork for success. The Browns had experienced employees and associates in Washington, D.C., reporting back regularly on possible opportunities, and these men systematically canvassed federal agencies for coming opportunities.[107] Personal connections with mid-level officials in the navy or the Corps of Engineers could be just as useful at times as contacts with a senate majority leader.

The Browns created and aggressively used circles of influence at the local, state, and national levels. Their role in the 8-F crowd, their aggressive lobbying of the Texas state legislature, and their very close relationship with Lyndon Johnson were much in the traditions of American political history. The Watergate scandals of the early 1970s brought a wave of reforms in the campaign financing laws. After that time, many of the practices followed by businessmen such as the Browns became illegal, but the controversies over the flow of money into politics obviously have not ceased. In the years before Watergate, there was an expansive gray area between what was clearly legal and what was clearly illegal. The Browns pushed aggressively into this gray area in support of Johnson and others. At the risk of rationalizing their behavior, it seems appropriate to observe that on close examination, most modern presidents and would-be presidents probably will be found to have a Brown or two in their campaign finance closets.

The Browns are best seen as mid-twentieth-century Texas variants of the aggressive entrepreneurs who placed their mark on the nation's economic and political life during the rise of big business in the late nineteenth century. What was unique about the Browns' influence was the place and time in which they lived. As they went about the business of building a modern Texas and helping Houston become a major city, their paths crossed with a young politician also intent on making a difference in his home state. The fact that Johnson rose to the presidency has opened for close scrutiny the political activities of his chief supporters, Herman and George Brown. In this sense, they left their mark on the broadest stage offered by American history, presidential politics and power.

The influence they wielded made many observers uncomfortable, since it reflected a concentration of power at odds with the commonly held vision of representative democracy. The Browns proved very adept at exerting well-organized political influence in concert with other businessmen with similar political goals. That this influence reinforced their economic success is beyond question. But how it also was used in the pursuit of a broader civic and even social vision is still only vaguely appreciated. A fuller understanding of the Browns' circles of influence in Houston suggests that their power extended far beyond the city's borders and far beyond simple economic self-interest.

CHAPTER EIGHT

Building Universities

O nce established as successful businessmen, the Browns turned more of their energies to civic endeavors. Higher education became a special concern of the Brown brothers and their wives. In the postwar years they devoted increasing amounts of their money, energy, and time to leadership in this area. George and Alice focused their efforts on Rice University in Houston; Herman and Margarett, on Southwestern University in the Central Texas town of Georgetown. The brothers' business skills, sense of mission, and connections had a substantial and lasting impact on both universities. From their time spent working on the issues facing Rice and Southwestern, the Browns took great personal pride and enjoyment.

In such civic endeavors, the Browns' business experience proved critical. The most obvious connection between success in business and successful involvement with civic institutions was money. Universities and cultural organizations have perpetual needs for funds to attract the best educators and upgrade their academic programs and facilities. But dollars alone did not register the depth of the Browns' commitment to their favored causes. They brought an entrepreneurial spirit to the task of building stronger organizations. The Browns succeeded in business by aggressively pursuing promising, though at times risky, opportunities. The same approach made them forces for change on the normally conservative boards of universities, and they helped make both Rice and Southwestern more aggressive in the pursuit and creative spending of funds. The Browns could be counted on to bring to the table more than their own money, time, and energies. Their enthusiasm and commitment helped convince others to support higher education. In the cases of Rice and Southwestern Universities, the Browns' financial support and leadership proved decisive in propelling the institutions to new levels of achievement.

The brothers went somewhat separate ways in supporting higher education. Herman, who briefly attended the University of Texas before engaging full-time in road building, worked most closely with Southwestern. His wife and his sister-in-law had graduated from there, and the university was just down the road from Belton in Central Texas, where the Browns had been raised. George focused his efforts on Rice Institute (whose name was changed to Rice University in 1963) in his adopted home-town of Houston. Yet, though several hundred miles separated these two quite differ-ent institutions, the brothers strongly supported each other's activities. When the needs of either school demanded their cooperation, they pulled together, either in-formally or through the Brown Foundation, which became an increasingly signifi-cant vehicle for supporting the two universities and other philanthropic activities.

Rice was the magnet that first attracted George Brown to Houston. When he en-tered the recently opened Rice Institute as a student in September of 1916, George found an ambitious new venture made possible through the generous endowment of William Marsh Rice. The original Rice Institute board of trustees was organized in 1891, but only after Rice's death in 1900 and a prolonged legal dispute over his will did the new institute open its doors to its first class in 1912. With an original endow-ment of about $5 million, Rice Institute held great promise of excellence.[1]

The young George Brown's decision to attend Rice in 1916 reflected the rapid establishment of the new school's reputation. Brown had graduated from Temple High School earlier in that year, and his choice of Rice over the nearer and more established state universities at Austin and College Station suggests the high esteem in which the new school was held. Brown had never before made the 170-mile trek from Temple to Houston, and his enrollment at Rice introduced him to the city that became his home.[2] His response to the lure of both Rice and the booming city of Houston mirrored that of several generations of ambitious young people who mi-grated from small towns in Texas to the urban-industrial center on the Gulf Coast. Rice Institute provided George Brown with a springboard into a broader world of opportunity, and later in life he would repay this favor many times over by assisting Rice in its own quest for national recognition.

Almost a quarter century passed between Brown's departure from Rice in 1917 and his reentry into the affairs of the school in 1942. During that time, George and Herman built the foundation for the business empire and the fortune that would allow them to become significant benefactors of higher education. While the Browns built Brown & Root, the Rice Institute prospered under the leadership of President Edgar Odell Lovett and Chairman of the Board "Captain" James A. Baker. The two led the way in establishing the institute's reputation as a regional center of high-quality education. They kept the student body small and highly selective while recruiting top-notch faculty who concentrated on research and undergraduate teaching. Houstonians spoke proudly of "the Princeton of the South," and graduates became increasingly evident in the leadership ranks of the region's businesses, law firms, and other professions.

As a private institution that did not charge tuition until 1965, the continued success of Rice rested squarely on its endowment. Recognizing that its large endowment gave Rice a unique position among regional universities, its leaders took a conservative approach in managing its financial affairs. They invested the endowment in the safest and most predictable places: local real estate and government bonds were the particular favorites. The endowment grew slowly, from $5.1 million in 1911 to $6.7 million in 1941, during which time the institute's total assets almost doubled from $9.4 million to about $18 million. During these formative years, the institute prospered under the firm leadership of Chairman Baker and President Lovett, who led the way in building an excellent modern university from the ground up in Houston, far from the traditional centers of higher education. Baker's death in 1941 and Lovett's impending retirement marked the end of the institute's formative years, causing all who were concerned with Rice to reflect back on the institution's past and to ponder its future.[3]

George Brown stepped forward in the 1940s to help mold that future. As the Browns' businesses prospered and George Brown entered his forties, he had more time and resources to devote to higher education. Both Brown & Root and Texas Eastern had headquarters in Houston and drew staff from the area, making the Rice Institute, with its strong commitment to engineering, a natural target for George Brown's support. Although his own engineering degree came from the Colorado School of Mines, it was only a matter of time before George became more deeply involved in the affairs of his first university.

The time was right in the early 1940s, when Rice faced the need for new leadership and new funding sources. George Brown helped provide both in a pivotal episode that greatly improved the institute's financial standing. A group of Rice supporters recognized an excellent opportunity to acquire a valuable interest in a proven oil field—the Rincon Field in South Texas. Brown took the lead in organizing, financing, and selling a "deal" acceptable to the traditionally conservative Rice board. The successful completion of this transaction required an entrepreneur's instinct to move quickly and decisively in seizing opportunity; Brown's two decades of experience in construction prepared him for this moment. It required the capacity to create a consensus in "the Rice community" for a departure from the safe patterns of the past.[4]

The opportunity was originally identified by Harris County judge and Rice alumni Roy Hofheinz. The Rincon Oil Field had been developed by W. R. Davis, who purchased in 1938 a lease of more than twenty thousand acres from a small exploration company that had drilled a wildcat well on the property. Davis moved quickly to develop this promising property in Starr County in the southernmost tip of the Rio Grande Valley. Excellent early results encouraged Davis to raise needed developmental capital by selling a half-interest to the larger Continental Oil Company and borrowing heavily from major banks using the proved reserves of the field as collateral. By 1941 Davis and Continental Oil had invested heavily in drilling 128 producing wells with an allowable production of about 100,000 barrels of oil per month and estimated reserves of 50 million barrels. When Davis died suddenly in a Houston hotel in August of 1941, the Rincon Field was on the threshold of becoming a very profitable venture—but he remained $5.5 million in debt from expenditures required to develop the property. Although many oil men recognized the long-term value of the Rincon Field, Davis's debts, potential tax liabilities, and the uncertainties of World War II all limited the attractiveness of this property to other oil companies.[5]

Judge Hofheinz learned of the Rincon properties and the peculiar details that made them of special interest to a nonprofit organization such as Rice Institute. He approached other Rice boosters for advice and support, calling on the expert evaluation of several Rice alumni active in the oil industry to assure him of the Rincon Field's potential value. Brown and H. C. Wiess, one of the founders of Humble Oil & Refining and a contributor to Rice, quickly joined Hofheinz in studying the opportunity and in trying to arrange the acquisition. In September of 1942, this threesome "approached the Trustees" of Rice with the "proposition that the Institute become the owners of certain properties then owned by the estate of W. R. Davis."[6]

The proposition was straightforward: for $1 million down and the promise to pay off about $5 million in existing debts using future revenues from the sale of oil, Rice would acquire properties conservatively valued at $10 to $20 million. If the estimates of the best geologists and petroleum engineers of the major Houston-based oil companies proved to be correct and if the price of oil did not decline sharply in the future, the Rincon Oil Field would be a bonanza for Rice, generating perhaps as much as $1 million per year for the institute once the debts had been paid off in the late 1940s. For its investment, Rice would receive twenty-nine sixty-fourths of the Rincon Field, in which Continental Oil would retain one-half interest and Davis heirs another three sixty-fourths interest. Continental would continue to operate the entire field. Brown considered this an offer the Rice board could not refuse; using the most cautious estimates of the reserves in the group at the Rincon Field, the institute stood to more than triple its existing endowment and approximately double its total assets in one decisive move. But to Brown's astonishment "they [the Rice board] didn't want any part of it." He recalled later that "they felt the whole thing was a gamble and Rice didn't have any business with an oil field."[7]

To convince the board otherwise, Brown, Wiess, and Hofheinz moved on several fronts. Expert opinions bolstered their argument that this particular deal was as close to a sure thing as was possible in the oil industry. The reserves were well established; the risks were minimal. Brown, Hofheinz, and a representative of the board journeyed to Chicago for a conference with representatives of the four major banks with claims against the properties. In the 1940s, most large banks still had qualms against making loans secured by oil in the ground, and the conservative bankers' enthusiasm for the Rincon Field and Rice's proposed purchase reassured the institute's board. Harris County's Eleventh District Court erased all doubts about the institute's legal authority under its existing charter to invest its funds in such a venture when it ruled in November of 1942 that the trustees had "full power" to "manage, control, and invest the funds of the Institute." This decision cleared the way for the trustees to make a simple choice: was the investment of $1 million in the Rincon deal a wise use of the funds available to them?[8]

Such an investment would require them to move a long step away from the institute's traditional investment policies. However safe Brown and his friends considered the Rincon Field, the risks involved were far greater than the risks of making very conservative investments in Houston real estate and in government bonds. And $1 million was, after all, some 6 percent of Rice's total assets. Yet Brown had a ready answer to such speculation. He responded in a way that the Browns found effective again and again in similar situations. He wrote a personal check and matched it with one from his brother Herman. As finally presented to the board, the Rincon deal required Rice Institute to put up $1 million to secure ownership of the properties. Brown, Wiess, and Hofheinz proposed that one-half of that amount come from the Rice endowment and one-half from donors. Thus before the institute had to commit its own funds, it would already be assured of $500,000 in new contributions. The Browns quickly broke through any remaining barriers of uncertainties by immediately pledging $100,000—$50,000 from George and $50,000 from Herman. Other donors also stepped forward, led by Wiess and several other prominent members of the Humble Oil family. The remainder of the "private half" of the needed funds came from a $300,000 contribution from the M. D. Anderson Foundation, a philanthropic organization created by one of the founders of the Anderson Clayton Company, a Houston-based cotton specialist with strong Rice connections. With such financial support, the Rice board closed the transaction on December 18, 1942.[9]

The payoff came steadily over subsequent decades. The field's revenues retired the original debts by 1947, and the Rincon oil flow continued for decades. By the 1960s, Rice had received some $30 million from the field, and money continued to flow to Rice into the 1990s.[10] This single decision proved as important to the long-term well-being of Rice as any other decision since the original choice by William Marsh Rice to endow an institute in Houston. The funds from the Rincon Field gave Rice extraordinary new flexibility to expand creatively in the boom years after World War II. The example of this bold break with past practices became a symbol of the institute's capacity to adapt and to find revenues in addition to the original endowment. Successful adaptation required new leadership as well as dollars, and the Rincon deal brought to the fore two such leaders who would play prominent parts in the next generation of Rice's evolution, George Brown and Harry C. Wiess.

George Brown came to the attention of the Rice trustees even before the Rincon negotiations. In February, 1942, the executive board of the Association of Rice Alumni urged the trustees to "consider seriously" naming an alumni of the institute to the vacancy on the board created by the death of Captain Baker. The association put forward a list of six alumni "on the basis of interest in the welfare of the Rice Institute, personal integrity, business acumen, and availability." Included on this list along with other rising young Houston businessmen from Humble Oil, Cameron Iron Works, and Reed Roller Bit was George Brown, who was identified as a member of the class of 1920.[11] In May, 1942, the board ignored this list and named Humble Oil executive Harry C. Hanszen to fill the vacancy. During the Rincon negotiations, another longtime member of the seven-person Rice board died, and George Brown accepted the board's unanimous offer to become a trustee on January 13, 1943. The next spot that opened was filled by Wiess in July of 1944.

Thus Rice approached the postwar era with an important new addition to its endowment and three new trustees on a board of seven. The board faced the necessity of searching for a new president for only the second time in its thirty years of operations. Amid a growing sense that the institute had reached a crossroads in its development, the new board members—Hanszen, Brown, and Wiess—asserted a strong voice in decision making. Hanszen served as chairman from 1945 through 1950, when he passed the reins to George Brown, who had been named vice chairman in 1946. Brown remained chairman until 1967. Wiess was very active in Rice affairs in the years after joining the board in 1944, and he served with Brown as vice chairman after 1946. When Wiess died suddenly in 1948, George Brown lost both a friend and a tireless coworker at Rice.

Because of changes in the selection process begun in 1946, by the time of Wiess's death the board had a completely new composition. Changes in the bylaws of the institute in 1946 created a new emeritus position for trustees over seventy years of age; the four "holdover" trustees from the Baker-Lovett era took advantage of this option, thereby opening the way for the appointment of four new trustees. The four included Gus Wortham, president of American General Insurance; William A. Kirkland, an officer with the First National Bank of Houston; Frederick Lummis, physician in chief at Hermann Hospital; and Lamar Fleming, president of Anderson, Clayton & Company. Two of the new trustees, Gus Wortham of American General Insurance and William Kirkland of the First National Bank of Houston, had close business ties with the Browns. From 1942 through 1946, the board at Rice underwent a complete transformation in membership. The seven new members represented the business and professional elite that had emerged in Houston between the wars, and they looked forward to building on the tradition of excellence established by their predecessors at Rice.[12]

With Wiess's death, George Brown emerged as the group's leader. While voicing respect for the traditions at Rice, he was impatient to push the institute toward greater national prominence. To do this required more money and a persuasive vision of the future backed by a practical plan of action. Brown quickly made an impact on the institute's financial affairs. One week after he took his seat on the Rice board, Brown became a member of a newly created investment committee charged with preparing "a list of high grade stocks that they recommend be considered for the account of the Institute." In less than six months, Brown, Hanszen, and banker A. S. Cleveland had directed the investment of $500,000 into blue-chip stocks. By March of 1944, the institute held $2.3 million in stocks out of total assets of about $16 million.[13] This represented a sharp departure from past investment policies, and Brown continued to lead the way toward a "more modern" investment policy throughout his years at Rice. He helped recruit to the board several Houston businessmen with special financial talents, notably Gus Wortham, a financial specialist who had built one of the largest insurance companies in the Southwest, American General Insurance Company. Brown quietly arranged for Wortham's selection as a Rice trustee in 1946, and the two close friends devoted much time and effort to Rice affairs during their fifteen years of shared board service.[14] In the 1950s Brown also recruited Fayez Sarofim as an investment advisor to Rice. Sarofim, who served in a similar capacity for the Brown Foundation, brought a new professionalism to investing at both institutions while helping to build their endowments. Together with the representatives of local banks who regularly held positions on the Rice board, Brown, Wortham, and Sarofim played key roles in guiding the investments of.the growing Rice endowment in the postwar years.

Although Brown had an aptitude for finance, he brought more to the Rice board than his investing skills and fund-raising ability. He had a deep commitment to Rice; those closest to him noted that Rice "was a mission with George" and that it could be called "the love of his life."[15] Kenneth Pitzer, who served as the institute's president for a portion of George Brown's tenure as chairman of the board, recalled that Brown "personally had a vision for Rice University as an institution, though small in size, but of national and international stature." Of course, this had been the goal of Rice's leadership from its earliest days, but Brown was in an excellent position to provide the leadership needed to push Rice forward in the postwar years.

Pitzer noted that Brown "was operating in that national circle" of influence and power to the benefit of Rice. As a businessman, a member of numerous corporate boards, and an active participant in governmental advisory committees and in electoral politics, Brown maintained close contacts with leaders in business, government, and education throughout the nation. On vital questions that came before the Rice board, Brown "had the contacts within the course of a few telephone calls or a few meetings or trips to contact people who were trustees of Cornell or Columbia or Princeton and ask them what their policy was."[16] On issue after issue, Brown and other members of the Rice board studied the policies of other leading universities across the nation before adapting "current best practice" to Rice's particular needs. Once a choice had been made—on issues as diverse as faculty-student ratio, tenure policy, and the size of the football stadium—Brown had the standing and the personality to maintain a consensus within the board and to gain the support of the broader Rice community.

The board charted the general direction to be taken by Rice after World War II in a comprehensive plan in 1945. After a thorough study of the institute's past and its current strengths and weaknesses, the board, led by Wiess, Hanszen, and Brown, set forth an ambitious "Long Range Program for Rice Institute." They stressed the "need for Rice to continue to provide good training for a limited number of students." This required a basic change in board governance, with the creation of "essential committees" on vital issues to assist the overworked board members. A top priority was a "substantial building program" to be carried forward "with vigor during the next ten years." Another pressing concern was the need to lower dramatically the existing faculty-student ratio comparable with that "at good schools" surveyed by the board. Recruiting and retaining highly qualified faculty was a high priority. The board acknowledged that this program might tax Rice's existing resources, but it voiced the belief that any required funds would be forthcoming once the Rice supporters understood its needs.[17]

One of the key steps in implementing this program was the selection of a new president to take the place of Edgar Odell Lovett. Rice's first president retired in 1941, but he agreed to remain in office until a suitable replacement was found. Upon joining the board in 1943, Brown became active in the ongoing presidential search, which examined the credentials of potential candidates from leading universities. Brown and Harry Wiess actively sought and recruited highly qualified candidates who seemed capable of leading Rice's ongoing quest for national prominence. William V. Houston, a physicist from the California Institute of Technology, became the target of their recruitment efforts in 1945. As he contemplated a move to Rice, Houston registered several concerns with the board in a letter to Wiess. He wanted the Rice business office moved on campus, perhaps to establish more clearly the lines of authority of the university president; he wanted to limit the growth of physical education, so that athletics would be "a physical training activity rather than essentially a means of advertising"; and he hoped to focus the institute's resources on the development of physics, chemistry, and engineering "somewhat to the exclusion of other fields."[18] Houston's background and his concerns seemed appropriate for a school of Rice's aspirations, and the board made him the new president. He took office on March 1,

1946, and remained for fifteen years. During this time he became a close friend of George Brown and worked closely with him and other members of the board to realize the program set forth in 1945.

The partnership between Houston and Brown was somewhat strained in one highly visible area: big-time college football. Where the president saw athletics as "physical activity" whose growth should be strictly limited, the chairman was a dedicated football fan who saw success in intercollegiate sports as one measure of a university's national stature. Along with his brother Herman and other friends who frequented 8-F, George Brown took great pleasure in attending college football games, particularly those involving Rice and other Southwest Conference teams. Indeed, the Brown & Root company airplane enabled the brothers at times to attend more than one game on a given Saturday in the fall, with George rooting for Rice while Herman cheered the University of Texas. By the 1940s, Rice football had become an important source of entertainment for many Houstonians and a symbol of the institute's growing quality to George Brown and others active in Rice's affairs, and Brown worked hard in his early years on the board to bolster football at Rice.[19]

As in many other aspects of the institute, Brown found a traditional commitment to excellence in football when he joined the Rice board, and he aggressively built on this tradition. In the 1920s Rice had hired John W. Heisman, who had established a reputation for coaching excellence at Georgia Tech, Clemson, and Auburn, to build a powerhouse football program. After his departure in the late 1920s, Rice enjoyed unprecedented success in sports during the 1930s. Before the 1940 football season, Jess Neely became athletic director and head football coach and subsequently spent a quarter century seeking to balance the demands of building a winning major college sports program at a small university with a highly competitive academic environment. George Brown backed Neely's efforts as board chairman driving Neely's long tenure at Rice.[20]

Soon after joining the institute's board, Brown became deeply involved in the affairs of the football program. He regularly volunteered for service on trustee committees related to athletics, and he served as a liaison between the board and a group of Houstonians attempting to establish a postseason bowl game at Rice Stadium. This proposed "Oil Bowl" promised to generate both revenues and prestige for Rice, and Brown was instrumental in arranging several Oil Bowls on the Rice campus in the late 1940s. The negotiations for this bowl and the growing popularity of Rice's successful football teams in the late 1940s called into question the continued adequacy of the school's football stadium, and Brown became a leading advocate of the construction of a bigger and better facility.[21]

Early in these discussions, several competing options emerged. The existing stadium had been expanded to seat more than thirty thousand by the late 1940s, and one option was the further expansion of this facility. Others argued for constructing a new stadium. Discussion of this approach was sidetracked for a time by extended negotiations with representatives of the city government and crosstown rival University of Houston to build a mammoth, 110,000-seat municipal stadium for the joint use of both universities and other potential citywide functions. George Brown and Gus Wortham represented Rice at most of the meetings to negotiate an acceptable plan for the financing and construction of this "Houston Stadium." But after more than a year of negotiations failed to produce a viable plan, Brown and Wortham returned to the board with a renewed plea for Rice to consider its options apart from this futile cooperative endeavor.[22]

Before the discussions regarding a possible municipal stadium, Brown had been named chairman of a board committee "to study the needs of the Athletic Department and the Oil Bowl situation." Brown, Wiess, and Hanszen collected information on existing stadiums at other universities before recommending to the board that a new stadium with a seating capacity of forty thousand be constructed if Rice could raise the $2 million needed to construct the facility. In January of 1948 the three trustees had formed a committee to "work out the plan" for the new stadium; such planning had been suspended for more than a year amid negotiations over the cooperative venture with the University of Houston. Finally, in the fall of 1949, the Rice board returned to the debate concerning a solely owned facility. Events then moved quickly. Engineering studies suggested only minimal savings in expanding the existing stadium as opposed to building a completely new one, and the board rapidly reached a consensus in favor of a new stadium. In November of 1949 the board voted unanimously to build an approximately 50,000-seat facility that would be completed if possible for the opening of the 1950 football season; it would cost about $2 million.[23]

Gus Wortham carefully guided the financing for this project. Borrowing from a plan previously used to expand the stadium at Texas A&M University and then discussed as a way to secure financing for the proposed municipal stadium, Wortham proposed selling twenty-year options to buy season tickets for $100 to $200 to raise a portion of the funds needed. George Brown later recalled his initial skepticism about this approach; he wondered if many people would be willing to pay several hundred dollars simply to assure the right to purchase season tickets. While discussing his doubts with Wortham over lunch at the Houston Club, Brown and his good friend realized that they had a ready-made market survey group, the numerous businessmen and professionals eating lunch around them. After a quick survey revealed an overwhelming support, the season ticket option plan went forward quickly under Wortham's direction, ultimately producing about $1.6 million of the estimated $2.1 million needed to build the stadium.[24]

Once the board voted to build the new stadium on November 17, 1949, construction plans went forward quickly. At that time, the board named George Brown as chair of the committee in charge of "Construction and Engineering and Letting of Contracts." Having talked about a new stadium for several years, the board was in no mood for further delays. Its members agreed on an ambitious goal: the completion of the new structure in time for the beginning of the next football season. Since the Owls were scheduled to play their first home game on the night of September 30, 1950, there was no time for delay if this timetable was to be met. At the November 17 meeting to approve the stadium, the board discussed how best to proceed with planning and construction. All members agreed that Brown & Root was "one of the best qualified to do this type of work" and that the choice of Brown & Root would save valuable time in getting the project started. Despite the fact that George Brown took part in these discussions, the board formed a committee "to contact Herman Brown" regarding Brown & Root's availability to complete the project at cost. Other suppliers and subcontractors subsequently followed Brown & Root's lead in donating their services without profit. By January of 1950 the previously vacant west end of the Rice campus hummed with activity as construction began on a state-of-the-art facility.[25]

The September 30 deadline quickly became a test of Brown & Root's ability. Both Herman and George Brown remained personally involved in driving the work toward completion. For much of the project, large crews worked two ten-hour-per-day shifts, causing complaints from neighbors of the Rice campus while making rapid progress in building the giant complex. George Brown made the task somewhat more difficult early on by returning to the Rice board with a proposal to expand the seating capacity from fifty-four thousand to seventy thousand. His argument that the expansion would be needed sooner or later and that it could be accomplished at significantly lower costs during the initial construction than at a later date convinced the board to authorize the change, which added some $400,000 to the stadium's cost. Brown & Root's crews pushed on through weather delays, Houston's summer heat, and the regular pickets of union workers protesting the company's open-shop policies. The race to finish the stadium became a topic of conversation among Houstonians and the source of some concern for President Houston. In response to a question about the likelihood that the stadium would be available for the first home game, Herman Brown replied by asking his questioner if the first game had been scheduled for day or night. When reminded that it was a night game, he feigned relief and observed slyly that in that case, they should be able to finish on time. This often-repeated comment came to symbolize the boldness of the entire project.[26]

Rice now had a "world-class" stadium. George Brown no doubt felt a special pride. "The house that Brown built" was a dramatic statement of the determination of Rice to compete on equal terms with the best universities in the nation. To Brown and his

Rice Stadium on opening day, Sept., 1950. *Courtesy Brown Family Archives.*

contemporaries on the board, a first-class football program was a highly visible symbol of Rice's aspirations to develop into a first-class university. Many questioned the capacity of a small, highly selective school such as Rice to compete successfully in major college athletics. The giant stadium dominating one end of the campus was a clear expression of Brown's determination to have Rice do so. This was an important part of his vision for Rice, and the stadium was the concrete embodiment of his commitment to see Rice excel in all areas. The fact that it would have taken the first 150 graduating classes from Rice to fill the stadium suggested that it also embodied Brown's vision for Houston as a major league city.

As the stadium moved toward completion, George Brown became chairman of the Rice Board of Trustees in February of 1950. The stadium was only the most visible structure in an ongoing building boom that altered the face of Rice during Brown's years on the board. During the late 1940s and 1950s, work went forward on a wide array of buildings on the campus. The new library defined a new center of the campus, and new buildings arranged themselves around the distinctive quadrangle connecting the library to the original administration building. George Brown remained quite active in overseeing the planning, financing, and construction of new buildings. In so doing, he helped guide the institution toward the fulfillment of one of the principle goals of the 1945 plan, the vigorous pursuit of a building program that would give Rice the first-class facilities it needed for research and teaching.[27]

A second key component of the 1945 plan was recruiting new faculty to lower dramatically the faculty to student ratio. Brown's primary contribution in recruitment was his involvement in hiring two presidents during his board tenure. He believed strongly in allowing campus officials near autonomy in faculty governance, including the recruitment and retention of new faculty. Brown saw his role as the provider of resources and facilities needed to attract the very best faculty and students. He sought to use his powers on the board to build the best possible academic environment, and he tended to listen closely to faculty members, administrators, and colleagues associated with other outstanding universities to educate himself about the university's needs.[28]

As would be expected from any experienced businessman, Brown relied heavily on other board members to take the lead in those aspects of Rice's affairs for which they had special expertise. Thus Gus Wortham became especially important in designing a new retirement plan for Rice. Experienced bankers on the board helped evaluate investment opportunities; Brown himself took a special part in construction and in fund-raising. Rice's rapid post–World War II expansion placed heavy demands on its board members. When rising obligations threatened to overwhelm the board's time and energy, Brown and other members pushed for a reorganization and expansion of the trustees. The new board that emerged contained the traditional seven "life trustees," but it also included eight new members appointed to four-year terms. This innovative arrangement gave Brown the option to develop a strong core group of trustees whose efforts could be temporarily supplemented by those of others with special skills needed by the board. Brown took a strong, though low-key, role in recruiting new trustees, and he drew on his contacts within 8-F and the broader Houston professional community to maintain a high-profile board.[29]

This high profile proved especially helpful in fund-raising. One traditional function of university board members has been to contribute money to the institution, and Brown and many of his colleagues on the board took this responsibility most seriously. A related function has been garnering support for the institution from others in the community; this, too, was done by Brown's board. The reputation of Rice as a well-endowed institution with abundant resources made Brown's fund-raising job difficult. He put forward several innovative initiatives in the 1950s. After sending a representative of the board on a trip "to the Pacific Coast, the Mid West, and the East to study plans being used by other schools," the board encouraged the creation of three groups of potential contributors. In 1953 the board organized a select group of "Rice Institute Associates" made up of substantial contributors to Rice. This group grew steadily in the 1950s, with the list of annual contributors giving at least $1,000 sprinkled with names from the 8-F crowd as well as from Brown's associates at Texas Eastern, Vinson & Elkins law firm, and First City National Bank. Brown took a strong interest in a second group of "Industrial Associates," which consisted of companies each willing to contribute $10,000 for three years to be used in research partnerships between Rice faculty members and the company. Despite Brown's enthusiasm for this program, it languished in the 1950s because of the lack of sponsors. A final group of potential contributors was the Rice Alumni Association, which grew slowly in the 1950s before emerging decades later as an important source of contributions.[30]

Brown understood that Rice had to develop much greater fund-raising capacities to continue along the path to national prominence. This became quite clear in 1960 when William Houston stepped down from the presidency because of health problems, and Rice faced a search for its third president. Brown knew that candidates for the position would seek reassurances about the institute's future prospects. One promising nominee was Kenneth Pitzer, a chemist at the University of California, Berkeley, and a former director of research at the Atomic Energy Commission. Brown made the trip to San Francisco to talk with Pitzer about the opportunity at Rice. Pitzer had only limited knowledge about Rice, but George Brown's vision of Rice's future impressed him. While negotiating with the board for the job, Pitzer requested and received assurances of support for his efforts to raise outside funding and to remedy a problem that had become increasingly embarrassing to Rice, its racial segregation policy.[31]

Pitzer's arrival as the new president in 1961 came at a time of great challenges. The year before, the trustees had decided to change the name from Rice Institute to Rice University, which they thought better captured the breadth and quality of the institution's varied programs. Pitzer challenged the university to live up to its new name by accelerating its efforts to build a national reputation for excellence. Standing in the way of further progress toward this goal were several formidable obstacles. The first was the lack of funds; the "little rich school" could no longer pay for its high aspirations out of its endowment. New ways would have to be found to raise substantial new resources.

Two related problems stemmed from the original 1891 charter, which called for the creation of a "free" institution for the education of "white" students (both male and female). George Brown discovered the costs of these restrictions when he sought funding from such national sources as the Ford Foundation: "They said if we're too rich to charge tuition, then we are too rich to give money to." On the issue of the charter's creation of a segregated institution, the world had changed dramatically since 1891. What had seemed appropriate to white southerners at the turn of the century in a relatively small southern city was no longer practical or defensible in a major metropolitan area almost a decade after *Brown v Board of Education* had begun the deseg-

George Brown, William V. Houston, Kenneth Pitzer at Rice Commencement.
Courtesy Brown Family Archives.

regation of public schools in the South. The civil rights movement made segregation at Rice increasingly untenable. Promises to Pitzer to address this issue during his recruitment, and his assertion that he would not remain at Rice if it was not addressed, placed the board on notice that change was necessary. From his position as chair of the Rice board, Brown reduced this issue to basics: "No first class university in the world has that restriction [racial segregation]." For Rice to continue its drive for recognition as a national leader in higher education, its charter would have to be reinterpreted to allow for tuition and desegregation.[32]

In the early 1960s, George Brown thus faced a decisive moment in his tenure as board chairman. He approached it as he had other important choices facing the board. First, he sorted out the key priorities of the university; then he sought to build a consensus for unified action. To Brown, the key issue facing Rice in its quest for excellence was the need to raise additional funds. Brown's personal trips to discuss grants with national foundations left him with the impression that substantial funding would not be forthcoming to a well-endowed school that did not charge tuition, since the foundations would assume that the school had not yet exhausted its internal options in raising funds. He also knew enough about government contracts to understand that federal funds were destined to dry up for institutions that continued to discriminate against minorities. Brown's passion was national prominence for Rice, and he would not allow free tuition and segregation to hamper the fund-raising needed to attain this overriding goal.[33]

Desegregation of higher education in Texas began in earnest in the 1950s, when a series of court cases opened the doors of many public universities, colleges, and junior colleges across the state.[34] Rice and other private universities in Texas and throughout the South were less vulnerable to legal challenges to their segregated admissions policies, but they were not immune to the broad changes in race relations that were gradually dismantling the southern Jim Crow system. Most private universities, including Rice, stood aside from desegregation in the 1950s before being forced to confront this volatile issue in the next decade.

Brown spent considerable effort building a consensus on the board to challenge the charter's exclusion of blacks. A board committee reported back to the entire board on "the colored question" in January of 1962. The report advised the board to continue careful consideration of this issue. It suggested that individual members take every possible opportunity to educate the community about the problems facing Rice and to seek sympathy and understanding from the university's friends. More deliberations followed; by May of 1962 there was some sentiment that desegregation should be deferred until after a major fund-raising campaign to avoid alienating potential contributors. Brown responded forcefully that he would be willing to wait another month to foster support for challenging the charter, but he would not let the issue be further postponed. After sending a team of lawyers to consult with the Texas attorney general regarding the legal issues raised by their desire to reinterpret the Rice charter, the board took action in September of 1962.[35]

On September 26, 1962, the board passed a lengthy resolution justifying its decision to file suit seeking legal ratification of their reinterpretation of the Rice charter to allow for charging tuition and admitting qualified students regardless of race. The resolution argued that the stipulations for a "free" and "white" institution reflected the customs of a time long past, and that they were always considered secondary to the charter's primary goal, the creation of "an educational institution in Houston of the highest quality." By the 1960s, the resolution argued, these secondary goals—the lack of tuition and the exclusion of black students—were no longer compatible with the primary goal of the charter. Thus the trustees' commitment to the primary goal of William Marsh Rice required them to alter the university's policies on tuition and segregation.[36]

As the resulting trial and subsequent appeal made its way through the courts, opponents voiced strong criticism of desegregation. The court case generated by the board's action riled the passions of those who opposed changes in the Jim Crow order. As chairman of the board, George Brown was a lightning rod attracting racist invective. Even his family was not immune; his daughter Isabel recalls leaving a fashionable gathering after being reviled by a drunken opponent of desegregation at Rice as a "nigger lover."[37] Although George Brown was far from a leader in the broader struggle for civil rights in Houston, his leadership at Rice required a quiet resolve in the face of vocal and determined opposition to change.

Brown's board remained firmly united behind the goal of reinterpreting the charter. Indeed, the board, the campus community, and Rice alumni showed admirable resolve on this controversial issue. As the case moved toward its ultimate resolution in favor of the board's position, Rice moved quietly to admit its first black student in the fall of 1964. A dramatic change had taken place at Rice without the violence and confrontation that accompanied this change at many other southern institutions. Brown's desire to do what he thought was best for Rice led him to assert leadership on this potentially volatile issue, for which his natural gift for salesmanship proved most useful. His success in building a consensus within the Rice community on the need for change helped guide the institution away from its Jim Crow past. The removal of the segregation barrier smoothed the way for the transformation of Rice Institute, an excellent southern regional school, into Rice University, an institution that could aspire toward a national reputation for excellence.

George Brown's activities at Rice benefited the city of Houston as well as the university. As chairman of the Rice board, Brown played a pivotal role in a decision that gave a sharp boost to the Houston economy, the decision to locate the Manned Spacecraft Center (MSC) on the southern outskirts of the city. The announcement that Houston had been chosen to house the MSC (later renamed the Johnson Spacecraft Center, or JSC) came in September of 1961, and it symbolized the coming of age of the city and of Rice University. In leading the successful lobbying effort for the selection of Houston, George Brown made good use of a nexus of power and influence constructed over a lifetime of business and political dealings.

The key actors in this drama all had close ties to the Browns and to the broader Houston business community. Albert Thomas, who long represented the Houston area in the U.S. House of Representatives, chaired the House Independent Appropriations Committee, which approved funding for NASA. He had been a friend of George Brown since 1916, when they met as students at Rice. Lyndon Johnson had chaired the Senate's Space Committee in the late 1950s, where he helped write the Space Act of 1958. As vice president, Johnson served as chair of the National Aeronautics and Space Council (NASC), which advised the president on all aspects of the development of the space program. Morgan Davis, the chairman of Houston-based Humble Oil & Refining Company, was a friend and associate of the Browns, and at least one representative of his company traditionally had served on the Rice board.

James Webb, who became director of NASA in February of 1961, was not at the time tied directly to Brown, but he had strong ties to the southwestern petroleum industry and to Lyndon Johnson's fellow "oil senator," Robert Kerr of Oklahoma. Kenneth Pitzer had ties to the Washington establishment in general and to Albert Thomas in particular because of his work as director of research on the Atomic Energy Commission in the early 1950s.[38]

Albert Thomas was perpetually on the lookout for government programs that might hold some benefit for his home district. He had campaigned hard to convince NASA to give Houston serious consideration as the site for its facilities. In 1958 he failed to persuade NASA to locate its Goddard Space Flight Center in Houston, and he vowed to use his power over the agency's budget to assure that Houston not "get passed by again." He knew that a major NASA scientific facility would mean an economic bonanza for the Houston area, and he led the way in alerting others interested in the growth of the city to the prospects for attracting a NASA lab. Thomas felt strongly that his alma mater should play a central part in any NASA-related research lab, and he turned instinctively to his old college classmate George Brown for assistance in organizing a Rice/Houston appeal to NASA as the agency began to explore its options for the location of the MSC.[39]

By April of 1961 the chairman of the Rice board wore another hat as a civilian member of the Space Council. The council's chairman, Lyndon Johnson, announced Brown's appointment as the addition of a "distinguished American from the private sector."[40] This appointment made George Brown an official participant in the oversight of the space program, including NASA site selections. It gave Brown direct and early access to information concerning NASA plans for future expansion, placing him in an excellent position from which to guide Houston's efforts to obtain a major research facility.

Early in 1961, Thomas and Brown met to compare notes on Houston's prospects. As Brown later recalled this meeting, Thomas told him that "if we could get some 1,000 acres of land . . . we could put this Manned Spacecraft Center here." Thomas noted that "everybody wanted it, and he couldn't get it for Houston or Harris County unless we got some land."[41] Brown knew the likely location of such a large tract of land, since several years before he had gone in with his brother Herman and several other investors in a futile attempt to purchase from Humble Oil & Refining a large plot of land near Clear Lake, some twenty miles south of the city. Humble had acquired the land as a future oil prospect relatively near a major field discovered in the 1930s, and it had previously donated a portion of this land to the government for the construction of Ellington Field, which became an active air corps training site in World War II. Humble saw its Clear Lake property as a potential future site for real estate development, and the oil company was immediately interested when George

Brown broached the idea of using the land to lure the MSC. Humble had strong ties with Rice, and company chairman Morgan Davis quickly saw the advantages of Brown's proposal to donate the land to Rice, which would then offer it to NASA. An original one-thousand-acre parcel was later expanded to sixteen hundred acres at the request of NASA, and Houston's bid for the MSC began in earnest.[42]

The offer of a gift of land on a site with most of the characteristics identified by NASA's site selection team as essential caught the attention of James Webb as he struggled to make one of his first major decisions as the head of NASA amid a swirl of intense political pressures from various regions interested in attracting the MSC. Webb found much to support Houston's case. In a detailed memorandum to Vice President Johnson in May of 1961, Webb noted that Congressman Thomas and "George Brown were extremely interested in having Rice University make a real contribution to the effort," before acknowledging that Brown "has been extremely helpful" in helping NASA look closely at the prospects of Rice and Houston. Webb felt that "it is going to be of great importance to develop the intellectual and other resources of the Southwest in connection with the new programs which the Government is undertaking."[43] He envisioned the emergence of a cooperative partnership between NASA, the educational institutions in the Houston area, and the technical experts from the industries located in and around Houston. The active roles played by Rice University and Humble Oil in Houston's bid pleased Webb, who looked most favorably on the development of a cluster of technical specialists drawn from the local universities, oil companies, and the space program.[44]

Events moved quite rapidly after May of 1961. George Brown worked quickly and efficiently to secure the cooperation of the Rice board, which granted Brown and President Pitzer authority to "take all such actions as they deem appropriate" to persuade NASA to accept Rice's gift of land and offer of future cooperation. Brown worked with Thomas and Johnson to sustain the political pressures on NASA, which was also being lobbied by at least seventeen other sites, including one in President Kennedy's home state of Massachusetts.[45] Brown also pulled together and focused the support of the Houston business community, which rallied to the cause of diversifying the region's economy. All of these individuals and groups played significant parts in building a persuasive case for Houston, but George Brown was the glue that held the coalition together. By moving decisively early on in the selection process, Brown and his compatriots stole a march on groups lobbying from competing cities. The announcement of Houston's selection as the site of the MSC on September 19, 1961, came before most potential competitors had put together a complete bid.

This episode presented a classic case study of the use of influence in a fluid, democratic system. NASA officials including Webb hurried to assert that "the decision was not based on political considerations or on the activities of the local business community, but on NASA's requirements."[46] But other sites undoubtedly also could have fulfilled NASA's needs; the timely intervention of Brown and others helped tip the balance in Houston's favor. President Pitzer left no doubt about his opinion of the decisive variable: "The Manned Spacecraft Center probably wouldn't be in Houston if it weren't for George Brown."[47] He might have added that without his ties to Rice and his long-standing relationships with Thomas and Johnson, Brown would not have been in a position to influence this decision.

The coming of the MSC to Houston proved a tremendous boon to the local economy and to local pride; it also marked Rice University for national attention and for substantial new federal funds to build space-related programs. Already, by December of 1961, Brown & Root had received a $1.5 million contract for design work on the MSC. Humble Oil's Friendswood Development subsidiary subsequently built a planned community in the area surrounding the land donated for the MSC. In time, the impetus provided by the space center gave the Houston area a major new industry while also encouraging the growth of a highly populated area between Houston and Galveston. In this case the Browns helped mobilize a coalition of business and political leaders in pursuit of a broader civic interest and in competition with other cities and other influential groups. In the case of the siting of the Manned Spacecraft Center, the Browns and their influential circle of friends at both the local and national levels cooperated to win a prize of great value to Houston.

The process of university building went forward somewhat differently in Georgetown, Texas, a small city some forty miles north of Austin that had been the home of Southwestern University since 1873. Southwestern was one of the oldest universities in Texas, dating the charter of one of its original schools to 1840 before Texas became a state. As a small liberal arts–oriented college, Southwestern relied heavily on its ties to the Methodist church for funds, students, and leadership. Despite its long history, Southwestern encountered hard times in the early twentieth century, amid the increasing costs of higher education and the intensifying competition for students and funds. During the 1920s and 1930s, Southwestern struggled to survive as it sought to define a mission for itself that would attract the support and enrollments needed to maintain a viable institution of higher learning.[48]

Herman Brown had strong geographical and personal ties to the university in Georgetown. Early in his business career, he maintained Brown & Root's headquarters at Georgetown as he oversaw road building in Central Texas. Although Herman never attended classes at Southwestern, he became quite well acquainted with several people who did. After graduating from the university, the woman who became his wife, Margarett Root, taught school in Georgetown when Herman courted her. Alice Pratt, George Brown's future wife, graduated from the university with a bachelor of arts in 1922; she had attended school there with Ed Clark, who became the Browns' lobbyist in Austin. Margarett's sister married Claude Cody Jr., a doctor who practiced in Houston but had ties to Southwestern. Cody's father, Claude Cody Sr., had taught mathematics at the university for thirty-six years, in the process earning the sobriquet "The Grand Old Man of Southwestern."[49] Cody Jr. maintained the family's strong ties by serving as chairman of the board of trustees from 1934 until his death in 1959. Such deep personal connections to Southwestern gave Herman Brown an ongoing interest in affairs at Georgetown, even after he moved to Houston during World War II. Indeed, Southwestern University came to play an important role in the last decades of his life.

Claude Cody Jr. kept his brother-in-law informed of happenings at Georgetown, and well before Herman went on the university's board, Cody sought his assistance for the financially strapped university. In 1939 Cody wrote to "my dear Herman" about "the present financial distress of Southwestern University and other denominational colleges and a proposal for their relief. If you will recall, I have discussed this with you a number of times in the past several years." Cody then went on to describe his efforts to gain some relief in Washington before inquiring of his brother-in-law: "Could you take up this matter with Mr. Lyndon Johnson? Also, could he find out Mr. Cooper's and Mr. Doughton's [other members of the House] reaction to the proposal. I realize that there will be no opportunity for anything to be done before Congress adjourns, but it would be desirable to have the matter presented to these men now in a general way."[50] Representative Johnson came through in helping his constituents at Georgetown gain WPA funds for the construction of the new Cody Memorial Library and a portion of a new gymnasium.[51]

Several years later, J. N. R. Score, the president of Southwestern, solicited Herman Brown's assistance in fund-raising in Houston. In launching a campaign to raise "a minimum of $3,500,000," Score stressed that "during the more than one hundred years of Southwestern's life, Houston has been intimately connected with the college. Trustees, members of the Faculty, and a large percentage of our students have come from Houston. In addition, citizens of Houston have provided a large percentage of our financial support." In acknowledging the city's importance to the university, "a permanent committee was selected for the city of Houston not only for the purpose of the solicitation of funds, but also for advice and counsel and assistance in our total university program. You were chosen as a member of this permanent committee."[52] It was a short step from such a committee to membership on the board of trustees, and in November of 1948, Herman Brown was elected as a trustee-at-large for a term of two years. In his letter accepting this position, Herman noted his "pleasant surprise" at his election while acknowledging that "I, of course, have always had considerable interest in Southwestern on account of my wife and her family and Dr. Cody."[53] Herman held this position as Southwestern trustee until his death in 1962. During these fourteen years he served as vice chairman of the board and as a member of the executive committee. Although his work in Houston and his poor health required frequent absences from the two board meetings per year, Herman Brown became one of Southwestern's staunchest and most active supporters.

He joined the board at a critical time in the university's evolution. Soon after Herman attended his first board meeting in 1949, President Score died after a brief illness, leaving the Southwestern board with the task of recruiting a new president while also continuing Score's initial forays into fund-raising. The board's choice for the presidency was William Finch, a professor of religion and philosophy and the director of student religious activities at Southwestern since 1941. Finch resigned his presidency to become dean of the divinity school at Vanderbilt University in 1961, his administration having lasted for most of Herman Brown's tenure on the Southwestern board.[54] The two men worked closely with other board members to push the university to a new level of financial stability and academic quality.

As was the case with George Brown's early involvement with Rice University, Herman Brown played a significant role in helping President Finch enhance an endowment that was much lower than required to support the university's aspirations. In 1950 the board announced an ambitious goal to expand the endowment from approximately $1 million to $5 million by 1955. Herman Brown became the point man in this quest and in related efforts to find additional funding for faculty salaries and new buildings on a campus that had run a deficit in 1950.[55] Throughout the 1950s, Herman used his business experience, his own resources, and the assistance of many of his friends to help Southwestern accumulate the money needed for improvements.

His business ties proved valuable in an important transaction that made an impact at Southwestern similar to that of the Rincon oil deal at Rice. As Herman settled into his new board position in 1949, the university was presented with an unusual opportunity to increase greatly its endowment. Fred McManis, a friend of President Score and president of a Houston-based oil tool manufacturer—the W-K-M Company—approached Southwestern and four other universities with a proposal to transfer ownership of the company to the five schools. Each would acquire a share of the company's stock without any investment of funds; 70 percent of the company's annual profits would be used to purchase the stock, with initial projections of a ten- to fifteen-year period required to complete the purchase. During this interim period, 10 percent of the company's profits would be available to the schools. Once the stock purchase was completed, the universities owned the company outright and were free to dispose of it as they chose. This certainly was not a standard university bequest. Indeed, three of the five institutions approached by McManis refused to participate in the proposal, which some considered a tax evasion scheme.[56]

In evaluating the potential risks and rewards associated with this unusual opportunity, Southwestern's board called on Herman Brown's special expertise and contacts. Grogan Lord, another board member, recalled Herman's role as follows: "Mr. Herman and Mr. George handled the evaluation and eventually the sale of that valve company and that was when Southwestern started bending up its finances."[57] In addition to more than $380,000 in profits from W-K-M's operations, Southwestern ultimately received $3 million from the sale of the company. Because of unexpectedly high profits, the university's share of the stock purchase was paid off in only five years instead of the ten to fifteen years originally projected. The proceeds from this single transaction made up most of the increase in endowment that had been called for in 1950. Although the Browns obviously did not create this opportunity for Southwestern, their business experience allowed the university to analyze the situation and to move quickly in a way that would assure the most profit.

Herman Brown played much the same role in advising the university on investments for its endowment funds. He recommended the same general investment strategy that George followed at Rice: make long-term investments in blue-chip stocks. Herman went one step further in 1950, when he received permission from the Southwestern board to buy stocks for the university through the City National Bank in Houston. He agreed to handle the transactions at the bank and personally pay any interest charges that accrued from the transactions.[58] Such arrangements allowed Herman Brown and other members of the board's "Investment Committee-Securities" to act quickly and decisively in taking advantage of favorable investment opportunities. Herman personally assured a flow of Texas Eastern stock into Southwestern's coffers.

The board also relied heavily on Brown as a fund-raiser. His most important on-going source of funds was his own pocketbook. Herman and Margarett Brown found numerous ways to assist Southwestern. They donated funds for a variety of purposes. Margarett took a special interest in the Browns' support of foreign students at Southwestern. She kept up with the scholarship students' progress and read their letters seeking support. Margaret also took a strong personal interest in the school's acquisition of artwork for the campus. During the mid-1950s, the Browns decided that the Southwestern campus needed trees and landscaping, and they led the campaign for campus beautification. Extensive landscaping planned by professionals and supervised by a gardener they hired dramatically improved the campus's appearance.

But perhaps the best symbol of the Browns' approach to giving at Southwestern was the construction of a swimming pool in 1955. After much discussion among the faculty and administration about the need for a pool on their campus, Herman Brown simply announced that he would build a large pool at his own expense, estimated at $50,000. In an episode reminiscent of the construction of Rice Stadium, Brown had bulldozers at work almost before announcing his intentions. By the end of the spring semester of 1955, the pool was ready for use. Herman took great pride in the pool and in the grateful response of the faculty, staff, and students for its construction.[59]

By convincing friends and business acquaintances to join him in supporting Southwestern, Herman Brown multiplied his own contributions. The first and most obvious place he turned was to brother George. The university's president in the 1960s, Durwood Fleming, recalled the brothers' mutual loyalty: "These brothers had such great ties and great relationships with each other that what one did, the other supported. This went without saying. George would say to Herman, 'Whenever you're ready to do something for Southwestern, count me in.' And Herman always supported him in what he wanted to do at Rice."[60]

This cooperation was most evident in the creation of the Brown Foundation and the gradual expansion of its donations to both Southwestern and Rice. The Brown brothers chartered their foundation on July 2, 1951, giving them a vehicle to organize and manage their philanthropic activities. The first resolution for a gift was passed on January 2, 1952, and that resolution called for a contribution of $150,000 for the creation of the Lucy King Brown Chair of History at Southwestern University. This memorial for their mother was one of the first endowed chairs at Southwestern, and it provided a fitting symbol of the strong ties between the Browns and the university. Throughout the 1950s, the Brown Foundation made a series of contributions to Southwestern for scholarships; building funds for the swimming pool, a student union building, and a dormitory; and a variety of other activities ranging from art exhibits to support for the library.[61] The Brown Foundation was not yet well-funded, but even in these early days of its existence, Southwestern was one of its special concerns.

In addition to his own contributions and those of the Brown Foundation, Herman Brown proved very adept at convincing friends to contribute their time and money to Southwestern. Businessmen such as Herbert Frensley, Bill Smith, and Bob Smith all joined Herman Brown in his efforts to boost Southwestern's fortunes. When Southwestern conveyed an honorary degree to Herman Brown in 1952, many 8-F regulars made the long trek from Houston to Georgetown. The possible long-term benefits to Southwestern were suggested in the recollections of another board member, who recounted Herman's offer to Bill Finch, the president of Southwestern in the 1950s, to come to Houston in search of funds: "Bill . . . come on over to Houston and we'll get old Abercrombie [Jim Abercrombie, the head of Cameron Iron Works and a frequent guest at 8-F] over a good steak and a good bottle of wine and some soft candlelight and I think we can get a hundred thousand dollars from him."[62] Some such "prospects," including Abercrombie, did not pan out; others, such as the ARMCO Foundation, became contributors to Southwestern after Herman Brown recommended how to approach them. Southwestern had always maintained strong ties to Houston, but Herman Brown helped strengthen these ties and assure that much-needed funds would continue to flow from the city to Georgetown.

Herman Brown and Durwood Fleming at Southwestern University.
Courtesy Brown Family Archives.

Such funds were essential for Southwestern to fulfill its ambitions of steady improvement and enhanced prestige. The university had moved beyond basic concerns
about its continued survival; instead, it was seeking to build a reputation as an excellent regional liberal arts college. In statements about its aspirations, the university at
times placed itself in a peer group of schools such as Davidson and Swarthmore. But
more often, it compared itself to Rice: "In short," stated the university's president in
1951, "Southwestern should do in the area of liberal arts and fine arts what Rice is
doing in the area of the sciences."[63] Herman Brown encouraged this comparison, as
when he announced the planned construction of the Brown swimming pool by noting that it was the same size as the pool at Rice. Unfortunately, despite the best efforts
of this generation of leadership, Southwestern's endowment remained much smaller
than that of Rice or the other schools most frequently used as points of comparison
by those with high aspirations for Southwestern.

Yet progress was made in the 1950s, and Herman Brown had a significant impact.
In presenting Herman with an honorary degree in 1952, President Finch noted that

"from the days when he was a resident of this community, Mr. Brown has been a true friend of Southwestern University."[64] Similar sentiments were voiced when George Brown received an honorary degree in 1958, and again at commencement in 1962, when Herman Brown was presented the Algernon Sydney Sullivan Award for "distinguished service as a leading businessman in Texas and in the nation, as well as to Southwestern University." Perhaps the clearest statement of appreciation for Herman's work at Southwestern came in the board resolution in 1963 after his death: "On this past November 15th Southwestern University lost one of its most devoted friends . . . Though his chair may be empty, his spirit of leadership will always be with us and will be a permanent contribution to the future of Southwestern."[65]

George Brown used the resources of the Brown Foundation to provide a commensurate tribute to his brother's commitment to Southwestern. The year after Herman's death, the Brown Foundation made its most significant grant up to that date in the form of funds for three new chaired professorships at Southwestern, the Lillian Nelson Pratt Professorship in Science (in honor of Alice's mother), the Margarett Root Brown Professorship in Fine Arts, and the Herman Brown Professorship in

Dwight D. Eisenhower, George Brown, and Oveta Culp Hobby at Rice.
Courtesy Brown Family Archives.

English. In addition to funding for these chairs, the foundation provided $500,000 as a matching grant for permanent improvements. A decade after these gifts, Southwestern's president summed up the impact of these and subsequent contributions from the Brown Foundation as follows: "These endowed funds, when translated, are at the heart and represent the core strength of our present academic program. They have lifted us from the plane of mediocrity and placed us in a position to attain the goal of excellence."[66]

The Brown brothers' leadership and money had far-reaching impacts on Rice and Southwestern. The Browns took their commitments very seriously; they brought to the task of building universities the same passion they brought to their work at Brown & Root and Texas Eastern. They also brought to the traditionally sedate world of the academy a healthy dose of innovativeness and a tendency to think creatively about problems of finance. Their willingness to use personal funds and Brown & Root's construction capacities to make things happen quickly helped rouse these universities into eras of sustained expansion and improvement. The Browns also helped mobilize the talents and resources of others in the service of the universities. Rice and Southwestern both benefited greatly from their involvement. In turn, Houston, Georgetown, and the state and region reaped rewards from the sustained growth in size and quality of the two universities.

Two of the primary beneficiaries of the Browns' activities at Rice and Southwestern were the brothers themselves. Both greatly enjoyed their work at the universities. They took pride in the obvious advances made by the universities during their board service. After his brother's death, George Brown gradually turned more attention to the Brown Foundation, which continued supporting Rice and Southwestern. In the two decades between the death of his brother and his own death, George greatly expanded work begun in the 1940s and 1950s at the two universities. As clearly as any of their business endeavors, Rice and Southwestern bear the marks of the long and deep commitment to community service of the Brown brothers.

Part Four

George Alone, 1962-83

Alice Brown, Oveta Culp Hobby, George Brown at Rice.
Courtesy Brown Family Archives.

CHAPTER NINE

Business after Herman's Death

On November 15, 1962, Herman Brown died, leaving George Brown alone in the business world for the first time since his brief stint at Anaconda Copper in the early 1920s. For forty years, the two brothers worked side by side; during the last seventeen of those years, they lived side by side. People had become accustomed to speaking of "the Browns" or "George and Herman" almost as if the two were one person. Of course, each had a distinctive personality and each brought different skills to their jointly owned enterprises. But the two men working together were stronger than the sum of their individual efforts. Their complementary skills made them a successful business team. Now half of that team was gone. At age sixty-four, George Brown suddenly found himself without the older brother who had been so much a part of his life.

Though sudden, Herman Brown's death was not unexpected. Herman Brown began his last day like he had many others. He flew from Houston to Austin and then drove the forty miles north to Georgetown, where he attended a board meeting at Southwestern University. After returning to Houston later that day, he ate lunch with Judge Elkins and then went to Suite 8-F to rest. After later meeting with George Darneille, he asked Darneille to drive him to the Ramada Club for a reception. Darneille dropped Herman off at the First City National Bank Building and returned home, where his children rushed out to tell him the news: "Herman Brown is dead." "He isn't dead," Darneille responded, "I just dropped him fifteen minutes ago at the Ramada Club."[1]

After leaving Darneille's car, Herman walked into the First City bank building and took the elevator to the Ramada Club on the twenty-ninth floor. While talking with a group of people shortly after his arrival, Herman suddenly gasped for air and collapsed to the floor. A doctor at the reception attempted to revive him but could not. Soon, an ambulance arrived and transported Herman to Methodist Hospital, where he was pronounced dead at 6:35 P.M.[2] Initially, it appeared that Herman Brown had died of a heart attack, but an autopsy revealed that a ruptured aneurysm of the thoracic aorta had caused massive internal bleeding.

After more than sixty years of relatively good health, Herman Brown's health had deteriorated rapidly in the last decade of his life. A lifetime of extremely hard work and long years of heavy smoking and drinking had finally caught up with him. In 1955, noted surgeon Michael DeBakey successfully operated on Herman Brown to remove cancer from his lung. During that initial operation, Lady Bird Johnson recalled that Isabel Brown, George Brown's youngest daughter, "was out for dinner at our house [in Washington] and we were all just sitting around tensely waiting for the news from Herman. . . . The operation turned out successfully."[3] In 1960, DeBakey again operated on Herman, this time performing a resection of a large aneurysm on Herman's abdominal aorta. Two years later, Herman died as a result of another aneurysm.

George Brown was in New York when his brother died. He had a dinner appointment with IT&T chief Harold Geneen. In the evening, just prior to dinner, George received a phone call and learned of his brother's death. Geneen recalled that "he was shaken but completely in control. He was intent on getting back to Houston."[4]

The funeral took place on Saturday morning, November 17, 1962, with Dr. Durwood Fleming, pastor of St. Paul's Methodist Church, officiating. Some fifteen hundred mourners attended, including Vice President and Mrs. Lyndon Johnson and Texas governor-elect John Connally and his wife, Nellie. Johnson's eulogy captured the essence of his friend's life: "Herman Brown was a builder of his community, his country and his world. He came from the bloodstream of America and poured back into it more than his share to help others realize the great American dream, which his life symbolizes so well."[5] Brown was buried at Glenwood Cemetery in Houston. Unfortunately, Margarett's health was so bad that she was unable to make the drive to the cemetery. Herman's pallbearers were all Brown & Root directors: M. P. Anderson, Harry G. Austin, L. T. Bolin, L. H. Durst, H. J. Frensley, Foster Parker, and Ben H. Powell.[6] Fittingly, honorary pallbearers included Vice President Lyndon Johnson and Otis Kerr, Herman Brown's longtime driver and friend.

After the funeral, several close friends visited Herman Brown's home for a reception. Margarett had been forced by her breathing problems to retreat upstairs to a room that contained a high concentration of oxygen. The Johnsons were about to leave when Margarett sent a request downstairs for the Johnsons and the few remaining guests to come up to her room for a visit. In a weak voice, Margarett told her friends that "I want to talk about Herman. . . . And the first thing you know," recalled Lady Bird, "we were laughing and everybody remembered him and it was really wonderful, loving, lighthearted goodbye. . . . I mean, it just doesn't do to let a man of that stature go out of everybody's lives without just having a kind of an Irish wake, in recalling some of these things."[7] Less than three months later, the emphysema that had so severely limited the activities of Margarett Brown in her final years finally claimed her life.[8]

After providing trust funds for his two children and funds to help look after his three surviving sisters and members of his household staff, Herman Brown left the bulk of his estate, valued at approximately $100 million, to the Brown Foundation. In death as in life, he trusted his brother's judgment. As the head of the brothers' charitable foundation, George would now have the resources to expand the foundation's activities.[9]

After Herman Brown's heart surgery in 1960, the brothers recognized that it was time to think about the long-term future of their economic interests. In addition to Herman's serious health problems, George was in his early sixties and a long·battle with ulcers and problems with his eyes had taken their toll on his health. George recalled their thinking in 1960: "It became apparent to Herman and me that we couldn't live forever, and the company had gotten so large that death tax would curtail operations pretty badly."[10]

More than bad health contributed to the decision to sell Brown & Root.[11] The Browns realized they had to loosen their grip on their firm, and they decided to sell rather than simply pass the reins onto existing management. They decided reluctantly that a sale to existing management would cause divisiveness among the strong-willed division chiefs.[12] In late 1962, *Engineering News-Record* reported that "rumors have circulated among financial circles for some months that, due in part to reasons of health, the Browns are considering a sale or merger." A company official responded, "If they are, they certainly aren't talking."[13]

As usual with the Browns, they went forward quietly, taking great care not to create unwanted speculation about their intentions. In the spring of 1962, they consulted about the company's future with their trusted aides Herbert Frensley, then secretary of Brown & Root, and Foster Parker, the firm's chief financial officer. The brothers sought a buyer with similar corporate values to Brown & Root. They hoped to sell Brown & Root to a company that was not a direct competitor and one that could pay much of the purchase price in cash. They also wanted assurances that Brown & Root's workers would continue to have secure jobs, and they preferred a company with a strong commitment to the open shop.[14]

Perhaps at the suggestion of Harold Decker, an executive with Highland Oil Resources and a director of Halliburton, the Browns approached Halliburton. In one early negotiating session, Halliburton's chief financial officer, Jack Harbin, met with Herman Brown, George Brown, and Herbert Frensley over lunch at the Ramada Club in downtown Houston. Subsequent meetings included Decker and several other Halliburton directors. Negotiations moved smoothly toward a deal acceptable to both sides. Halliburton quickly agreed in principle to a price of $46 million, which Herman Brown originally proposed based on the current book value of Brown & Root. L. B. Meadors, the head of Halliburton at the time, announced to his employees that a merger had been successfully negotiated. The negotiations then paused briefly as each side reexamined its position.[15]

The Browns felt that Halliburton was a good fit for Brown & Root in many respects. The company's history of rapid growth was quite similar to that of Brown & Root's. Earl P. Halliburton had created the firm in 1919 to develop a particularly popular cementing process for which he owned the patent. In 1924, after several large oil companies were accused of infringing on Halliburton's patents, the firm incorporated, and in a compromise settlement, eight major oil companies invested $388,000 for a 48 percent interest in Halliburton and representation on the board. At this point, Earl Halliburton retained 52 percent of the firm. Over time the oil companies sold their interests back to the company and withdrew their board representation. As an independent company once more, Halliburton gradually evolved into a diversified firm offering a broad range of services to oil companies.[16] In 1961, years before acquiring Brown & Root, Halliburton moved its headquarters from Duncan, Oklahoma, to Dallas, Texas. Halliburton was growing rapidly, often through acquisition, and it had overseas operations in as many as eighty countries. Brown & Root was an engineering and construction firm, and Halliburton was an oil field service company. Their work was complementary, and they were not direct competitors.[17]

Halliburton had a history of acquiring firms and then allowing them to operate independently. In fact, Halliburton Corporation was a holding company that contained the original oil service business in a division called Halliburton Oil Services. In the 1950s, Halliburton's acquisitions included Welex, an oil field logging company, purchased in 1957, and Otis Engineering. Both firms remained virtually separate companies operating with their own name, employees, and management structure. It was vitally important to both Brown brothers that Brown & Root remain as independent and autonomous as possible, and a merger with Halliburton seemed to offer this prospect.[18]

After Herman's death, George reopened negotiations for the sale of Brown & Root to Halliburton, and the companies quickly reached an agreement based on the price previously suggested by Herman Brown. In early 1962 the Brown brothers still owned 95 percent of Brown & Root's stock. L. T. Bolin, the firm's executive vice president had acquired a 5 percent interest in Brown & Root in 1939 for $7,500 and retained it throughout his life.[19] As always, each brother held one-half of their joint interest. Before Herman's death, they gave this stock to the Brown Foundation. Thus Halliburton's purchase of Brown & Root was made from the Brown Foundation. The Houston law firm of Vinson & Elkins drew up the contract and also provided the tax opinion for the Brown & Root side. Halliburton purchased the 95 percent of Brown & Root and its subsidiaries (including Southwestern Pipe, Joe D. Hughes, and Highland Insurance) that belonged to the Brown Foundation for $36,745,000. In the deal, Brown & Root borrowed $10 million and bought part of its stock as treasury stock. "We did that," recalled John Harbin, "so Brown & Root earnings could go to pay off debt rather than be paid to the parent company and we'd have a tax on that . . . company dividend."[20] The following year, Halliburton completed the acquisition by acquiring all of the outstanding shares of Southwestern Pipe, Inc., Joe D. Hughes, Inc., and Highlands Insurance Company for more than $600,000 in cash.

After the merger, Brown & Root became a subsidiary of Halliburton, although as had been discussed in the merger negotiations, it was managed as a "self-contained operating unit." George Brown took a seat on the board of Halliburton, where he enjoyed considerable influence as the chairman of Brown & Root and an accomplished businessman. As an outsider at Halliburton, Brown provided a welcome new perspective on the company's diversified business operations. His position at Brown & Root gave him considerable independence on the board.

George Brown served briefly as the president of Brown & Root in 1962 and 1963, but once the sale of the company had been completed, he oversaw a far-reaching reorganization of the company. For leadership, Brown turned to two old and trusted colleagues. As the new president in 1963, he named Herbert Frensley, who had joined Brown & Root in 1942. Frensley was a trusted friend and advisor who had been invaluable for his knowledge of financial affairs. Brown chose as senior executive vice president L. T. Bolin, who had remained with Brown & Root after directing the construction of the shipyard during World War II. Bolin was sixty-two and nearing retirement at the time of the reorganization, and he preferred not to take on the company's presidency. But his vast knowledge of the construction business made him a logical choice to provide continuity and leadership. Brown knew that he could depend on the team of Frensley and Bolin to help guide Brown & Root through a difficult transition.

Before Herman Brown's death, Brown & Root had no formal board of directors. The tightly held company was simply managed by the Browns with the advice of an operating committee. Now George had to create a more traditional management system, beginning with naming a board of directors. He became chairman of the newly created board, which contained two Halliburton employees and several representatives of Brown & Root. Brown in turn created a formal operating committee headed by Bolin, who became a sort of chief operating officer without the official title. Halliburton's willingness to grant managerial autonomy to Brown & Root was evident in the fact that George Brown took control of this reorganization.[21]

Brown & Root headquarters on Clinton Drive, Houston.
Courtesy Brown & Root

Halliburton's acquisition of Brown & Root worked out well. Brown & Root remained more highly publicized than its parent company, but Halliburton traditionally kept a relatively low profile. In Brown & Root, Halliburton acquired a firm with an outstanding history of success and excellent prospects for the future. Of particular interest to both companies were two quite high-profile contracts already under way at Brown & Root—the design contracts for the recently proposed Manned Spacecraft Center and Project Mohole, an ambitious, highly publicized initiative to drill deep into the core of the earth.

Both of these projects were not only high profile but highly political. Herman and

George Brown had learned during their road building days that success in gaining county road contracts depended upon knowing which counties were spending money on roads and then getting to know county officials involved in this business. In this way, the brothers could market their firm's capabilities through an expanding network of contacts. By the late 1950s, Brown & Root was operating on a national and international scale. Its web of connections to both government officials and fellow business executives was extensive and impressive. Herman and George Brown's relationship with Lyndon Johnson and Albert Thomas, in particular, was by then public knowledge, and a high-profile project typically meant high-profile public scrutiny not easy to control. But the brothers had close connections with highly influential business leaders as well, who could be counted on to help in business matters.

Well before Herman Brown's death, the Brown brothers helped Houston win the highly political contest with numerous other cities for the right to house the Manned Spacecraft Center. Four men played crucial roles: George Brown, chairman of the Rice University board of trustees and the only businessman on the National Aeronautics and Space Council; Albert Thomas, longtime Brown friend, representative from Houston, and chairman of the House Independent Appropriations Committee (which approved funding for NASA); Lyndon Johnson, vice president and chair of the National Aeronautics and Space Council; and Morgan Davis, chairman of Humble Oil & Refining Company and collaborator with Brown and Rice University in the acquisition of the land needed for the NASA site. As previously discussed in chapter 8, Davis arranged for Humble Oil to donate a sixteen-hundred-acre tract of land out of its thirty-thousand-acre site in southeast Houston. Rice then donated that land to NASA for the MSC.

Once Houston had been announced as the site for NASA in September of 1961, events moved very quickly. On September 22, 1961, James Webb, NASA administrator, formally assigned the Corps of Engineers to design and construct the spacecraft center. Lieutenant General W. K. Wilson Jr., chief of engineers, in turn selected the Fort Worth District (FWD) of the Corps to manage the job. NASA and FWD officials met on October 3, 1961, to begin the process of choosing an architect-engineer (A-E) contractor to produce the architectural plans and construct some of the buildings. NASA had already compiled a listing of 175 A-E firms. During a two- to three-week period, a selection committee composed of NASA and FWD officials sifted through the firm resumes. Finally, they decided upon three firms: Kaiser-Warnecke of Oakland, California; Parsons-Becket-Johnson of Houston; and Brown & Root. In part because of NASA's preference for using local talent and labor, Brown & Root received the contract.[22]

This contract received a mixed reception in the engineering and architectural community. One disappointed Houston consultant said, "Frankly, we were horrified when they got the engineering contract on the Manned Spacecraft Center because we consider them contractors, not engineers. The fact that politics had a lot to do with this award is too well known to warrant comment. George and Herman Brown and Mr. Lyndon Johnson are close and longtime friends."[23] The intensity and harsh tone of such criticism gave Brown & Root a sense of the sort of probing scrutiny it could expect now that Johnson occupied high national office. Other competitors noted that "there is some politics in any large project like that, but I certainly would not downrate them on their ability to get things done."[24] But the company's competitors conceded that Brown & Root's management talents offered it a legitimate competitive strength it could use in the consulting business.

Brown & Root quickly assembled a team of architects to begin designing the Manned Spacecraft Center. This team included five architectural firms and one mechanical engineering firm. The prominent Charles Luckman Associates of Los Angeles led the architects, which also included Brooks and Barr of Boston; Harvin C. Moore of Houston; Mackie and Kamrath of Houston; and Wirtz, Calhoun, Tungate and Jackson of Houston. Barnard Johnson Engineers, Inc., rounded out the team, which was managed by Brown & Root and project manager William Rice. Brown & Root's own architectural staff included eight men, led by Albert Sheppard.[25]

Speed was of the essence in the atmosphere generated by the "space race" with the Soviet Union. Project manager Rice and his assistant Harry Hutchens assured Vice President Johnson with "their personal word . . . that they would meet all schedules."[26] According to G. A. Dobelman, a top administrative engineer for Brown & Root, 120 of the 200 men working on the project were Brown & Root employees, who integrated the services of the various subcontractors and provided auxiliary services including "coordination, scheduling, accounting and all management functions." In addition, Brown & Root "had a large group of . . . men doing mechanical and structural design and detail work for each of the buildings and designing the central utility and air conditioning plant. In terms of manpower uses, B&R did about two-thirds of the work." For this work, Brown & Root ultimately received about $2.3 million. On September 20, 1962, Brown & Root submitted the master plans and specifications to the Corps of Engineers and retained the responsibility for providing interpretation and consultation services to the firms contracted to build the "campus-like facility."[27] At about the same time, President John F. Kennedy made a major speech in Houston on space exploration; fittingly, he delivered the speech at Rice Stadium with George Brown on the speaker's platform. Houston was to be the center of the nation's endeavor to send men into space, and Brown & Root and Rice were closely involved in launching this effort.[28]

After the completion of the Manned Spacecraft Center, Brown & Root continued its services to NASA under another contract in a joint venture with Northrop Corporation. Brown & Root–Northrop provided nearly one thousand employees at NASA for routine operational and maintenance services in NASA's experimental laboratories. This project included ongoing work on vibration and acoustic energy, shielding materials, a Space Environment Simulation laboratory, and numerous other testing laboratories and facilities. None of the company's work at the Manned Spacecraft Center (later renamed the Johnson Spacecraft Center) was particularly lucrative. But Brown & Root's involvement in this prestigious, cutting-edge endeavor was a reminder of the firm's growing technical expertise—as well as its ties to Lyndon Johnson.

Brown & Root's second high-profile contract was Project Mohole, an ambitious government-sponsored project to design special apparatus to drill through the earth's crust into the mantle. "Mohole" referred to the Mohorovicic crust, which is the boundary between the earth's crust and the mantle. On land, this crust is as much as 125,000 feet below the surface, but it is accessible at depths between 20,000 and 35,000 feet below the ocean floor. The project sought to recover and analyze samples of the mantle surrounding the earth's core.[29] Proponents expected great advances in the earth sciences and in the practical application of drilling technology.

The Mohole project originated in 1957 during a meeting of the Earth Sciences Panel of the National Science Foundation (NSF), which had been created in 1950 to initiate and support basic scientific research. This panel advocated the proposition to drill through the Mohorovicic Discontinuity in order to enhance earth science research in the United States. Just as one federal agency, NASA, began planning voyages into space, another federal agency, the NSF, began plans to explore the inner earth. As the United States and the Soviet Union competed in the "space race," they also competed in the "race for inner space." Ultimately, scientists hoped to gain a greater understanding of the origin and age of the earth and the moon, the causes of earthquakes and volcanoes, the origin of the solar system, and the composition of the earth's magnetic field.[30]

To launch the project, the NSF granted seed funds to the American Miscellaneous Society (AMSOC), an affiliate of the National Academy of Sciences. Chartered in 1952, the AMSOC included numerous renowned scientists, primarily geologists and oceanographers. It focused on "miscellaneous work" not in the mainstream of scientific activity. AMSOC initially received NSF funding for Phase I of the Mohole project, which was to be completed between 1958 and 1961. A unique funding arrangement placed responsibility on the NSF for both funding and administering the project while AMSOC provided advice on the scientific work.

In late 1960, AMSOC contracted a Global Marine Exploration Company drilling ship, CUSS I, to begin trial drilling off the La Jolla, California, coast in 3,000 feet of water. At this time, the deepest drilling on land had reached about 25,000 feet and only relatively shallow holes had been drilled on the ocean surface. Subsequent test drilling off the coast of Guadalupe, Mexico, reached an approximate depth of 12,000 feet below the subsea surface. These tests resulted in substantial new information about heat flow, ocean currents, and the performance of new drilling techniques and tools.[31]

After the completion of Phase I, AMSOC and its members, including the irascible Willard Bascom, agreed that they should continue with the Mohole Project. Since AMSOC's members had full-time jobs or research projects, they could not devote all their efforts to Mohole. Instead, the group made strong recommendations to the NSF that Phase II proceed with two ships. One ship, the "intermediate ship," would conduct preliminary scientific investigations of crustal material while developing techniques to be employed on a second ship responsible for drilling to the mantle. In addition, AMSOC proposed that it remain in control of the project's scientific policy and have the authority to review engineering and expenditure proposals.[32]

In the summer of 1961, the NSF began planning for Phase II of the project. At an initial briefing in July, eighty-four firms expressed interest in the project; the Browns were not among this group. The NSF called for proposals by September 11, 1961. During the meeting it became clear that the NSF expected the winning bidder to utilize AMSOC in the project. Also, the NSF preferred proposals that prioritized the scientific aspects of the project as well as proposals that were "no-fee." At some point after the July meeting, perhaps as late as early September, the Brown brothers learned of the project and quickly composed a proposal. A total of ten proposals ultimately came forth, and the NSF narrowed the field to five using a point system based on five criteria in descending order of importance: (1) management and policy, (2) technical and scientific experience and capability, (3) support facilities available or readily obtainable, (4) financial capability, and (5) comprehension and soundness of approach.

Of the firms chosen, Brown & Root placed fifth in this initial ranking. National Science Foundation director Alan T. Waterman then formed a second review panel that reduced this list to three finalists, including Brown & Root in third place based on a point grading system. The other two firms chosen were Global Marine Drilling Company of Los Angeles and a consortium headed by Socony Mobil Company, which included Standard Oil of California, General Motors, Texas Instruments, and Humble Oil. Global Marine had already done substantial work on the project, successfully completing some of the first borings through the ocean floor.

The NSF committee noted that the top two firms had great scientific capabilities and skills while Brown & Root had an impressive engineering record with special expertise in marine engineering. These bids from the three firms ranged from $23 to $44 million. Brown & Root also made contingency plans with Rice University's six-year-old geology department to act as consultant in carrying out the project.

The NSF committee proved unable to choose a winner from among the three finalists, leaving the final decision to Waterman. On February 28, 1962, he announced the choice of Brown & Root. Referring to the original list of criteria for evaluating proposals, he noted that his decision was swayed by Brown & Root's strength in the "management and policy" area rather than in the "technical and scientific experience and capability" area. George Brown's flair for salesmanship also helped secure the contract. He had ended his presentation with a forceful sales pitch: "Gentlemen, we can drill that hole." Brown & Root legend has it that one of the selection committee members was overheard to say, "Let's give it to that white-haired man. He is the only one that said, 'We can drill that hole.'"[33]

The day after the selection, Waterman and his staff presented the NSF's 1963 budget to the House Appropriations Subcommittee on Independent Offices, chaired by Albert Thomas. Waterman requested $5 million for Mohole, noting that "Brown & Root, Inc., of Houston, Texas, have been selected for negotiation of a prime contract." Thomas, who had a reputation for being "cool" to NSF proposals, replied that this budget was "a work of art."[34] Brown & Root had once again prevailed in winning a high-stakes government project. On March 17, 1962, the NSF announced that it had contracted with Brown & Root for a $1.2 million preliminary design and engineering studies for the Mohole project. The NSF noted that the firm that ultimately undertook the full-fledged project would receive a $35 to $50 million contract.

On June 11, 1962, Brown & Root officially received a cost-plus fixed fee contract for the project, $46,728,000 plus a fixed fee of $1.8 million. Initially, Brown & Root placed fifty employees on the Mohole project with the understanding that "we will need additional help on a lot of phases of the Mohole contract."[35] The company organized a separate Mohole division for the project and began lining up a long list of specialized subcontractors.

AMSOC continued to serve as a scientific advisory group under the name of Ocean Science and Engineering, Inc. (OSE), and it attempted to work with Brown & Root on the project. Problems soon erupted in this relationship. According to records, "Brown & Root did not welcome outsiders telling them how to run their business and OSE did not welcome Brown & Root; their association lasted two months. OSE returned to Washington and was employed by NSF to evaluate Brown & Root's performance for the next ten months." After the completion of its evaluation, the advisory group's association with Mohole ended, but intense political infighting continued.[36] Herman Brown, shortly before he died, explained the controversy this way: "You'll find a lot of experts will develop overnight in something no one has ever done before, and we're running into that here."[37]

In February, 1963, Brown & Root completed its own project study and recommended construction of a $40 million drilling platform with an estimated annual operating cost of $9 million.[38] The NSF reaffirmed its intention to carry out the Mohole project on January 21, 1964. Brown & Root continued planning for the project under a new project manager in William Rice, who had completed his work managing the design of the Manned Spacecraft Center.

As had been the case with the Manned Space Center engineering contract, the Browns found themselves again at the center of controversy after the Mohole award announcement. The *Los Angeles Times* report directly connected the two episodes: "It was the second time in recent months that eyebrows in the capital have been raised by the Texas firm snagging a spectacular government project [the manned space craft center]. . . . It is not lost on political observers that Vice-President Johnson . . . [has] long-standing loyalties to his native Texas."[39] Johnson was only one of two highly placed Texas politicians in a position to help the Brown brothers. As chair of the House Appropriations Committee, Albert Thomas exerted considerable influence on the approval of all funds for the NSF and NASA. But the ties to Johnson became the focus of criticism; the Browns discovered again that their long relationship with Johnson would be a two-edged sword now that he had ascended to the vice presidency. Whatever benefits came from connections to this powerful political figure were somewhat offset by the intense scrutiny and criticism trailing every move of those closely associated with one of the nation's most visible political positions.

The article ignited California senator Thomas H. Kuchel's anger. He and Colorado senator Gordon Allott, both Republicans, had "constituent" firms that lost the Mohole project, and they assailed Brown & Root while criticizing the politics of the bidding process. Kuchel went so far as to request that the comptroller general investigate the NSF's procedure for awarding the contract. Kuchel presented a speech on the Senate floor stating that Brown & Root's cost estimate for the project was $35 million while the California group provided an estimate of $23 million. Such charges of political favoritism played well, particularly in California papers.

The comptroller general's office reported that no worthwhile cost estimate could be made for such an experimental project, and the scientific community accepted the fact that cost estimates for this type of project could not be relied upon. In addition, the report noted that "we are unable to conclude that the award to Brown & Root was not in the public's interest." After Kuchel's initial outburst and the comptroller general's response, Senator Kuchel's criticism of Brown & Root subsided, but Senator Allot continued to attack the firm. Allot was not the project's only critic. As estimates for the project subsequently grew to a figure of $100 million, many congressmen became anxious about the final cost of this highly publicized project.

Senator Allott's attack on Mohole reached an almost fever pitch. On May 2, 1963, he presented a stern statement about the Mohole bid-letting process. Allott said that he had spoken with several prominent scientists who "unequivocally feel that this project is a fiasco. They laugh at the results obtained and they laugh at the direction in which it is going."[40] Allott promised to tell Congress more about Mohole soon. In the meantime, *Fortune* magazine printed an investigative story on Mohole, bringing the project to the national public's attention. In response, George Bush, then president of Zapata Off-shore Company (and one of the unsuccessful bidders) wrote a letter to *Fortune* magazine explaining how the politics of Mohole were threatening the entire project. "If the NSF would call off the dogs, cut out trying to throw a bone to the ex-AMSOC committee, then let Brown & Root get on with the job, maybe this difficult project can get done."[41]

Senator Allott, however, was only warming up. He made a long speech to Congress on May 8 in which he reviewed the entire history of the Mohole project. He also read into the record parts of an investigative article written by Herbert Solow that appeared in *Fortune Magazine*.[42] Allott assailed the NSF for accepting Brown & Root's original proposal even though "Of its 190 pages, all but 20 consisted of off-the-shelf promotional material. It was in marked contrast to some proposals that elaborated on possible solutions of Mohole's technical problems."[43] At the conclusion of Allott's speech, Senator Dominick thanked Allott for the presentation and indicated that the Mohole episode was not a unique situation. Dominick noted that a recently introduced congressional resolution covering contracting for defense and space projects be expanded to cover contracts let by all government agencies. The resolution was designed to shed light on "how a contract is negotiated and the conversations which go into it, with the necessity for memorandums on all executive communications with the contractor or other representatives who are trying to influence the selection of a contractor." Allott replied that he had uncovered no dishonesty on the part of Brown & Root; he simply believed the project—which he supported in theory—was not being managed well.[44]

It had become clear to George Brown that the Mohole Project had reached a level of political attention that made Brown & Root's project management role difficult. While Lyndon Johnson and Albert Thomas might have facilitated Brown & Root's entry into the project, the construction company now had other political obstacles to overcome. Brown's circle of influence was much wider than politicians and political institutions. As a well-respected member of the board of directors of IT&T and other large corporations, he had access to highly influential persons and resources. On May 9, Harold Geneen, the head of IT&T received an important memo, which he relayed to George Brown. The memo noted that Allott's attack on the NSF was being "researched by the Republican Policy Committee" and that Allott "is known as a patriot and a man who will go down the line for his constituents—particularly, those having any interest in oil shale." As a member of the Independent Office Sub-Committee of the Appropriations Committee and the ranking Republican on the Interior and Insular Affairs Committee, his opposition could be formidable. The memo suggested that "We can contact the [Republican] Policy Committee and, I believe, we may be able to convince them to omit references to B&R in the material that they prepare for the Senator." The memo also suggested a meeting between George Brown and Senator Allott, although the meeting was apparently quite difficult to arrange for a variety of reasons related to travel schedules and other commitments.[45]

At the advice of Joseph Thomas, a Halliburton director who was a partner at Lehman Brothers, Brown Booth wrote a letter to Robert Six, president of Continental Airlines and a friend of Allott's, asking for his help in the Mohole matter. Booth explained Brown & Root's position. "We think we are Senator Allott's 'kind of folks' and that if he knew us he would be with us."[46] Six later reported to Thomas, who passed the message on to George Brown: "I believe there is really no trouble with Allott once a meeting takes place."[47]

Criticism did not subside. But at the suggestion of the White House, a special committee of the National Science Board in August, 1963, did endorse Brown & Root's continued role in the project.[48] Later in the same month, Dr. Leland Haworth, director of the NSF, wrote a stern letter to George Brown that reflected his concern about the firm's commitment to the project. "I am deeply concerned," he wrote, "about a letter I have just received from the *Houston Chronicle*." A reporter apparently requested from Brown & Root information about the process of determining crew sizes and operating costs for the project. Allegedly, members of the project staff responded that the costs would "probably be much higher now that President Kennedy has another child. . . . I am sure you will agree that such an answer is not appropriate under any circumstances."[49] Fortunately for Brown & Root, the *Houston Chronicle* reported this incident to the NSF instead of to its readership.

While behind-the-scenes and more public attention buffeted Project Mohole, Brown & Root was working hard to make significant progress on the project. One Brown & Root employee said that the project was difficult and "mind-stretching." Bucking the AMSOC recommendation for a two-ship project, Brown & Root insisted on a one-ship program. Project engineers recommended a massive column stabilized semisubmersible drilling platform. Besides constructing a huge drilling platform, Brown & Root designed special drilling apparatus, including specialized drilling pipe. The project required a dynamic positioning system to keep the acre-sized platform within a maximum circumference of five hundred feet even in rough weather. The project also required sonar equipment capable of providing mechanical guidance for reentry into the hole. Brown & Root manager Joe Lockridge recalled that Honeywell and General Dynamics worked on dynamic positioning systems for the Mohole project. Their work, although never actually used on this project, made significant advances in existing technology.

In 1964, Brown & Root completed an extensive drilling test in which it drilled for eighty-two days straight using a drilling motor driven by fluid pumped down the pipe at high pressure. The turbocorer operated for 203 hours, twice its design maximum. It penetrated five and a half feet of rock an hour. The rock, located in a deep plug, or intrusion of basalt near Uvalde, Texas, was the hardest rock Brown & Root could find and most resembled the rock of the deep crust.[50]

As Brown & Root went about its work, the spiraling cost of Mohole forced Congress to reexamine the project. During this crucial period in the life of the project, Brown & Root and Project Mohole lost a critically important supporter when Albert Thomas died of cancer in early 1966. Thomas's successor, Democratic representative Joe L. Evins from Tennessee did not have Thomas's strong commitment to Mohole, and he asserted his independence from Thomas's long reign by attacking the project's funding.

On March 8, the NSF's new director, Leland Haworth, appeared before Evins's subcommittee to explain the agency's $450 million budget for 1967. Although Evins seemed largely satisfied with the overall budget, he indicated that the NSF might apply the Project Mohole funds to other scientific projects or even forgo them in light of costs associated with the Vietnam War. Despite the agency's continued support of Project Mohole, in one of his first major acts as chairman, Evins suspended the project.

Proponents of Mohole did not give up easily, and "a great deal of lobbying ensued on behalf of the stricken program. Many senators were approached, and [President] Johnson assisted in the effort. The fight to avoid termination, led by the technoscience presidency, seemed to be working. The Senate voted to reinstate Mohole."[51] Efforts in the Senate to restore funding for the project succeeded, but the House of Representatives voted 108 to 59 to reject the 1967 appropriation of $19.7 million necessary to carry the project further.[52] The House remained opposed even in conferences between the Senate and House appropriations committeemen.[53]

Project Mohole's downfall resulted from several factors, including the project's rising cost, the death of Albert Thomas, and the relationship between George Brown and Lyndon Johnson. During debate on funding for the project on August 19, 1966, Republican congressman Donald Rumsfeld from Illinois charged that George Brown and his family had contributed $25,000 to the president's club "between the time that the House had originally rejected the project and the time President Johnson asked the Senate to restore it."[54] Records showed that George Brown's three daughters and their husbands contributed $23,000 on May 13, apparently "just a few days after Johnson had contacted several legislators in a successful appeal to the Senate to reverse the House action," and the remaining $2,000 came personally from George Brown.[55] Brown told reporters that the money had been pledged "long before this thing came up" and called it "ridiculous" to link the contributions to the Mohole project.[56] Several days later, reporters questioned Johnson about the same allegations. He replied: "You can expect to have periodic political charges of this kind until November. They usually come from the party that has been rather strongly rejected by the people. [Contributions] do not influence the awards."[57] Owing at least in part to the public controversy surrounding the Brown family contributions, presidential and congressional support for Mohole waned, and the House voted not to restore funds for Mohole.

Criticism of the use of political influence to gain the contract for a major government-funded project next appeared in a *Life* magazine article, and the furor surrounding Project Mohole finally brought a strong response from Brown & Root. The company's president, Herbert Frensley, noted with disdain the attempts of an Illinois congressman to "identify Brown & Root's election as prime contractor with Mr. George Brown's personal political contributions. . . . The Illinois politician revealed a remarkable naivete by asserting that any national officeholder would practice the political idiocy of influencing the selection of an incompetent service firm to perform a highly complex undertaking."[58]

Such counterarguments were too little too late. Bowing to both public criticism and concern over escalating costs, the project director at the NSF recommended termination of all activities on October 31, 1966. Work on the project came to a halt, and in December, 1966, the advisory committee was dissolved. Project Mohole became a fading memory of an ambitious idea to push back the frontier of knowledge of the earth sciences—and an often-cited reminder of the complexities of government funding of science in a highly politicized process.

Although Brown & Root did not complete Project Mohole, the firm benefited from its involvement. Mohole enhanced the firm's capabilities in both engineering and design of deep-sea drilling barges.[59] It also added to the company's overall reputation as a "cutting-edge" engineering firm capable of undertaking one of the most advanced projects in the world. Unfortunately, Mohole also reminded the company of the possible costs of its reputation as a "well-connected" company adept at using political influence. In the case of Mohole, Brown & Root lost one of the most attractive projects in its history in part because of the success of political opponents in convincing a majority of Congress that the company enjoyed unfair advantages in the contracting process. One lesson was clear: the more powerful and controversial Lyndon Johnson became, the more public scrutiny Brown & Root could expect over its ties to him.

This new level of scrutiny became quite uncomfortable for the Browns and Johnson during the weeks leading up to the Democratic National Convention of 1964 as journalists began investigating Johnson's background. The president was concerned about stories suggesting wrongdoing, especially regarding his relationship with the Brown brothers. Reporters from *Life* magazine started researching Johnson's life for a two-part article.[60] George Brown apparently saw a draft of the article containing inaccuracies. He then complained to a former client, a director of Time-Life for whom Brown & Root built a paper mill in East Texas. Soon thereafter, *Life* sent a reporter to interview Brown. The resulting article nonetheless contained a brief overview of the Brown-Johnson relationship suggesting that Johnson was responsible for making Brown & Root rich in return for the Browns' patronage.

In the morning of July 29, George called Johnson to discuss the article. Brown complained to his friend that the news media had always credited either Jesse Jones or Lyndon Johnson for Brown & Root's success. "It's a little unflattering to me," George told Johnson, "to think that we didn't have any capability at all to get the work on our own."[61]

Johnson was concerned about possible negative political ramifications of the article. "All right now," he told George, "you told him we never heard of the Mohole thing." Brown replied, "Well, he never brought it up. He never brought Mohole up." Johnson, apparently informed of the article's content, replied, "Well, he's got it in the story." Brown told Johnson that the reporter was not interested in facts; he just wanted a story. He was simply rehashing newspaper articles printed during the last twenty-five years, according to George. Johnson believed that the inspiration for the story came from Bobby Kennedy, positioning himself for the vice presidential nomination at the convention. Brown said he would use his influence on the Time-Life director he knew to get the story "killed." "I probably won't be successful," he said, "but I'm going to try to do it this morning. I think you're going to have to try to get it killed up there somewhere. I don't know how you're going to do it."[62] Neither man could have been pleased about one "political fable" in the article: "Shortly after the 1960 election, the story goes, President-elect John Kennedy remarked dryly to his Vice-President-to-be, 'Now, Lyndon, I guess we can dig that tunnel to the Vatican.' And to this Johnson replied, 'Okay—so long as Brown & Root get the contract.'"[63]

Later the same day, a clearly agitated Johnson spoke with Clark Clifford. The president told Clifford about his concerns: "I have a friendship. I know a man. I had a heart attack at his house, therefore I'm responsible for all the work that he's done, all the contracts he's gotten, and I've never been responsible for one. I never discussed one with anyone and didn't even know he had a good many of them, and 90% of his stuff is all over the world and is private, not public. And there's not any evidence anywhere, anytime, that I had it." Clifford tried to soothe Lyndon's concerns. "I reiterate what we discussed briefly this morning," said Clifford. "I do not consider it one of the serious issues of the campaign. . . . it will get uglier and tougher than that, but the American people have a great understanding of the kind of crap that gets printed during an election campaign."[64]

Despite these denials, another telephone conversation between George Brown and Johnson suggests how these men made their interests known to each other. While discussing the political situation in Panama in 1964 and calls for U.S. surrender of the Canal Zone, Brown suggested a solution: "Competition is a great thing to make people come to their senses. I'm in the construction business, I know." Brown told Johnson that competition forces people to "straighten up and fly straight." In this situation, creating competition meant that the United States should suggest building a new canal in a friendly spot. "All you've got to do is let us go down there and core for that goddamn canal . . . and goddamn your trouble would be over in Panama, I'll tell you." Johnson paused and then asked, "How much is involved in coring . . . $10 million?" Brown replied that the entire project would cost no more than a half million dollars.[65]

The conversation turned to another article by Marcus Childs about the Brown-Johnson relationship. Johnson told Brown, "Well, I just told him I never had talked to you about NASA or Mohole either. It had never been mentioned to me. I didn't even know what Mohole was." Brown shared Johnson's frustration that while the media portrayed Brown & Root as involved in a "$500 million" contract with NASA, the reality was much different. "I've been raising hell with Jim Webb," George told Lyndon, "because he has been so goddamn mean to us."[66]

Johnson and George Brown discussed many issues during their telephone calls and visits. Johnson often seemed more engaged in those parts of the conversation that related to his own political strategies about which George Brown was quite interested himself. In February of 1964, Johnson and Brown discussed the "smear squads" that Barry Goldwater employed to get dirt on Johnson. Johnson told Brown he was simply trying to do a good job: "We got more wars now than we can win and Vietnam, I'm just not running around over the country with my shirt tail out trying to get us in more trouble. I'm trying to put out the fires we've got." Brown replied, "I think you're exactly right."[67] Six months later, Johnson told the nation about a North Vietnamese attack against American ships in the Gulf of Tonkin. Although more recent historical accounts proved that the Tonkin Gulf episode was highly exaggerated by the Johnson administration for political purposes, Congress quickly passed the Gulf of Tonkin Resolution, giving Johnson new powers that he later used to escalate the war.[68]

As the Vietnam war escalated, Brown & Root saw an excellent opportunity to expand one of its traditional strengths, the construction of military bases. Yet this work came with a high price tag, as the company—and, in particular its chairman, George Brown—encountered vocal public protests.

Brown & Root became a major contractor in Vietnam in 1965, when it joined a giant consortium of American construction companies known as RMK-BRJ.[69] To critics, Brown & Root emerged as part of the "military industrial complex" about which President Dwight Eisenhower warned the country in his farewell address years earlier. The "BRJ" group consisted of Brown & Root and J. A. Jones Corporation of Charlotte, North Carolina. In the summer of 1965, these two companies joined "RMK"—Raymond International of New York and Morrison-Knudsen of Boise— which had been active in Vietnam during the previous years of a gradually expanding American presence in the war. American military planners knew that a greatly enlarged American military commitment in Vietnam required extensive new support facilities, and RMK-BRJ was given primary responsibility for building these facilities.[70] The basic contract for this work provided cost plus 1.7 percent of the estimated cost. An additional 0.76 percent could be awarded to the contractors based on performance. From August 1, 1965, through the project's closeout in 1972, the contract totaled almost $2 billion; Brown & Root's 20 percent share equalled more than $380 million. The joint venture built ports, ammunition depots, fuel storage facilities, airfields and related maintenance and operational facilities, medical facilities, utilities, command and control functions, command functions, and ammunition storage.[71]

Brown & Root was deeply involved in expanding harbor facilities at Saigon, the capital of South Vietnam; Da Nang, the headquarters of the U.S. Marine Corps located far to the north; and Cam Ranh Bay, a strategic port also located to the north of Saigon. Before construction began, these ports were simply not capable of handling the vast shipments needed to support the war effort. According to H. William Reeves, who managed Brown & Root's work in Vietnam, in early 1965 "things were just in a horrible mess."[72] Later that year Secretary of Defense Robert McNamara visited Saigon to discuss "large-scale plans for building the war."[73] Reeves suggested using prefabricated techniques as opposed to on-site construction that was presenting many problems in Vietnam. His suggestions were accepted, and much of the work on the expansion of Vietnam's ports used prefabricated structures built in the Philippines.[74] Brown & Root crews also built bridges, including one half-mile span built across the fifty-foot deep Han River at Da Nang in thirty days.

At its peak, in July, 1966, RMK-BRJ had fifty-two thousand employees at fifty locations in Vietnam. During March, 1967, alone, the joint venture completed $67 million in work.[75] By 1969, RMK-BRJ began gradually phasing out its work in Vietnam, as its workers were slowly replaced by Army Engineers, Seabees, air force construction forces, and Vietnamese nationals. July 3, 1972, was the official closeout day for RMK-BRJ. At the ceremony held on that day, Ambassador Ellsworth Bunker noted, "I am pleased to have an opportunity to acknowledge the contribution of RMK-BRJ achievements. The ports that were built here to move the cargoes of war can in the years to come support the movement of cargoes of commerce and world trade. The airfields, roads, and bridges that now bear military traffic can serve as lifelines for the distribution of goods and services throughout the nation. At a time when all too many forces are bent on destruction, RMK-BRJ's ten years of accomplishment have been in my opinion one of the finest episodes in our nation's history."[76]

Other observers did not share this assessment. The RMK-BRJ contractors received criticism for their work and their accounting practices in Vietnam. In 1967, the U.S. General Accounting Office (GAO) found that RMK-BRJ lost accounting control of $120 million. The report also cited lax security that resulted in millions of dollars of losses from theft. It concluded that "normal management controls were virtually abandoned."[77] The companies excused such problems by noting the need to rush construction in support of the war effort. But critics were not convinced. Connecticut senator Abraham Ribicoff commented that the Senate Subcommittee on Investigations should look into this matter regarding "federal funds which are now being squandered because of inefficiency, dishonesty, corruption, and foolishness."[78]

Criticism of Brown & Root in Vietnam became more personal and much more discomforting for George Brown. As a frequent visitor to college campuses, this most visible symbol of Brown & Root and close friend of Lyndon Johnson was hounded by protestors. In October of 1971 protestors at the University of Texas embarrassed George Brown during a ceremony honoring distinguished alumni of the university. Although he had only briefly attended the University of Texas before earning his degree from the Colorado School of Mines, Brown felt honored by the award. But the event took an ugly turn. During the ceremony several hundred persons gathered at the entrance of the LBJ Library where the ceremony, chaired by Governor John Connally, was being held. The protesters angrily objected to honoring George Brown. They claimed that Brown & Root contracted with the Department of Defense to build "tiger cages" to house prisoners in South Vietnam. These cells had been used through the summer of 1970, when a contingent of U.S. congressman discovered them. In response, RMK-BRJ began building redesigned detention cells to replace them.[79] Brown was not entirely shielded from the protestors. The *Austin American-Statesman* reported that "during the ceremonies, a couple mounted the stage and announced a 'special award' to Brown. It was a photograph of a 'tiger cage.'" This award also included a letter denouncing Brown.[80]

The man who had enjoyed public acclaim for his defense-related work during World War II now faced public condemnation for his defense-related work in Vietnam. Brown could not escape public questions about his company's work in Vietnam even on the Rice campus, where students also picketed. Ironically, Brown was not a strong supporter of the continuation of the war in Vietnam. According to his old friend Gus Wortham, in the last years of Johnson's presidency, Brown and Wortham had decided that "if Lyndon did not get out of Vietnam, it would destroy him." They resolved to speak frankly and directly to their old friend about this problem, and they arranged to talk with the president during one of his frequent visits to the LBJ Ranch. Yet the two men could not bring themselves to reveal their feelings to Johnson. When Wortham later asked Brown why they had both hesitated, Brown replied, "Well, Gus, maybe you felt like I did. . . . it's not just Lyndon. He's the president of the United States. He has the intelligence community. He has the State Department. He has all of the generals and the Defense Department. Who are we to tell him what to do on a situation that we know so little about?"[81] A mutual friend of the two men later reflected how Herman Brown might have handled the same situation: "Herman Brown would have said, not 'Mr. President,' but 'Lyndon, this damn thing is going to eat you alive.'"[82]

In an interview in 1977, George Brown himself recalled a different kind of conversation with the president about Vietnam. Johnson was concerned, recalled Brown, about simply letting South Vietnam fall to the communists. But George advised that Johnson could not "help people who couldn't help themselves." This angered Johnson, who "got up from the table and walked around and put his finger in my face and wanted to know what I meant when I said he couldn't help people who couldn't help themselves. He said, 'You mean you wouldn't help Sam Houston, and he is an alcoholic.' I said, that doesn't have anything to do with this . . . you can help him, but you won't do him any good.'" Brown continued to tell Johnson to get out of Vietnam until the two old friends decided to drop the subject for the time being.[83]

The Vietnam War pulled the entire nation into turmoil. While receiving lucrative construction contracts in Vietnam, Brown & Root also received considerable and vocal criticism. In addition to antiwar protests, the company heard complaints from major competitors such as Bechtel, which charged that "Brown and Root . . . had gotten most of the choice projects during LBJ's administration, from constructing the Space Center in Houston to building the infrastructure for the Vietnam War. Bechtel, by contrast, had come away with only comparative crumbs."[84] Although Halliburton had purchased Brown & Root several years before the firm joined the RMK-BRJ joint venture, journalists and critics still singled out Brown & Root because of the long-standing connection between President Johnson and Herman and George Brown. Like its partners at RMK-BRJ, Brown & Root made money from its work in building infrastructure for the Vietnam War, but the personal cost to George Brown proved high.

The "tiger cage" episode highlighted a dilemma facing the aging George Brown. Although he was now more detached from his company's day-to-day operations, he remained a quite visible symbol and representative of Brown & Root. In 1971 at the age of seventy-three, George Brown had little to do with the company's activities in Vietnam, but public opinion nonetheless held him accountable for the company's actions. He continued to serve as chairman of the board at Brown & Root until his

George R. Brown under the Loop 610 bridge, 1972.
Courtesy Brown Family Archives.

retirement at age seventy-seven in 1975. He retired earlier as chairman at Texas Eastern in 1971, but he remained active as the chair of that company's executive committee. Even before these retirements, he began to ease away from direct involvement in
the company's normal operations, choosing instead to advise their top managers on
strategic issues.

Brown & Root enjoyed tremendous expansion during Brown's tenure as chairman of the board from 1963 to 1975. In addition to its work on the Manned Spacecraft Center, Mohole, and construction in Vietnam, the company continued to grow
in its traditional areas of strength, such as highway and dam construction, the building of power plants and other industrial plants, pipeline and pipeline-related con-

struction, and the design and building of offshore platforms and pipelines. George Brown served as a forceful advocate for Brown & Root on Halliburton's board, and he remained prominent in selling the company's services and in publicizing its achievements. Thus, when Brown & Root began construction in 1972 of a major bridge over the Houston Ship Channel on Loop 610 (the freeway loop around Houston), Brown appeared in publicity photos. The highly decentralized firm had projects spread all over the world, and George Brown remained largely in Houston overseeing work of special interest to him.

Statistics on Brown & Root's expansion during these years show the company taking advantage of a booming economy to record a dramatic period of sustained growth. Employment grew from about 10,000 in 1960 to almost 25,000 in 1970 to some 66,500 in 1975.[85] By the late 1960s, Brown & Root regularly placed high in the *Engineering News-Record*'s annual sales rankings of American engineering and construction companies. In 1966 the company showed sales of more than $1.2 billion; in 1969 its sales of $1.67 billion gave it the number one spot on the annual listing for the first time. During the next six years, it remained ranked number one, two, or three in the nation, with sales ranging from $2.35 billion in 1970 to $7.2 billion in 1974.[86] The company did not suffer under the somewhat detached leadership of George Brown during his years as chairman. (See table 9.1.)

The same was true at Texas Eastern, although here George Brown remained a bit more involved in key decisions. The company's core business, natural gas transmission from Texas to the East, grew steadily and profitably in the years from the early 1960s through the 1970s, as George Brown gradually withdrew from the firm's management.

TABLE 9.1. RANKING OF BROWN & ROOT, INC.

Year of Contract	Ranking
1957	*
1959	47
1961	*
1963	8
1964	7
1965	2
1966	3
1967	2
1968	2
1969	1

* Not listed in Engineering and News Record list of top 40 construction companies ranked by Gross Revenues.

Source: Various years of *Engineering and News-Record*.

At its origin in 1947, the company had less than 3,000 miles of pipeline. This figure grew to more than 5,600 miles in 1960 and to more than 12,000 miles in 1975. During this time, its total assets increased from $156 million in 1947 to $897 million in 1960 to $2.5 billion in 1975. For a time in the early 1970s, Texas Eastern could claim the only gas pipeline system that reached both coasts of the United States. During these same years, the company diversified into several related industries, notably petroleum products pipelines, liquefied natural gas, and oil and gas production.[87]

As at Brown & Root, George Brown relied on a seasoned group of longtime employees to manage Texas Eastern's day-to-day operations. But the extent to which he could become involved in the company's strategic decision making is illustrated by one of the clearest departures in the firm's history, the diversification into Houston real estate development on a grand scale in the 1970s. Texas Eastern previously experimented with the idea of entering this new field, but it had not found an appropriate project that made business sense but was also large enough to "make a difference on the bottom line" of a giant, capital-intensive gas transmission company.[88]

According to company legend, George Brown was responsible for leading the company into such a project, the giant Houston Center proposal that originally sought to develop some thirty-two blocks on the east side of Houston. In 1970, the seventy-one-year-old Brown and Baxter Goodrich, a longtime Texas Eastern engineer then serving as the company's president, talked about real estate ventures while looking out of Goodrich's office window. Houston's downtown had been growing to the north, south, and west, but not to the east, of the traditional downtown area. Looking out the window to the east from the Texas Eastern office building, Brown "remarked that the property could be put to better use. 'Let's see if it can be bought.' They agreed and went to the Texas Eastern board and got approval."[89]

The property in question spread over thirty-two acres, and the company quietly went about acquiring it under the direction of George R. Bolin, a Houston real estate specialist who was the son of a former Brown & Root executive and the namesake of George Brown. When Goodrich formally announced the purchase of most of the land in April of 1970, local and national newspapers voiced astonishment at the magnitude of the proposed project. The *Houston Post* correctly noted that this was the "largest single deal in the city's history," while the *Cincinnati Enquirer* reported simply "Gas Firm Buys Downtown Houston."[90] Texas Eastern's October, 1978, announcement for "Houston Center" created another wave of excitement. This $1.5 billion development called for a futuristic city within a city, complete with large sections of air conditioned mall with distinct levels to separate automobile traffic from pedestrians. In addition to ample new office space, stores, and hotels, the proposal called for downtown housing for all income levels and "people movers" to help pedestrians move through the thirty-two-block area. In effect, this development would double the size of downtown Houston while pointing the city's future expansion decidedly toward the east.

After basking in the glow of this early publicity, Texas Eastern faced the somewhat daunting task of beginning work on a giant real estate development in a city already becoming saturated with available office space by a wave of newly completed office buildings. Over the next decade, the company gradually retreated from its initial master plan, settling instead for the construction of a series of modern office buildings and a modern hotel. Although Houston Center as actually constructed fell far short of the city of the future promised in 1970, the new buildings did make a dramatic difference on the east side of Houston. Texas Eastern established its new headquarters in Houston Center. In 1979 it entered a joint venture with Cadillac Fairview, an experienced urban real estate developer, to help salvage as much of the project as possible. Later, it was appropriate that a large new convention center built on part of the original thirty-two-block site purchased for Houston Center became the George R. Brown Convention Center in memory of the man who envisioned new development on Houston's east side.

In the years after Herman Brown's death, George retained a particular interest in one area of importance to both Brown & Root and Texas Eastern, the involvement of both companies in the development of North Sea oil and gas reserves. The North Sea was a sort of "last frontier" in oil exploration and production, and Brown remained fascinated by the extraordinary advances in engineering required to find and produce oil in this hostile environment. Since his personal participation in the design of the earliest offshore drilling rigs at Brown & Root, George Brown had followed the evolution of this booming industry.

During the time George Brown served as chairman of Brown & Root and Texas Eastern, both firms became involved in North Sea work. Texas Eastern reaped great rewards as an investor in the North Sea while also maintaining a presence as an operator of gas-related facilities in the field. Brown & Root became one of the major North Sea contractors, building some of the massive platforms required to drill in the rough waters of the North Sea and laying much of the underwater pipeline needed to transport oil and gas to shore. Both companies remained significant players in the area during George Brown's years as chairman.

The North Sea, covering 185,000 square miles, provided great challenges to any firm. Severe storms often arose quickly out of calm seas and clear skies, especially in the fall and winter. Gales and other storms sweeping down from the north literally piled up in the North Sea, as the waters ricocheted off the English, Norwegian, and northern European coasts causing huge waves. Offshore drilling equipment had to be designed to withstand such severe conditions, and it had to reach extreme depths to tap the oil and gas deposits in parts of the North Sea. The hard ocean floor composed of clay, shale, sand, and even boulders posed still another challenge, as did tidal movements of up to eighteen feet.[91]

Texas Eastern's initial involvement in the North Sea developed through industry contacts. In the early 1960s, the British Gas Council invited Amoco, Amerada Petroleum, and Mobil to form a group to search for oil and gas in the North Sea. Amoco had a working relationship with Texas Eastern, and it sought the company's involvement because of its expertise in the natural gas industry.[92] George Brown's close relationship with Amoco's president and CEO, John Swearingen, facilitated cooperation between the companies. Swearingen had assumed the top position at Amoco in 1960, but he had known George Brown since the late 1940s. He recalled that "one of the reasons for choosing Texas Eastern, of course, was that they were very active in the pipeline business and they were also active in the construction business and they were also active in the construction business of offshore platforms and undersea pipelines."[93] Of course, Texas Eastern's work involving platform and undersea pipeline construction was done by Brown & Root.

Texas Eastern joined the Gas Council–Amoco Group in 1965, participating in the United Kingdom's first licensing round for exploration concessions in the North Sea. The British government awarded the group exploration licenses in approximately 3,600 square miles, or approximately 2.3 million acres, off the coasts of Scotland and England. These particular sectors turned out to contain substantial gas fields. With this auspicious start, Texas Eastern organized a new wholly owned subsidiary, Texas Eastern (U.K.) Ltd., for the management of its North Sea operations. Texas Eastern then began exploration in the Norwegian sector of the North Sea as a member of a group of private companies including Amerada Petroleum Corporation, American International Oil Company (a subsidiary of Amoco), and the NOCO group, a consortium of Norwegian industrial companies. The Norwegian government awarded this group of American and Norwegian interests licenses on 1.2 million acres off the Norwegian coast.

George Brown's view of North Sea opportunities through Texas Eastern's early participation underscored to him the possibilities for construction work. Brown & Root's long experience in offshore pipeline projects and offshore drilling rigs made it a strong candidate to develop North Sea oil and gas facilities. The company's earliest work in the North Sea was for the German North Sea Consortium and for Conoco in 1965.[94] From these projects through 1970, according to one estimate, Brown & Root participated in the construction of 70 percent of the structures installed in the North Sea.[95] A London office offered the firm a certain early logistical advantage, but the firm continued to keep its engineering staff in Houston during the 1960s. Throughout the history of North Sea oil development, Brown & Root remained a major contractor for constructing pipelines and platforms. One researcher described the firm as having done "the lion's share of North Sea pipe-laying."[96]

George Brown once presented a vivid description of these efforts. He noted that

North Sea platform. *Courtesy Brown & Root.*

"when the winds blow for twelve hours running, it's hard to dig a ditch in a hundred feet of water and lay your pipeline and bury it, too. But that's what we have to do in the North Sea. The pipelines must be buried to avoid ocean travel in that shallow water." This was a difficult job made much worse by unpredictable conditions: "The work is so tough that our competitors don't want to bid on it. Some months we are able to work only three or four days. Any sea can get rough, but the North Sea lays it on. During one period we expected to lay forty miles of pipeline. We laid only nine."[97]

George Brown was not deeply involved in the day-to-day design and construction of these advanced offshore structures. For that, Brown & Root had a growing cadre of engineers equipped with the latest in computer-assisted design technology. But he nonetheless remained an important resource for those seeking to expand the company's lucrative offshore business. He was an excellent, well-connected salesman. He also had the stature and authority needed to cut through the red tape and bring decisions to a head. Perhaps the most prominent example of this came in Brown & Root's early years in the North Sea, where the company had formed a joint venture with the Dutch entrepreneur Pieter Heerema. With a blunt, full-speed-ahead personality somewhat similar to that of Herman Brown, Heerema could be difficult to "manage" when he was upset. Tensions between Heerema and Brown & Root's managers rose over the authority of each partner in the joint venture to make decisions, and Heerema finally threatened to scuttle the venture's working vessels in Rotterdam harbor if agreements could not be reached over the relative authority of each partner. When things had reached a seemingly hopeless deadlock, Heerema flew to Houston and met with George Brown. The two quickly agreed to a mutually acceptable parting of the ways, and Brown & Root was freed from the impossible task of conquering Pieter Heerema so that it could return to the difficult task of conquering the North Sea.[98]

Brown & Root's work in the North Sea required constant innovation in offshore technology for the production of both oil and gas and the transportation of these products to shore. In December, 1969, Phillips Petroleum discovered a massive oil and gas field in southern Norwegian territorial waters at a field called Ekofisk. The field was located 150 to 160 miles from the nearest Norwegian shores and was separated from Norway by a trench that varied in depth from 1,200 to 1,500 feet.[99] Brown & Root participated in several phases of the development of Ekofisk. During Phase I, the company assisted in installing a converted jackup drilling rig, the "Gulf Tide," into a temporary production platform. This 252-foot-high, 2,200-ton rig had to be towed 5,000 miles from Houston to Ekofisk for installation.

The Ekofisk discovery moved North Sea drilling operations from 100 feet of water in previous operations to 230 feet of water and more hostile weather conditions. This necessitated building platforms that would stand not only in deeper water but higher above sea level as well. Other installation work done at Ekofisk by Brown & Root included "the field terminal production jackets and decks, the flare structure and deck, numerous bridges and connecting platforms, the entire pipelines system, and two thirty-inch 'hydrocouples' used to connect the pipelines to pre-installed risers."[100]

More large oil and gas discoveries in the North Sea soon followed, bringing additional construction work. In October, 1970, British Petroleum (BP) announced the discovery of the massive Forties field, the first giant oil field located in U.K. waters.

BP announced a $400 million plan—a figure that would more than double over the next several years—to develop the Forties field to the point that it would be capable of producing 400,000 barrels of oil per day.[101] BP intended to build four drilling and production platforms to sit in 420 feet of water. This platform would have to withstand winds of 130 miles per hour and waves as high as 94 feet.[102] Brown & Root joined other major contractors in pursuing work on this expensive and innovative project. Despite political objections to foreign companies such as Brown & Root, the company ultimately received the contract as the managing contractor with responsibilities to plan and design the Forties field installation and to construct much of the equipment.[103] Brown & Root also was project manager of a joint venture that built two of four huge platforms scheduled for Forties,[104] and it received a portion of an estimated $39 million contract to lay a 110-mile pipeline from the Forties field to Cruden Bay, Aberdeen, on the Scottish coast.

The new surge of platform construction and pipe laying in the North Sea prompted Brown & Root to enlarge its North Sea engineering staff at its London office in 1972. This move enabled the firm to work more quickly and efficiently on its North Sea design and engineering tasks. The staff grew quickly from forty to four hundred persons. Brown & Root also augmented its construction capabilities by building new pipe-laying barges. But the firm seemed to gravitate more toward pipeline work on its own. "We tended to specialize more in the pipeline side than the offshore platform side," recalled Hugh R. Gordon, one of the key figures in the company's North Sea work. "We did a lot of that but we'd pushed the pipeline harder just because I liked it more."[105]

Throughout the 1960s and 1970s, Brown & Root remained a leader in most phases of pipeline and platform construction in the North Sea's hostile environment, where the company claimed an impressive series of innovations in building large facilities under very harsh conditions. At the same time, the company continued to build state-of-the-art platforms and pipelines in the Gulf of Mexico, the Middle East, and other areas with offshore petroleum developments throughout the world. Until the oil price collapse of the mid-1980s, the company maintained one of the largest fleets of specialized offshore construction vessels in the world. Its leadership in offshore construction brought Brown & Root lucrative contracts, enhanced technical capabilities, and considerable publicity as a high-profile engineering and construction company. It was an activity in which George Brown and the technical staff of Brown & Root took great and justified pride.

It also brought one of the bleakest moments in the modern history of the firm. In 1977, after George Brown had retired from Brown & Root, the company was rocked by a federal grand jury investigation of its pricing and marketing practices during the sixteen-year period before 1977.[106] This investigation resulted in an indictment of both Brown & Root and its primary rival in offshore construction, J. Ray McDermott and Company, charging that the two firms "got together to decide which firm was to win specific jobs, particularly in the Gulf of Mexico, and arranged for the firm not assigned the job to submit 'intentionally high' bids."[107] In his deposition regarding this matter, George Brown recalled little about specific events. The U.S. Department of Justice fined each firm $1 million after pleas of "no contest" to these antitrust charges. Numerous civil lawsuits followed.[108] The impact of this suit on Brown & Root's offshore work is difficult to assess, since a precipitous drop in oil prices in the early 1980s dramatically altered the prevailing conditions in the industry, calling into question continued investments in high-cost areas such as the North Sea.

In marine construction as in most other aspects of Brown & Root's operations in the years of George Brown's tenure as chairman, close public scrutiny followed the company's success. Perhaps attitudes toward competition and contracting molded during the early days of the company's history were no longer appropriate. More likely, the company was simply having problems adjusting to the new regulatory demands and public scrutiny that faced all American industry after the 1960s. Herman Brown had died in 1962, before his approach to business was challenged by the changed political demands of the Kennedy-Johnson years. Brother George lived twenty-one years longer, and during this time, regulatory and legal constraints on American business behavior greatly increased. He confronted more than the normal challenges facing all American companies in these years. He also had to adjust to the bright spotlight of public scrutiny that was aimed at Lyndon Johnson but also often landed on Brown & Root and Texas Eastern. Although the years after Herman's death were often trying times for George Brown, they were also quite good years for his two major companies, which grew and prospered in the 1960s and 1970s.

As George Brown's responsibilities at Brown & Root and Texas Eastern eased after the late 1970s, he turned more of his attentions to the smaller enterprises that he and Herman had organized. In the late 1960s, Brown moved with the assistance of Ralph O'Connor to consolidate all of his personal holdings. These included Brown Securities, Highland Oil Company, Frio Pipe Line Company, the Herman and George R. Brown Partnership, the Driskill Hotel, the Greens Bayou Terminal, Brown Engineering, Victoria Gravel Company, properties in East Kansas and Arkansas, and other small properties. Brown Securities was the oldest of these concerns, having been created in 1932 to acquire paving certificates for Brown & Root. Over the years this company had purchased other interests. In 1967 Brown decided to amend Brown Securities' charter, changing its name to Brownco, Inc., with the expressed purpose of consolidating all of the enterprises owned by Brown and his family into a single concern. Alfred Negley, who had been president of Brown Securities, became Brownco's first president. Upon its creation, the new company had about five hundred employees. Its corporate logo was a large "B" with profiles of Herman and George Brown inside the letter.

An important part of Brownco was the Highland Oil Company, which became the new company's oil and gas division under the leadership of Ralph O'Connor. Highland concentrated primarily on oil and gas and with Brown's backing was able to participate in some large ventures both on and offshore.[109] For a time the company also had the Houston-area Coors distributorship. Brown knew Adolf Coors, who was a trustee of the Colorado School of Mines.

Brown's lifelong fascination with the oil industry at times led him into questionable ventures. At one point, he decided to invest heavily in a drilling company started by a friend who had retired from Brown & Root. O'Connor, in charge of Highland, prodded Brown about his interest in the venture. Brown responded, "Well, you run your business and I'll run mine." Later, one of Brown's close friends, John Swearingen, head of Standard Oil of Indiana (Amoco), told him, "George, my people say you're getting in over your head. You ought to protect yourself." Realizing the dangers of a venture being run by persons not familiar with the operations of the oil and gas business, Brown decided to get out. Only then did O'Connor learn of Brown's investment. Brown had financed two rigs by using as collateral all of his Texas Eastern stock, which was then worth about $160 million.[110] Although Brown ended up taking a loss to get out of this investment, his Texas Eastern stock was freed up.

Another unpleasant venture was Brown's attempt to drill for oil in Memorial Park, the major park in Houston, and give the proceeds to the city of Houston for the development of more parks. The land originally had been sold to the city in the 1920s by the family of James S. Hogg, a former governor of Texas, on the condition that it be used only as a public park. George Brown sought to satisfy this stipulation by obtaining the written permission of the remaining members of the Hogg estate, including Governor Hogg's ninety-three-year-old daughter, Ima. Environmentalist groups protested, and after the Daughters of the Texas Revolution objected adamantly that this activity would destroy historical property, Brown decided to give up on the project.[111]

In the years after Herman's death, George was alone in business, and he had his hands full. His brother's absence forced him to assert a stronger influence within their companies, and he responded by showing a decisiveness that he had not often been asked to show in his previous years at Brown & Root. After Herman's death, George took the lead in completing the sale of Brown & Root to Halliburton. He then took a strong role on the Halliburton board while overseeing a fundamental reorganization that transformed Brown & Root's management system from the private domain of the Brown brothers into a modern corporate structure. Brown & Root remained highly successful and expansive during his years as chairman, as did Texas Eastern.

Although George became increasingly withdrawn from the day-to-day management of these companies, he nonetheless remained a potent force on crucial strategic choices. By the time he retired as Brown & Root's chairman in 1975, he was seventy-six years old. George Brown filled his days productively, looking after his Brownco affairs while turning more and more personal attention to the Brown Foundation's programs.

CHAPTER TEN

Innovations in Philanthropy

Whlie forcing greater business responsibilities onto George Brown, his older brother's death prompted George to pause and reflect about life's meaning. In the words of a newspaper reporter who interviewed him about these years, "Brown looked at his life, reassessed his priorities, and decided to live for himself instead of his business."[1] During his remaining years, George spent increasing amounts of his time and energy on philanthropy. The Brown brothers had always been builders; now George sought creative ways to use philanthropic gifts to help build stronger educational and cultural institutions. During his years without Herman, George struck a balance between business and philanthropy, which he summarized as follows: "You know, in the mornings, I work to make money, and in the afternoons, I work to give it away."[2]

George Brown soon discovered a talent for philanthropy. His natural salesmanship gave him the skills to solicit cooperation from others in the community to support worthy causes such as Rice University. His creative bent helped him imagine new and innovative ways to leverage charitable contributions so that they could have the greatest possible effect. Finally, his engineer's mentality helped him construct well-designed systems for giving that proved efficient and durable. George Brown experienced great success in his business career, but some of his greatest achievements came as a philanthropist.

George Brown's major interests included education, and philanthropy allowed him to pursue his passion for educational improvement in a practical way. Brown felt that "I've always been in agreement about having mass education in the United States. I always thought there ought to be some way any boy who wanted to go to college could go."[3] One way to do this was to provide assistance to private universities. By focusing on private instead of public universities, Brown avoided "being in competition with the government. . . . They're so much bigger and can do much better than the individual can so you leave them alone."[4] He also felt strongly that no single individual "can . . . have much influence on the welfare and betterment of mankind. But you can act like a catalyst and get other people to interact, and then you may have some influence. . . . That's what philanthropy is—trying to make a better world."[5]

Over time, he spread out into numerous other areas, notably the arts and medicine. The first reflected the influence of his wife Alice and his sister-in-law Margarett Brown. The second, his growing fascination with the medical center that grew up across Main Street from Rice University in the postwar years. Indeed, many of the major gifts of the Brown Foundation during George Brown's leadership can be understood best as long-term contributions to worthy institutions deemed vital to Houston's emergence as a "major league city," complete with excellent universities, cultural attractions, and medical facilities.

The vehicle for most of George Brown's philanthropy was the Brown Foundation, which he and Herman had created in 1951. After Herman's death, the Brown Foundation used the proceeds from the sale of both Brown & Root and Herman and Margarett Brown's estate to grow quickly into one of the most significant philanthropic institutions in Houston and the state of Texas. Other major foundations in Houston included Jesse Jones's Houston Endowment and Gus Wortham's Wortham Foundation. Indeed, these three major Houston-based foundations supported many of the same educational and cultural endeavors. Their activities allowed the influence of the 8-F crowd to extend long after the deaths of its members.

Charitable foundations are uniquely American institutions that have gone hand-in-hand with great estates acquired through entrepreneurship. The first generation of large foundations were created in the early twentieth century to manage the distribution of charitable contributions from the fortunes amassed by the first wave of big businessmen in America. The Carnegie, Rockefeller, and Ford Foundations had substantial endowments, and they helped establish the rules for "big philanthropy" in America. Based in the older industrial sections of the nation, these first giant foundations provided models for those in other parts of the country who wanted to create institutions capable of effectively managing the flow of large gifts to charitable causes.

Although smaller than the largest of these older foundations, the Brown Foundation, by the time of George Brown's death in 1983, ranked as the thirty-fifth largest out of approximately twenty-two thousand independent foundations operating in the United States.[6] Much of the credit for its rise to prominence in its home region as well as in the state and nation must go to George Brown. He outlived Herman by twenty years and two months. During that time, he built a fitting memorial to his brother and, ultimately, to himself and their wives. Herman provided much of the money, although the Brown & Root stock that the foundation sold to enhance its endowment in 1963 was jointly owned by the brothers. George provided much of the creativity and energy needed to find ways to make this money accomplish as much as possible.

The brothers created the Brown Foundation on July 2, 1951. In the last year of his life, Alvin Wirtz advised Herman Brown on the best possible way to organize the new foundation. The incorporators and officers of the original organization included Herman Brown as president, George Brown as vice president, and Herbert Frensley as secretary-treasurer. The three officers joined Margarett Brown and Alice Brown on a five-member board of trustees. Although formed during the summer of 1951, the Brown Foundation did not receive IRS approval as a tax-exempt charitable and educational corporation until August 28, 1952.[7] At that time, the foundation could begin to distribute gifts from a relatively small endowment supplied by the Browns.

The first gift—a $150,000 donation to establish the Lucy King Brown Chair of History at Southwestern University—was approved on January 2, 1952, to be distributed upon official approval of the foundation's tax-exempt status. It is likely that the Browns organized the Brown Foundation at the time they did in order to go forward with this specific bequest, in honor of their mother, as well as to prepare for future giving. Such future giving did not, however, come quickly or in large amounts. During the 1950s, the Brown Foundation distributed a total of only about $700,000 in grants. Southwestern University remained the most frequent recipient of its grants, which included numerous small gifts targeted for specific needs of the university and other organizations.

In the late 1950s, however, the foundation made its first step to becoming a national philanthropic institution when it amended its charter, allowing it to make gifts outside of the state of Texas.[8] This allowed the Brown Foundation to arrange gifts to institutions in other states; it could cooperate with other foundations in agreeing to support each others' favored causes. Rice University was one beneficiary of this change in policy.

The early sixties marked a turning point in the Brown Foundation's history. On June 30, 1962, the Brown brothers donated to the foundation their Brown & Root stock, which represented approximately 95 percent of the company's value. Herman died five months later, and a special meeting of the foundation elected George Brown president, a position he held until his own death.[9] George resumed negotiations with Halliburton to sell Brown & Root. On December 13, 1962, the Brown Foundation agreed to sell 52,578 shares of its Brown & Root stock to Halliburton for $32,659,912 in cash. Several days later, the foundation sold its remaining 16,098 Brown & Root shares back to Brown & Root for $10 million.[10]

The sale of Brown & Root brought the foundation its first great influx of money. At about the same time, Herman and Margarett Brown left much of their estates to the foundation. In the mid-1960s, George and Alice Brown gave large gifts to the foundation. Together, the two Brown brothers and their wives created a foundation that grew into one of the largest in Houston and Texas. Careful management helped increase this endowment, giving the Brown Foundation a growing source of funds for charitable gifts.

During George Brown's life, the foundation remained tightly controlled by family members. In January of 1963 Herman and Margarett Brown were replaced on the board by James R. Paden, a nephew of Margarett Brown, and Alfred W. Negley, the husband of George and Alice's daughter Nancy.[11] During the summer of 1976 the trustees voted to increase the number of board members from five to nine. At this time, Alfred Negley resigned and six additional trustees were elected to join George and Alice Brown and Herbert Frensley on the board. These included James R. Paden, who was reelected; George Brown's three daughters, Nancy Brown Negley, Maconda Brown O'Connor, and Isabel Brown Wilson; as well as Herman Brown's two children, Louisa Stude Sarofim and Mike Stude.[12] The foundation's chief financial advisor during these years was also a family member and professional investor, Fayez Sarofim, husband of Louisa Stude. Thus the Brown Foundation remained an organization controlled by the Brown family. George Brown was the president and the unquestioned leader on the board. All trustees shared an unspoken commitment to support worthy causes important to the family, which was understood to include Herman and Margarett Brown in the sense that their special interests were remembered by their survivors.

The Brown Foundation held its early meetings at Brown & Root's offices on Clinton Drive in Houston. Later, it occupied offices in downtown Houston adjoining those of the George R. Brown Partnership at the San Jacinto Building, which was owned by Texas Eastern. At times, the foundation met at the Lamar Hotel in Suite 8-F. However, new rules governing the management of foundations passed by the IRS in the 1969 required the foundation to occupy offices totally separate from the family's business interests. The foundation then moved into a small building on Welch Street on the southwest side of Houston.

In the early 1960s, Brown Foundation trustees began putting to good use their newly enlarged endowment. The 1960s and the 1970s witnessed an impressive wave of giving. In the 1960s, the foundation disbursed approximately $20 million in gifts; in the 1970s, about $75 million; and another $75 million from 1980 to 1983, when George Brown died.

These gifts went to a wide array of educational, cultural, and medical institutions, but the Brown Foundation targeted a majority of its grants to three major institutions: Rice University, Southwestern University, and the Museum of Fine Arts, Houston. Other universities receiving substantial grants in these years included the Colorado School of Mines (George Brown's alma mater), the University of Houston, Southern

Methodist University (Dallas), Texas A&M University (College Station), the University of Texas, Texas Christian University (Fort Worth), Abilene Christian College (Abilene, Texas), St. Thomas University (Houston), and Carnegie-Mellon Institute (Pittsburgh). The Brown Foundation also supported numerous cultural organizations in addition to the Museum of Fine Arts. A final area of special interest to George Brown and the Brown Foundation was the medical center in Houston, and a steady stream of grants went to programs and institutions there.

Yet despite the Brown Foundation's diverse list of grants to numerous institutions, George Brown believed strongly in focusing gifts so that they could have a significant impact on the recipients. To this end, he paid particular attention to the needs of his three favored institutions, Rice University, Southwestern University, and the Museum of Fine Arts, Houston. These three institutions received the bulk of the foundation's gifts during George Brown's life. In 1976, the foundation recognized its special commitment to these three institutions when it announced that for the next ten years it would give them an estimated $40 million in matching grants. Reflecting George Brown's intention to focus his giving and to leverage it as well, these bequests were in the form of "matching grants" that required the recipient to raise additional funds in order to receive the Brown Foundation grants.

Rice University remained George Brown's special favorite. Indeed, in the 1960s and 1970s, affairs at Rice took increasing amounts of his personal time and energy. As a member of the Rice board since 1943 and chairman from 1950 to 1967, Brown oversaw tremendous growth of the university's physical plant and endowment, as well as in its national reputation. In the 1950s Brown had played a crucial role in helping Rice find the funds to construct a series of much-needed buildings. In the early 1960s, he led the board through the often contentious process of reinterpreting the university's charter to allow for desegregation and charging tuition. Then in the remainder of 1960s, he took personal charge of the university's first major capital campaign.

As the suit to reinterpret the charter moved forward, Brown prepared the board for what he considered its most significant challenge of the 1960s, the first major fund-raising campaign in the history of Rice. As early as 1945 the board discussed the inadequacy of the existing endowment to sustain needed expansion and improvement. By the early 1960s, it became increasingly clear that the university suffered from a "lack of sufficient funds . . . to provide for the expanding program recommended by President Pitzer." In a ten-year plan put forth in the early 1960s, Rice asserted its continued determination to be "a university of the highest quality" that was both "an educational center of excellence" and "a center of creativity where new knowledge and new ideas result from research and other scholarly-creative activities." Initial discussions concluded that the university needed to increase its endowment by $20 million, and this figure quickly grew to $33 million. After studying the annual fund drives of other schools, the board hired a professional consultant to help conduct a preliminary survey for the proposed fund drive.[13]

By November of 1962, the month of Herman Brown's death, this preliminary report had been completed and the board began planning the campaign.[14] George Brown again stepped up to the front of the line to contribute both money and time. From 1962 through the formal conclusion of the $33 million campaign in 1968, Brown and the Brown Foundation contributed more than $8 million to Rice. Of course, trustees traditionally led the way in support of university fund-raising efforts, but Brown's generosity to Rice was unmatched since William Marsh Rice's original grant of approximately $5 million to found Rice Institute at the turn of the century.

In 1963, the Brown Foundation provided its first major grant to Rice. It was a $1 million contribution for the construction of the Margarett Root Brown College, a new on-campus facility for housing women students. In the 1950s Rice had adopted a "college system." Throughout their stay at Rice, students remained in a single college, which served as a combination dormitory, cafeteria, and social club. The Brown Foundation agreed to provide $1 million over three years for construction costs, conditioned only on the foundation's approval of the college's design. Two years later, the foundation agreed to give an additional $30,000 needed to complete Brown College, which opened to residents in the fall of 1966.

This small grant for Brown College was only part of a much larger package. In October of 1965, "Mr. Brown stated that Rice University has started their campaign to raise $33 million, and in order to get certain foundation grants, Rice University itself must raise $9 million, and he would like to see the Brown Foundation make a pledge of $2.5 million toward this amount payable in not less than $250,000 per year."[15]

At the same meeting, the foundation's board approved a $500,000 grant for the creation of the Albert Thomas Chair of Political Science. This grant no doubt filled George Brown with pride and even nostalgia, since he and Thomas had arrived together at Rice as freshmen in 1916. The chair was a fitting tribute to the lifelong friendship between the businessmen and the Houston area representative. It symbolized the increasing commitment of the former Rice Institute to the new social sciences.

Later in the same year, the Brown Foundation announced what was then the second largest grant to Rice in its history, $2.5 million to be used in part to fund the Edgar Odell Lovett Chair of Mathematics. Lovett, an astronomer, mathematician, and the original president of Rice, set the young institute's tone and aspirations during George Brown's student days. This chair absorbed a portion of the earlier $2.5 million pledge the Brown Foundation had made to the Rice capital campaign. George Brown tied the grant to the development of the Manned Spacecraft Center: "Rice University is at the forefront in space research and is making a major contribution toward bringing to fruition the dreams of space conquest and exploration cherished

Mayor Fred Hofheinz, George Brown, Alice Brown, and Alexander McLanahan
at opening of Brown Pavilion. *Courtesy Brown Family Archives.*

for centuries by astronomers. But none of this would have been possible without the
revolutionary progress and achievements in the science of mathematics."[16]

In 1967 the foundation made a $500,000 grant to establish an endowment for the
George R. Brown Program for Excellence in Teaching. This creative program sought
to encourage commitment to classroom teaching at the research-oriented university
by creating large cash rewards for those selected annually as the best teachers on campus. These teaching awards were important to George Brown, who spoke often of the
teachers who had made a strong impact on him during his brief stay on campus.
With a large endowment to fund these awards, Rice developed a thorough and innovative system of teacher evaluation that included questionnaires to former graduates,
who were asked to evaluate their teachers' courses in relation to the off-campus world.

This extraordinary period of largesse was capped off in 1968 by a $3.05 million
grant to establish the Brown Engineering Fund and a $750,000 contribution for the
Herman and George R. Brown Chair of Civil Engineering. Rice had always been strong
in engineering, but this large grant helped provide the funding needed to pursue continued excellence in a field characterized by high equipment and laboratory costs.

During the years he helped organize and carry forward Rice's $33 million capital campaign, George Brown engaged in a sort of one-man campaign of his own, directing substantial gifts toward areas in which he had a special interest.[17] As campaign chairman, Brown mobilized the steering committee and directed its three-year effort. At 8:30 each Monday morning, Brown presided over steering committee meetings designed to define the best possible strategy for approaching old friends and potential new friends of Rice for gifts. In this endeavor, Brown called on Houston's business elite; he also had good luck in persuading nationally active corporations with local operations to contribute. Assisting Brown was a steering committee composed of administrative leaders at Rice and prominent Houston area businessmen and professionals. Rice's board of governors helped approach potential contributors while also giving generously from their own pockets. Houston-based foundations and corporations joined alumni in pushing the campaign more than $10 million over its original goal. When the campaign formally ended in December of 1968, Rice had a significant new source of funds targeted for further building, endowed professorships and scholarships, and operating capital. All agreed that George Brown had been the glue that held the campaign together. He took great pride in the success of the campaign, which owed much to Brown's money, his organizational skills, and his friendships and business ties with other potential contributors.[18]

Although he retired from board chairman as the capital campaign was coming to a close, George Brown remained active in Rice University's affairs. He continued to direct large Brown Foundation contributions to meet Rice's needs. The foundation helped fund a second residential college, this time for men. It assisted the library, underwrote a lectureship, created a fund in honor of a longtime philosophy professor, and helped endow a chair in geology. Then in 1974 and 1975, the foundation announced two more large grants, a million-dollar donation to develop fine arts at Rice and a $750,000 contribution to establish a chair in engineering in honor of 8-F member Jim Abercrombie. At this time, George Brown received what he considered one of the greatest honors of his life when Rice dedicated the George R. Brown School of Engineering. George Brown, who had dropped out of Rice almost sixty years earlier, was rightly hailed as one of the most important men in the history of the university.

But he was far from finished. In 1976, the Brown Foundation announced a challenge to the Rice community. In what was initially announced as a ten-year program, the Brown Foundation offered almost $20 million in grants if others interested in the future of Rice would raise about $24 million. This "challenge grant" program represented the culmination of George Brown's efforts to leverage the foundation's grants by involving other contributors. Before 1976, he had used the carrot of "matching grants" to stimulate contributions for specific projects. Now he applied this concept on a much larger level and over a much longer time span. His experience on the capital campaign obviously shaped his thinking on this grant, which challenged different targeted groups of potential Rice supporters to meet specified annual fund-raising goals. The foundation promised different matching grant levels for each of these groups, which included members of the board of governors, alumni, nonalumni friends, and corporations. By making the foundation's gifts contingent on other fund-raising, he assured a spirited effort to meet these goals. By extending the challenge over a decade, he assured a well-planned approach to fund-raising.[19]

The foundation subsequently extended the challenge to cover the period from 1976 through 1995, and the total amount committed to Rice reached more than $53 million. This grant paid Rice between $2 million and $3 million a year for twenty years, and it was matched by an even larger sum raised from other sources. This was George Brown's most creative contribution to Rice and to philanthropy. The challenge grant concept perfectly married his desire to focus the Brown Foundation's contributions on a few institutions and his belief that the philanthropist should serve as a catalyst for giving by others. In the case of Rice University, Brown clearly succeeded in making a difference. The steady influx of millions of dollars helped propel Rice up the hierarchy of the nation's elite universities.

The Brown Foundation had a similar impact on Southwestern University. The foundation's first major contribution after Herman's death was a $750,000 gift in 1964 that established three chairs in honor of Herman and Margarett Brown and Lillian Nelson Pratt, Alice Brown's mother. The Margarett Root Brown Chair in Fine Arts was particularly fitting, since Margarett had devoted a lifetime to studying literature and fine arts as well as acquiring and donating art to Southwestern. The Herman Brown Chair in English also seemed fitting in a less obvious way. Perhaps it was simply a way to put her husband's name on a chair that would have been most interesting to Margarett Brown. Or perhaps the chair holder could succeed in conveying to students a sense of the forceful, direct use of the English language that Herman Brown had practiced throughout his life. The Lillian Nelson Pratt Chair was in Science, matching in a sense the chair in history that Herman and George had earlier established in honor of their mother at Southwestern. Whatever the reasoning behind the naming of the chairs, the addition of three major chairs was a landmark event in the life of a university the size of Southwestern. Several years later, in 1968, the Brown Foundation contributed $1.5 million to increase the endowment of each chair to $500,000 while also creating an additional chair, the Elizabeth Paden Chair in Sociology, in honor of Margarett Brown's younger sister.

The Brown Foundation continued to meet Southwestern's financial needs by contributing building and operating funds and $2 million in 1970 for to the university's $10 million capital campaign. In what amounted to a precursor of the challenge grants of 1976, the foundation conditioned its large contribution on the university's ability to raise four times as much money from other sources. In 1974 the foundation again increased the endowment of the existing "Brown chairs" while also creating another endowed position, the Brown Visiting Professor Chair.

To thank the Brown family for its long support, Southwestern University held a ceremony in April of 1975 to honor the Browns with the naming of the Brown College of Arts and Sciences. Durwood Fleming, Southwestern's president, acknowledged the Browns' extraordinary support; they had given more than $5 million to the university from 1960 to 1975. After naming all the projects supported by Brown contributions, he concluded, "We are forever in debt to Mr. and Mrs. Herman Brown, Mr. and Mrs. George Brown and the Brown Foundation, Inc."[20]

That debt deepened dramatically in 1976 with the announcement of the Brown Challenge Program at Southwestern University. As at Rice, the original grant extending over ten years was later lengthened to twenty years. More than $28 million was made available to Southwestern over the two decades, contingent on its ability to raise a slightly higher sum from other sources. Again, this dramatic, long-term challenge galvanized the fund-raisers at Southwestern, giving them a clear planning horizon and a pressing reason to approach potential donors. The Brown grants flowed to Southwestern in sums of $1 million to $2 million per year. These funds may have been more important to Southwestern than the Brown Challenge funds were to Rice, since Southwestern had a much smaller endowment and fewer large donors than its "sister" school in Houston. A long article in the *Bulletin of Southwestern University* published after George Brown's death summarized the feelings of many: "There can be no doubt in the minds of anyone who knows the history of the university that one name, one family, stands out in philanthropy and personal interest. That name is Brown."[21]

The Brown Foundation targeted for special consideration Rice and Southwestern, but it also proved generous to other schools, especially George Brown's alma mater, the Colorado School of Mines. The Brown Foundation's significant contributions to the Colorado School of Mines began in the mid-1970s. In 1974, the Brown Foundation initiated a five-year grant of $25,000 per year to fund a professorship or bring in a visiting lecturer.[22] On April 9, 1981, Dr. Guy T. McBride Jr., president of the Colorado School of Mines, began ceremonies to dedicate the George R. Brown Hall, a $5 million, 85,000-square-foot building, designed to house the School of Mines and Mining, and Basic Engineering. Previously, these schools were housed in two separate buildings built in 1900 and 1905. Construction was made possible by a $4.4 million gift from the Brown Foundation in 1971 to the college.[23] The Foundation contributed to the construction of other campus buildings as well.

George Brown's concern for the Colorado School of Mines was illustrated by his agreement to give a rare public speech there. He delivered the commencement address to the graduating CSM seniors in June, 1962. He told the class that engineers must learn to communicate better: "The one thing which fails to improve throughout the years is man's ability to communicate with his fellow man. Pay heed to the usefulness of communication, or your fine education and natural talents will never bear their full fruit for you or for the nation."[24] As persuasive as he was in person-to-person negotiations, George Brown never felt comfortable in addressing large audiences. But the contributions of his foundation spoke quite clearly of his commitment to higher education.

The Brown Foundation and the Brown brothers also provided philanthropic support for a unique educational institute located near Camden, Arkansas, the Southwest Vocational Technical Institute. During April, 1961, the Brown Engineering Company, owned by the Brown brothers, purchased at a government auction a fifteen-thousand-acre naval ammunition/industrial area in Arkansas. The naval depot had been in operation since World War II and remained operational through the late 1950s. However, in the late 1950s, the navy closed it because the rockets housed there had become obsolete. Although the Brown brothers purchased the property for its salvage value, Democratic senator John L. McClellan of Arkansas persuaded the brothers to delay their demolition efforts for three years while he worked to attract industry to the site. Then, after Herman Brown's death in 1962, the plans to demolish the site were further delayed.[25]

George Darneille was then vice president of Brown Engineering. He had assisted the Browns with other salvage operations, notably the purchase of Fort Clark and the Aircraft Reconversion Company. Darneille and Roy Ledbetter worked with local community leaders and developed a master plan to transform the area into an industrial park. George Brown also assisted the effort. His board membership at IT&T gave him the opportunity to suggest to IT&T chief Harold Geneen that the company locate one of its technical schools there. Darneille contacted other firms, including Aerojet General, Baldwin Electronics, Dixie Chemical Company, and International Paper Company. Although interested, these firms said they needed employees with a technical background. At this point, the group decided to establish a vocational technical school to provide trained personnel to work at the plants that were built at the transformed industrial park. The Brown Foundation donated seventy acres for the Southwest Vocational Technical Institute. Senator McClellan then attracted state money to continue developing the project. What began as just another salvage operation turned into a technical institute and industrial park located in a depressed area of Arkansas.

George Brown favored educational institutions for his charitable giving, but his wife, Alice, developed an equally strong commitment to cultural institutions, particularly the Museum of Fine Arts, Houston. The Brown Foundation made its first significant gift to the museum in 1964, when it established the Alice Pratt Brown

Museum Fund for purchasing works of art by Bertel Thorvaldsen, Frank Stella, Phillip Guston, and Adolph Gottlieb. While the Brown Foundation provided funds for the museum to purchase art, Brown family members, particularly Alice Pratt Brown, single-handedly purchased a large number of museum pieces for the museum.

The Brown Foundation not only helped the museum's art collection grow, it also purchased land for the museum's auxiliary facilities and financed much of the museum's major expansion in 1974. In October of 1968 the Brown Foundation provided the museum $1 million to purchase land on which to build the new Cullen Sculpture Garden. The land purchase included the aging five-story Montrose Apartments, the Central Presbyterian Church, and land occupied by a clinic as well as parking lots. At approximately the same time, the foundation gave a $150,000 matching grant toward a projected $500,000 goal needed to purchase land for the proposed Contemporary Arts Museum to be located across the street from the Museum of Fine Arts. The foundation also gave a $150,000 contribution to the Houston Museum of Natural Science for the Herman Brown auditorium.[26]

Alice Brown's lifelong interest in art originated during her youth. After her father died when she was only six years old, Alice moved in her teens to Dallas to live with her mother's sister, whose husband, Leslie Waggener, was a banker who collected art. An undergraduate degree from Southwestern University also contributed to her interest in art. Throughout her life, she took fairly frequent "art trips" around the world, helped support favored artists, and became acquainted with some of the country's most well-known art collectors. President Johnson also recognized Alice's interest in art. Perhaps the highlight of her long involvement with the arts was her service on the original board of trustees of the prestigious John F. Kennedy Center for the Performing Arts when the center opened in the early 1960s.

But her most significant service to the arts was in Houston, where she a driving force in the life of the Museum of Fine Arts, Houston. Although her husband did not strongly share her interest in art, he did support her initiatives to improve Houston-area culture with funds from the Brown Foundation.

The Museum of Fine Arts opened in 1924 and remained in the original building on the same site. However, in 1958, the museum arranged for Ludwig Mies van der Rohe to design an expansion to the museum's existing building. This 1958 design for the Cullinan Hall included a provision for a later expansion.[27] Mies completed his drawings for this expansion, the Brown Pavilion, only months before his death. His plan nearly doubled the museum's existing space to 133,750 square feet, with the addition of a 70-by-300-foot upper story gallery space, rest rooms, a library, and administrative offices, as well as a 450-seat auditorium in the basement level of the three-story expansion. The museum instituted a $15 million capital funds drive for this expansion, which would cost $4 million, and to pay for additional projects. The

Brown Foundation provided $4 million for the construction of the Brown Pavilion.[28]

Groundbreaking was held in December, 1971, with George and Alice Brown participating in the ceremonies. The groundbreaking ceremonies were part of an even larger effort to enhance the museum's national reputation. In September of 1969 the museum hired its fourth director, Phillipe de Montebello, who had previously been the associate curator of European painting at the Metropolitan Museum of Art in New York City. Montebello had set certain conditions for accepting the job, including the stipulation that the museum begin a major fund drive. This request was fulfilled by museum board president Alexander K. McLanahan's announcement of a $15 million capital funds drive to pay for the expansion, building a new art school building, setting up an endowment for art purchases, and providing money for the operation of its Bayou Bend museum.

The dedication of the Brown Pavilion was in January of 1974, with an exhibition titled "The Great Decade of American Abstraction: Modernist Art 1960 to 1970." Museum director Montebello remarked that the exhibition "was thematically appropriate for the Brown Pavilion because like Mies' late architecture, the exhibition represented twentieth-century modernism in its 'classical' phase rather than at its vanguard origin."[29] Other remarks at the dedication made an interesting connection: "It is appropriate that the Brown Pavilion, named for an old friend, should be the last work of another dynamic builder, Ludwig Mies van der Rohe."[30] This was a variant of a comment that Margarett Brown had made decades earlier, when she acknowledged that Herman Brown's construction jobs were his art.

The Brown Pavilion attracted both praise and criticism in the national press. Art historian Robert Hughes writing in *Time* magazine said: "Apart from the Houston Astrodome, one could barely imagine a less sympathetic space for showing art than Mies' vast curving hall. . . . Every grand old man has a prescriptive right to his cliches." However, Paul Goldberg wrote in the *New York Times* that "Mies' design is so refined—and simply so beautiful—as to rise above the limitations of an all too familiar idiom."[31]

After the Brown Foundation's generous gifts to the museum, Alice continued the family's involvement with the Museum of Fine Arts. At various times between 1957 and her death she served as an elective trustee, advisor, vice president and vice chairman, and secretary.[32]

In 1976 the Brown Foundation established a ten-year challenge grant for operations and acquisitions. As with the foundation's similar programs at Rice and Southwestern Universities, the grant required the museum to meet specific annual goals. In 1981, the grant was extended to 1995. The grant "matches one-to-one every gift that the museum receives for accessions, membership, and the general operating budget and matches two-to-one all new and increased gifts up to the limit of the grant." In the late 1980s, museum director Peter Marzio stated, "The Brown Foundation has provided not only the endowment growth that makes the institution stable and capable of growth, but also the incentive for expanding the museum's base of support."[33] (See table 10.1.)

In 1989, the Museum of Fine Arts reflected on the importance of Alice and George Brown and the Brown Foundation in its development. Although both of the Brown brothers and their wives were by then dead, the museum acknowledged the debt that it owed to them. "The Brown Foundation, Inc., remains the preeminent supporter of the Museum of Fine Arts, Houston. . . . Over the past four decades, the foundation has provided funds that have formed the backbone of the museum's operation and expansion."[34] The Museum of Fine Arts was greatly aided in its own quest for "major league" status by the generosity of the Browns and their foundation.

One other area of giving grew in importance to George Brown as he became older. The health problems that afflicted both Herman and George Brown brought them into close contact with some of city's most prominent surgeons and doctors, many of whom worked in Houston's large medical center located across Main Street from Rice University. George Brown also carried on Herman's relationship with surgeon Michael DeBakey. DeBakey recognized the Brown Foundation as an important source of funding to help build the Houston Medical Center, which was gaining a national and international reputation based in no small part on DeBakey's own participation in important cardiovascular developments. DeBakey wanted to expand both the cardiovascular and orthopedic pediatric surgical departments at the Methodist Hospital.

One morning shortly after Herman Brown had died, Dr. E. L. Wagner visited George Brown at his home. Wagner had served both Herman and George Brown's general medical needs. Wagner was surprised to see DeBakey drinking coffee that morning with George Brown. DeBakey was describing to Brown his efforts to fund the expansion of cardiovascular and orthopedic surgical wings at the Methodist Hospital in Houston. DeBakey asked Brown to help by providing $1 million in seed

10.1. THREE MATCHING GRANTS (as of March 15, 1998)

Name	Grant No.	Total Paid in Matching Funds	Total Paid to Date
Rice University	76-082	$52,851,128.00	$75,529,135.00
Southwestern University	76-083	$28,206,802.00	$40,637,901.00
Museum of Fine Arts	76-084	$35,161,631.00	$61,346,544.00

Note: Original grants awarded at April 23, 1976, meeting for ten-year period. Period was extended to an additional ten years at the November 23, 1981, meeting.

money. At the time, recalled Wagner, Brown said, "I think we'll think about it."[35]

DeBakey had already approached Ella Fondren, wife of the deceased Humble Oil president, regarding a contribution from the Fondren Foundation. Both the Brown Foundation and the Fondren Foundation ultimately agreed to provide funds for the proposed expansion. The Brown Foundation dedicated monies to a five-floor expansion for cardiovascular research and surgery while the Fondren Foundation committed funds for a six-floor expansion for pediatric orthopedics. Albert Alkek provided additional funds to enlarge the annex. The Brown Foundation gave the hospital $1 million to help build the annex, which was called the Fondren-Brown building.[36] Soon after agreeing to give seed money, George Brown discussed related matters with Michael DeBakey. In order to have the finest facility, DeBakey wanted to inspect top-notch cardiovascular units around the country. George Brown agreed to allow a group of architects to use a Brown & Root plane to visit other sites. DeBakey also requested from Brown another $1 million to purchase two highly sophisticated and expensive operating room lights. Brown agreed.[37]

George Brown and Michael DeBakey teamed up once again several years later to lead a rebellion of sorts within the Houston Medical Center. The Baylor College of Medicine had originally been located in Dallas, but Baylor College of Waco moved the medical college to Houston in 1943 as part of a larger effort to develop a medical center in Houston comprised of various hospitals and colleges.[38] Although Baylor College of Waco and Baylor College of Medicine had separate boards of trustees, both boards were comprised solely of Baptists. This raised questions about state financial support for denominational schools.

In 1968, DeBakey consulted with George Brown about a problem brewing at the Baylor College of Medicine. In the rapidly growing medical center, Baylor College of Medicine became increasingly uncompetitive and required increased funding to expand its operations. Baylor College, in Waco, controlled the medical school and proscribed that only Baptists could be trustees, a restriction, DeBakey recalled, "that made it difficult to get additional funds." DeBakey "found [himself] in the position of having to assume leadership of the college, that is, to be the dean and vice president of the school."[39]

Seeking support, DeBakey contacted Oveta Culp Hobby and George Brown. DeBakey visited George Brown in his home. "It seems to me, George," DeBakey explained, "the only way the school is going to get out of this is for the school to separate itself from the Baptists, and to get a new charter." Brown replied, "There's no way you're going to continue this way. You've got to get a new Board of Trustees representing the best people in the community, people who can get you money, no matter what their religion is. You're a secular organization anyway. We're a Baptist school only because we're connected with Baylor University."[40]

DeBakey went forward with his plans to separate the college from Baylor University. He then received from both George Brown and Oveta Hobby a list of names of potential board members. Brown and Hobby both declined memberships on the board themselves, telling DeBakey they could be more help to him off the board. Two of their mutual friends, L. F. McCollum and Herbert Frensley, joined the Baylor board. Frensley became the board's vice chairman.[41] These changes helped to prepare the Baylor School of Medicine for far-reaching changes that solidified its standing as one of the leading medical schools in the region and the nation.

George Brown recognized the growing importance of the medical center and medical technology as a significant new force in the Houston-area economy. He was particularly interested in what might be called "medical engineering," the application of basic engineering knowledge to such problems as the creation of an effective heart pump. He cooperated with other friends from 8-F such as Jim Abercrombie to build support for various institutions in the booming medical center. He also used Brown Foundation money to support a multimillion-dollar program in biomedical engineering at Rice University. Taken as a whole, the contributions of George Brown and the Brown Foundation were significant parts of a citywide effort to build a modern medical center in Houston, and this center grew into a significant contributor to the economic—as well as the medical—health of the region.

One special interest of George Brown was the Lyndon Baines Johnson School of Public Affairs at the University of Texas at Austin. Brown felt that the LBJ School was an appropriate tribute to his old friend, and he enjoyed helping the former president organize this school in the years between the time he left Washington in 1968 and died in 1973.

During Johnson's presidency, the Browns continued to see the Johnsons but not as regularly as they had in previous years. "Lyndon," recalled Lady Bird, "considered him [George Brown] one of the resources of his life for his wisdom and his judgement and his marvelous creative ability and his relationship with the business world for which Lyndon had a great deal of respect."[42] George served as both an official advisor to Johnson through the offices of several committees as well as close personal friend and advisor. The Browns "came to many, many dinners, usually spent the night at the White House."[43] Years later, George Brown even recalled a time when Johnson told him during a swim at the White House in 1967, long before he formally withdrew from the 1968 presidential campaign, that he planned to withdraw.[44] Brown remained skeptical of this claim, but he realized later that Johnson's health was deteriorating badly under the relentless demands of the presidency.

Alice Brown continued to enjoy a close friendship with Lady Bird Johnson in these years, and Lady Bird invited her to join the Committee for the Preservation of the White House originally created by Jacqueline Kennedy. Alice Brown also con-

George R. Brown, Lyndon Johnson, Alice Brown, and Lady Bird Johnson. *Courtesy Brown Family Archives.*

tributed art to the White House. One evening she and George Brown carried the Winslow Homer watercolor *Surf at Prout's Neck* "wrapped in a paper parcel and she herself is carrying it," Lady Bird remembered, "and we went walking around finding a place to hang it and actually did hang it right where she chose and there it stayed for a long time."[45]

George Brown was one of several persons instrumental in providing for the Lyndon Baines Johnson School of Public Affairs at the University of Texas at Austin. Shortly after Johnson left the presidency and retired to his ranch west of Austin, he set in motion the process through which he could create a graduate school of public affairs. Johnson also wanted his presidential library to be located at the University of Texas at Austin and to operate as a research archives. Johnson told his former undersecretary of the treasury, Henry H. Fowler, "I just wish that I'd had the opportunity to have a far deeper education in public affairs than I was able to get at San Marcos."[46]

In his pursuit of these two goals and in discussions with various associates including George Brown, Johnson decided to organize a foundation to provide funding for both the presidential library and the affiliated graduate school. Johnson asked Fowler to assist him in raising funds for the foundation and to work with Arthur Krim in national fund-raising. Johnson relied on George Brown to raise funds in Texas.[47]

The formal process of creating the foundation began in September, 1970, with the creation of the HEC Public Affairs Foundation; George Brown was a charter member of the board and remained on the board until his death. In February, 1973, the foundation's name was changed to the Lyndon Baines Johnson Foundation.[48] Early in Johnson's presidency, in February of 1965, the University of Texas at Austin had offered to locate and build the Johnson presidential library and a school of public affairs. The school, which originally offered a master's degree, moved into its own building beginning with the fall semester of 1970, and this helped fund-raising efforts for a larger new facility.[49]

The joint dedication for the Lyndon Baines Johnson Library and the Lyndon Baines Johnson School of Public Affairs was held on May 22, 1971. During the dedication, Johnson officially presented his papers to the United States government. After the ceremony, guests attended a barbecue lunch on the northwest side of the LBJ library.

In response to George Brown's fund-raising efforts, Johnson wrote George and Alice a letter in October, 1970, indicating the significance and longevity of their relationship: "You are the best friends I have ever had and the only ones left of the original group that started off with me to build the dams on the Colorado River. Since then you have had a part, always a very big part, in each of my endeavors, and now you have helped us to start to try to make the Lyndon Johnson School of Public Affairs the best in the land. . . . The only way I can thank you is to say that I hope the school will produce many men and women with your wisdom and ability, as well as your generosity in helping others make their dreams come true."[50]

In these years, George frequently visited former president Johnson at his ranch near Johnson City, Texas. Brown described Johnson as "like a caged lion, he couldn't slow down." After a few years passed, separating Johnson from the pressures of the presidency, Brown commented that "he's like a deep sea diver, he's been down pretty deep and it's going to take him a long time to decompress."[51]

During one of George Brown's last visits with Lyndon Johnson, he and Henry Fowler met Johnson at his ranch one morning during the fall of 1972. After Fowler arrived, Johnson said, "George Brown is flying over from Houston and the three of us are going to have a day together." Johnson greatly enjoyed taking his two friends through his birthplace and then to the Admiral Nimitz Museum in nearby Fredericksburg. Fowler recalled that Johnson was in "great spirits" that day. Johnson motioned his guests into one of his Cadillacs, which he drove while George Brown sat in the front seat. Johnson was aware that from his position in the back seat Fowler had taken notice of his long hair. Johnson asked, "What would you think of a guy who was the best friend you ever had in life and you had him over to dinner with a bunch of your friends and right in the middle of the evening sitting around after dinner, he said in a big, loud voice, 'Lyndon, why don't you get a haircut?'"[52]

When Johnson died in 1973, Lady Bird Johnson recalled that "George and Alice were some of the first people I phoned after his death. And a few hours later, with a lot of confusion and plans and military and staff and everything running around, George and Alice arrived in their plane . . . and George's comment was, 'You need a man in the house.' And so, he assumed the kind of 'I will help manage role,' which was just fitting and just fine."[53]

In memory of his longtime friend, George Brown actively supported the development of the LBJ School of Public Affairs until his own death a decade after that of Johnson. This was both a personal commitment and a philanthropic one. Johnson had been a friend for more than thirty-five years, and the two men had shared some extraordinary times in their quest to pull themselves up out of the world of Central Texas, and at the same time to pull Central Texas forward with them. The Johnson family had built close ties with the Herman and George Brown families, and George, Alice, and Lady Bird respected the memory of their old ties with the missing members of their families while remaining in contact whenever circumstances allowed. The LBJ School was a fitting symbol of the ties between George and Lyndon, since the earliest meeting of their minds on the need to improve education.

In pursuing a variety of philanthropic endeavors, George Brown found a new calling late in his life. From the time of Herman's death in 1962 until his own death in 1983, George derived more satisfaction from his work with the Brown Foundation than with any other aspect of his public life. He was well-suited to be an excellent philanthropist, with his natural gift for salesmanship and his capacity to organize. He had promised Herman that he would look after his business and family affairs for him after Herman's death, and he did so. Part of those affairs was the management of the Brown Foundation, and George and other family members managed it in such a way that Herman's interests continued to be looked after long after his death.

George Brown's innovative programs made dramatic contribution to the development of quality programs in a few selected institutions by leveraging the foundation's giving. The challenge grants to Rice, Southwestern, and the Museum of Fine Arts, Houston, proved to be successful, durable programs that gave direction to the fundraising efforts of those institutions for twenty years. In the process of embodying Brown's ideas about philanthropy in an innovative new approach to giving, the challenge grants also extended his influence over the direction taken by the Brown Foundation for years after his death. He was an entrepreneur in philanthropy, one with a good engineer's talents for designing effective and durable gifts.

CHAPTER ELEVEN

Legacies

In the late 1970s, as George Brown approached eighty, failing health severely limited his activities. Having retired from Brown & Root and Texas Eastern, he remained involved in Brownco. Although he had retired as chairman of the board at Rice University in 1968, he continued his work there as an emeritus board member. His abiding passions were Rice and the Brown Foundation, but he also remained interested in political and civic affairs. As he neared the end of his long and productive life, George Brown took great pleasure in his family. As the years passed, he outlived most of his closest Houston area friends and business associates. To many younger Houstonians, he became a living symbol of an older time in a city that had been transformed by the growth his generation had fostered.

Public recognition of a long life well-spent in business, engineering, and civic affairs was one benefit of age. Awards too numerous to list honored him in the 1970s and 1980s. He took special pride in the recognition of his standing in his chosen profession of engineering. In April, 1977, he received the John Fritz Medal, a prestigious honor awarded annually by representatives of five national engineering societies: the American Society of Civil Engineers, the American Institute of Mining Metallurgical and Petroleum Engineers, the American Society of Mechanical Engineers, the Institute of Electrical and Electronic Engineers, and the American Institute of Chemical Engineers. Joining such past winners as George Westinghouse, Alexander Graham Bell, Thomas Edison, and Orville Wright, he received the award in recognition of "his outstanding achievements as an engineer, as a builder and in management in many fields, for his public service and for his personal and material support of educational, medical and scientific institutions."[1] Five years later he received another coveted engineering-related award, the American Petroleum Institute's Gold Medal for Distinguished Achievement, which acknowledged his designs of early offshore platforms.[2]

Even more than these national awards, one "local" honor held a special place for George Brown. In 1975 Rice University dedicated the George R. Brown School of Engineering, comprising all engineering programs at the university. After his long service on the board and his key role in fund-raising, Brown took great pride in this symbol of his contributions to the school he had attended for three semesters when he first came to Houston in 1916. For almost sixty years he had grown with Rice, and he had helped foster the development of a major university with a rising national reputation for excellence in engineering.

The city of Houston and the state of Texas also acknowledged his contributions in business and civic affairs. In 1982, twenty years after the death of Herman Brown, the city of Houston proclaimed "Herman and George R. Brown Appreciation Day" in recognition of the contributions of the brothers to "the offshore oil, gas, and mineral industrial development boom which has played such a significant role in the development of our city."[3] That same year the Texas Business Hall of Fame Foundation announced the nine original members of the Texas Business Hall of Fame. The Brown brothers made the list, along with only one other Houstonian, their friend from 8-F, Jesse Jones.[4] No awards were handed out for "civic/political influence" of the sort exercised from Suite 8-F, but George Brown regularly appeared on lists of the most powerful individuals in Texas and in Houston.[5]

He retained his keen interest in politics. During the late 1960s, Brown and Walter Mischer had begun the Political Action Committee of Texas. This PAC was designed to provide funds as rewards for politicians "doing a good job" or to help fund the campaign of a politician running on a probusiness platform. "But the purpose," recalled Mischer, was to try to build greater awareness of "how important it was . . . that we had good people in office in politics and that we encouraged young people to get involved in politics and it was fairly successful in doing just that."[6] This effort reflected a lifelong commitment to a simple approach to government: "We always believed in good government and keeping good people in office. Those things go hand in hand. There's a place for both."[7]

Of course, George and his brother Herman had their own particular definitions of what constituted both "good people" and "good government." The brothers believed strongly in government committed to creating a healthy climate for business growth. Throughout his life, George felt deeply that education was a prime responsibility of government and that public policies at all levels should encourage educational opportunities. On this and other issues, he never lost interest in the game of politics, which he played throughout his life with the zest of one who enjoyed involvement and the pragmatism of one who understood the need to take a long view.

In George Brown's last decade of life, however, the rules of the game of politics changed dramatically. The Brown brothers and Brown & Root and Texas Eastern had thrived in the post–World War II era, when weakly drawn and poorly enforced campaign contribution laws allowed considerable flexibility in the exertion of political influence or lobbying. George had operated for most of his adult life in a political environment that allowed money and other sorts of "in-kind" contributions (such as use by politicians of privately owned airplanes) to flow easily and with little legal constraint from the private sector into the campaign coffers of politicians. Just before and after the Watergate scandal of the early 1970s, Congress passed new laws on campaign contributions, requiring much greater public disclosure.[8] George Brown felt that these news laws inhibited his right to take part in the political process, and he mounted a spirited legal challenge to them.[9]

But he did not shy away from political combat under the new laws. When he felt strongly about a candidate, he fought hard. For example, he helped fund an effort to keep Frank Church from winning reelection to the Senate from Idaho. John Bookout, a former CEO of Shell, recalled a discussion about Senator Church in which Brown expressed the fear that Church was "basically dismantling our CIA and our ability to attract intelligence around the world." Brown wanted to make sure that residents of Idaho understood Church's views, or at least Brown's perceptions of these views. Bookout recalled that "he did provide the leadership and provide a good bit of the funds and all to carry that message back to Idaho. . . . I know he brought to Idaho the contradictions in Church's position domestically or at home, say, in Idaho, as opposed to Washington. And not that these things were certainly not totally unknown but he provided the organization and funds and the efforts to publicize it widely and bring it to peoples' attention. And I guess I would have to say he played a major role in Church's defeat."[10] That Brown would become so active in a Senate race in a distant western state highlights the extent to which he had come to take a national view of politics.

Although George Brown was clearly a conservative, he contributed financially to a wide range of political candidates. Houston attorney Searcy Bracewell learned during an investigation of Brown's political contributions one year that he contributed to a wide range of political figures, "from Senator Goldwater on the far right to Bob Eckhardt . . . the real flaming liberal." Bracewell once helped Brown organize a reception for Eckhardt, a Democratic representative from the Houston area. Brown told Bracewell, "You know, that fellow is going to be there a long time. We just need to get to know him. There's no point in us just always talking about how bad he is. We don't know him, he doesn't know us. Let's get together and talk with him and why don't we have a little reception for him over at the Ramada Club? And you set it up." Brown, who never enjoyed talking at meetings, also asked Bracewell to contact John Connally and ask him to give a talk at the reception, which Connally did.[11]

Although George Brown, like Herman, had long remained a traditional southern Democrat, he favored Gerald Ford over Jimmy Carter in the 1976 presidential campaign. Brown telephoned Posh Oltorf, who had served as a type of lobbyist for Brown & Root in the 1940s and 1950s. Brown and Oltorf agreed that they would "come out of retirement" from political campaigning and help Ford. Oltorf was out of touch after fifteen years spent in retirement on a ranch in East Texas, and at seventy-eight years old, Brown was no longer actively associated with national political figures. Yet they decided to organize a letter-writing campaign in the conservative precincts in Harris County by soliciting the assistance of several leading law firms. Oltorf's description of their visits to the managing partners of the city's law firms shows that the two old friends had lost neither their zeal nor their sense of humor:

> George and I would go out and we would make, say, three of them a day. It took us only three days. We'd call on the managing partner, George would tell him what he wanted to do and in every instance but one, they agreed to do it. Such was the prestige of George. And I am completely deaf in this ear. And George was blind. . . . We would walk. And George walked at a great pace. And people would scatter. I said, "My God, George! Do you know how ridiculous it is, a blind man and a deaf man walking down the street! No wonder the people are scattering!" Horns were beeping! I couldn't hear them honk and George couldn't see the cars! . . . And consequently, I guess George put out, oh, maybe 200,000 letters urging people, without missing any names.[12]

By the late 1970s, George's infirmities were not laughing matters but rather serious constraints on his daily activities. Ulcers continued to plague him, as they had since his forties. But his most troublesome ailment was deteriorating eyesight. His eyes first began to bother him after a rapid descent in an unpressurized airplane, and his vision worsened over time owing to the combination of a detached retina in one eye and glaucoma. By the mid-1970s, his secretary, Doris Johnson, was forced to find ways to help him cope with his near blindness. She began a morning ritual of reading the *Wall Street Journal* to him. She also began to order "talking books" on records from a library in Austin for the visually handicapped. Brown favored business-related works and biographies. Among his favorites was *Mornings on Horseback,* which paints vivid scenes of Theodore Roosevelt's Washington, D.C. He would take these recorded books home at night to listen to when he could not fall asleep.[13] He also refused to give up one of his favorite hobbies, shooting. As his eyes weakened, he had an aide stand nearby and point him in the right direction to fire at the birds. After his eyes became so bad that he was legally blind, at times he stood out in the field with his aide, who fired his rifle for him. His doctor, E. L. Wagner, a general practitioner who treated both George and Herman beginning in the late 1950s, recalled that "he went to all the finest of the fine eye doctors not only in this country but in London and everywhere and there was nothing that could be done for him."[14]

The bouts with illness took their toll in 1982. In December George Brown suffered a heart attack. He was taken to St. Luke's Hospital, where he remained for the next ten days before being allowed to return home on December 13. At the turn of the new year, he seemed to lose interest in continuing his struggle to recover his health. He stopped eating for a time before returning to St. Luke's, where he remained until his death on January 22, 1983, at 3:00 P.M.[15]

His death was noted throughout the nation. The obituaries in the major Houston papers stressed his role as a builder and called him "one of the state's most generous philanthropists." The *Washington Post* focused on his relationship with Lyndon Johnson, asserting that "in effect, the Browns and Johnson made each others' careers possible." A combination of these views was captured in the *New York Times* headline: "George R. Brown, Industrialist, Dies," and the subhead "Benefactor of Lyndon Johnson Played Key Role in Texan's Rise to the Presidency."[16] On Tuesday, January 25, Senator Lloyd Bentsen, whom George Brown had personally encouraged to run against Ralph Yarborough for the seat, addressed the Senate. Bentsen stated: "Mr. President, it was with great sorrow that I learned of the death this past weekend of my good friend George Brown. . . . Be it making democracy work, funding and shepherding the education of young minds, or encouraging the healing of the ill, George Brown was there. . . . I consider myself fortunate to be among those who counted George Brown as a friend and I am deeply saddened by his passing."[17] Lady Bird Johnson voiced similar emotions: "George Brown had one of the most creative, wide-ranging minds I ever knew. He retained a youthful search for knowledge and, quite simply, all of our family loved him."[18]

Brown left his estate to his wife, Alice, and his children. Alice quickly followed him in death on April 14, 1984, after suffering an aneurysm at the age of eighty-two. Her funeral service was held at the Church of St. John the Divine in Houston, as had been her husband's service the year before. She received posthumous recognition in the spring of 1986, when the Rice University board of governors established the Alice Pratt Brown Library, an art library annex containing seventy thousand volumes. In the 1990s, Rice honored George and Alice again with the naming of the Alice Pratt Brown Music Building and the George R. Brown Engineering Building.

Such recognition reflected one of the primary legacies of the Brown brothers and their families, the impact of their sustained support on the development of Rice University, Southwestern University, the Museum of Fine Arts, Houston, and numerous other institutions. In all, George Brown helped give away about $150 million in the years from the creation of the Brown Foundation in 1951 until his death in 1983.[19] These dollars influenced the lives of institutions and the individuals who passed through them, in the process enriching the life of the city of Houston and the state of Texas. Herman's much earlier death meant that his primary contribution to the foundation was financial. His capacity to build a strong, profitable company such as Brown & Root helped establish the endowment needed to run a successful foundation. George also contributed resources to the foundation, but in addition he had both the privilege and the responsibility of distributing these funds. His creative use of charitable contributions stands as one of his outstanding achievements.

A less obvious legacy of the Browns was the companies they built. Brown & Root came to symbolize the aggressive international construction companies that surged forward in the post–World War II boom years. The can-do spirit instilled by Herman Brown and the pride in engineering and salesmanship contributed by George defined the company's culture long after the death of the older brother and founder and the withdrawal from active management of the younger brother. The Texas Eastern Corporation took its place as one of the most prosperous and diversified of the handful of major gas transmission companies that rose to dominance after 1947. As at Brown & Root, Texas Eastern established a tradition of tackling ambitious projects and thinking big about potential opportunities, a trait best exemplified by the ambitious, if ill-fated, Houston Center project to develop thirty-three blocks of downtown Houston. Both of the Browns' primary companies enjoyed decades of sustained expansion from the 1940s throughout the 1980s, and their success marked the Browns as successful men in economic, political, and civic affairs.

Herman worked hard to push these companies forward, always appreciating the wealth that came to him only in his forties and fifties after decades of hard work on the roads of Central Texas. His death in 1962 prevented his enjoyment of the surge in size and profitability of both companies during the heady boom years in Houston in the 1960s, 1970s, and early 1980s. At the same time, he avoided the heartbreak he no doubt would have felt had he lived through the wave of government activism in the 1960s and 1970s. George lived just long enough to witness the oil- and gas-led growth of the decade after the energy crisis of 1973, when higher petroleum prices fueled an extraordinary boom in industries served by Brown & Root and Texas Eastern. He also died soon enough to be spared the wrenching experience of the oil bust of the mid-1980s, when plummeting oil prices devastated most oil- and gas-related companies, including Brown & Root and Texas Eastern.

Both companies took hard blows in the aftermath of the oil price collapse. The companies into which the Browns had poured their energies for decades did not survive this onslaught intact. A part of the lasting legacy of an entrepreneur is the company he or she creates, and the events of the 1980s diminished the size and the visibility of the Browns' companies, somewhat diminishing the memory of their achievements. Brown & Root survived as a subsidiary of Halliburton, but like many companies in oil- and gas-related businesses, it underwent a drastic restructuring—complete with massive layoffs and the reduction of the construction capacities that had long been one of the company's great strengths. Texas Eastern went through the ordeal of a much-publicized takeover battle before being absorbed in 1989 by Panhandle Eastern, where it became an important part of a diversified, national gas transmission company, PanEnergy, which itself was acquired by Duke Energy in the 1990s.[20]

But even if the once familiar names of the Browns' companies moved out of the public eye, they continued to perform vital services as subsidiaries of larger companies. The Browns had built companies that carried their names and reputations around the globe. But the actual projects of these companies, not the companies themselves, were the great business contributions of the Brown brothers. The roads and bridges, dams and military bases, offshore platforms and pipelines, gas transmission facilities and gathering systems remained in use, durable legacies of the lives of builders.

The Browns did not build the skyscrapers that defined the urban landscape of modern Houston, but they nonetheless left a legacy in the city itself. Indeed, as members of the 8-F crowd, the Browns are remembered for their power to shape the political and civic life of their adopted city. Although it seems clear that contemporaries and historians alike have exaggerated the cohesiveness and the power of those who gathered regularly in the Browns' suite at the Lamar Hotel, the image of the 8-F crowd's power grows stronger as the passing years dim the memories of the Browns and their friends. The symbolic "fall" of 8-F came in 1985, with the demolition of the Lamar Hotel. Ironically, one of the Brown brothers' primary legacies in the 1990s appears to be a recurring question: where are the strong business leaders of the past now that we really need them? This question arises as business and civic leaders in the 1990s grapple with a variety of difficult issues, from financing proposed new sports stadiums to revitalizing downtown Houston.

The difference between the 8-F crowd's era and the modern era is immense. Houston has grown much larger, and all segments—male and female—of its increasingly diverse population are exerting voices in the city's affairs. Business leaders of the current generation tend to be more corporate-oriented than those of the Browns' generation. Whereas many of the most prominent leaders of the 1990s owe their primary allegiance to national and international corporations, the Browns and their friends tended to be small-town Texans who migrated to Houston in search of their fortunes and felt a strong sense of commitment to the future of their city. They generally owned the companies they built, and by the time they reached their forties and fifties, they were in a position to devote much of their time and personal wealth to issues broader than the fate of their particular company.

In contrast, corporate leaders currently face much more demanding economic pressures in a era of increasing international competition and deregulation. They are transferred more often, and they seldom possess the kind of personal fortunes required to make a dramatic difference in the life of a city. In addition, deregulation and growing international competition in the 1990s means that current business leaders have to work harder and longer to compete successfully, making it difficult for them to devote the kind of time and energy given by the 8-F generation to political, civic,

The Brown family, Christmas, 1971. *Standing from left:* Mike Stude, Walter Negley, Toby O'Connor, Lisa Negley, Reagan McAllister, Leslie Negley McAllister, Ralph O'Connor, George O'Connor, Fayez Sarofim, Alfred Negley. *Seated from left:* Isabel Wilson with son Will Mathis, Maconda O'Connor, George Brown, Alice Brown, Nancy Negley, Louisa Sarofim with daughter Allison, Anita Stude with son Herman. *Seated in front:* John O'Connor, Isabel Stude, Nancy O'Connor, Travis Mathis, Christopher Sarofim.
Courtesy Brown Family Archives.

and cultural affairs. When they turn to politics, they encounter much stricter regulations on campaign contributions and much greater public scrutiny of activities. When they turn to civic and cultural pursuits, they often find that those who went before them had the capacity to create the cultural and civic institution needed to build a "major league city," while they face the less exciting, more daunting task of sustaining existing institutions.

That said, one legacy of the Browns' lives is a simple message that speaks across the decades to today's leaders. To make a difference, you must roll up your sleeves and commit both time and money to causes you consider worthy. The 8-F crowd acted on the assumption that what was good for their city and region ultimately was good for them. They shied away from neither commitment nor controversy in pursuing their vision of Houston as a major metropolis. Focusing on issues of importance to them, they helped build a thriving economy and a major league city. They had an obvious blind spot on the important issue of race relations, leaving for those who came after them the challenge of embracing and including the talents of the city's rich racial and ethnic diversity. In a fitting extension of 8-F's influence, the philanthropic foundations created by these leaders have taken a strong role in funding educational opportunities for all Houstonians.

The controversy over building a civic center in downtown Houston in the 1980s illustrated how civic leadership has changed since the 1950s. This center directly involved Texas Eastern, since it became a part of the company's overall strategy to continue developing Houston's east side after the grandiose Houston Center plans had been whittled down to a more manageable size by the city's economic trials. In the 8-F days, one might have envisioned a small group of business and political leaders meeting quietly, working out the project's details, and moving on their own toward its swift completion. But in the more turbulent days of the 1980s, the debate over the civic center erupted into a hard-fought public controversy, as different factions of an increasingly diverse business environment pushed different plans and locations for the proposed facility. After much public rancor, the east side site was confirmed and the new "George R. Brown Convention Center" was finally constructed.[21]

The officers and directors of Texas Eastern Corporation decided to honor Brown by commissioning a statue of him and placing it in a small public park near the George R. Brown Convention Center. With the approval of family members, Texas Eastern officials commissioned the statue from sculptor Wei Li "Willy" Wang. The sculpture was unveiled in 1988 to a small private audience including family and close friends.[22]

The statue is a fitting tribute in an appropriate location to a man who did much to shape modern Houston. At the back of George Brown's statue stands the civic center he never saw but of which he would have been quite proud. His eyes look out toward his former offices and the former headquarters of Texas Eastern. Off to the right lies the parking lot where once rose the Lamar Hotel. Several miles behind his right side—beyond freeways built by his company—sits the Clinton Road headquarters of Brown & Root. Farther off to his left are the lights of the stadium he helped complete on the Rice University campus. On up the road in River Oaks sit the houses where he and Herman lived side by side during their years together in Houston.

Indeed, all that is missing to make the setting complete is a statue of Herman standing beside his younger brother. The two, after all, spent a lifetime together as builders of giant construction projects, large corporations, circles of civic and political influence, universities, and philanthropic initiatives. They traveled the roads together, sharing equally in all enterprises, on an extraordinary journey from young boys in Belton, Texas, to fortune and a measure of fame as builders active around the world.

NOTES

PREFACE

1. Lady Bird Johnson, interviewed by Christopher J. Castaneda, Austin, Nov. 15, 1990. Herman and George R. Brown Archives, the Brown Foundation, Inc., of Houston, hereafter HGRBA. (Unless otherwise noted, all interviews are part of this collection.)

2. Ibid.

3. Useful sources on Brown & Root and Texas Eastern include the following: Christopher J. Castaneda and Joseph A. Pratt, *From Texas to the East: A Strategic History of Texas Eastern Corporation* (College Station: Texas A&M University Press, 1993); *Brownbuilder: 75th Anniversary Issue, 1919–1994* (Houston: Brown & Root, Inc., 1995); Joseph A. Pratt, Tyler Priest, and Christopher J. Castaneda, *Offshore Pioneers: Brown & Root and the History of Offshore Oil and Gas* (Houston: Gulf Publishing Company, 1997); and Jeffrey L. Rodengen, *The Legend of Halliburton* (Fort Lauderdale, Fla.: Write Stuff Syndicate, 1996).

4. "Roadbuilders with a Flair for Other Jobs," *Business Week*, May 25, 1957.

5. Biographies of Lyndon Johnson that discuss his relationship with the Browns include Ronnie Dugger, *The Politician: The Drive for Power from the Frontier to Master of the Senate* (New York: Norton, 1982); Robert Dalleck, *Lone Star Rising: Lyndon Johnson and His Times* (New York: Oxford University Press, 1991); and Robert Caro, *The Years of Lyndon Johnson: The Path to Power* (New York: Knopf, 1982) and *Means of Ascent* (New York: Knopf, 1990).

6. One notable exception is the excellent biography of Henry Kaiser by Mark Foster. See Mark S. Foster, *Henry J. Kaiser: Builder of the Modern American West* (Austin: University of Texas Press, 1989). Interesting, but less useful as history, is Laton McCartney, *Friends in High Places, The Bechtel Story: The Most Secret Corporation and How It Engineered the World* (New York: Simon and Schuster, 1988). Also see Stephen Adams, *Mr. Kaiser Goes to Washington: The Rise of a Government Entrepreneur* (Chapel Hill: University of North Carolina Press, 1997).

7. "Roadbuilders with a Flair," 91.

8. Several histories of Houston and Houston-based businesses provide useful background on the business elite of the city. These include Walter Buenger and Joseph Pratt, *But Also Good Business: Texas Commerce Banks and the Financing of Houston and Texas, 1886–1986* (College Station: Texas A&M University Press, 1986); Kenneth Lipartito and Joseph Pratt, *Baker & Botts in the Development of Modern Houston* (Austin: University of Texas Press, 1991); Fran Dressman, *Gus Wortham: Portrait of a Leader* (College Station: Texas A&M University Press, 1994); Bascom N. Timmons, *Jesse Jones: The Man and the Statesman* (New York: Holt and Company, 1956); and Patrick J. Nicholson, *Mr. Jim: The Biography of James Smithers Abercrombie* (Houston: Gulf, 1983). See also the more general treatments: George

Green, *The Establishment in Texas Politics: The Primitive Years, 1938–1957* (Westport, Conn.: Greenwood, 1979), and Joe R. Feagin, *Free Enterprise City: Houston in Political and Economic Perspective* (New Brunswick, N.J.: Rutgers University Press, 1988).

9. Castaneda and Pratt, *From Texas to the East.*

CHAPTER ONE

1. The best source of information on early Belton is George W. Tyler, *History of Bell County* (San Antonio: Naylor, 1936).

2. Ibid., 309–12, 381–89.

3. Robert Wiebe, *The Search for Order* (New York: Hill and Wang, 1967), uses this phrase to describe the communities scattered out over the nation's countryside before the coming of modern communication and transportation.

4. Tyler, *History of Bell County,* 171–73, 309.

5. Ibid., 305 and 321.

6. Rupert N. Richardson, Ernest Wallace, and Adrian Anderson, *Texas: The Lone Star State,* 5th ed. (Englewood Cliffs, N.J.: Prentice Hall, 1988), 169. Also see Walter L. Buenger and Robert A. Calvert, *Texas through Time: Evolving Interpretations* (College Station: Texas A&M University Press, 1991). Also see Billy Mac Jones, *The Search for Maturity* (Austin: Steck-Vaughn, 1965); and G. Terry Jordan, *German Seed in Texas Soil: Immigrant Farmers in Nineteenth Century Texas* (Austin: University of Texas Press, 1966).

7. Richardson, Wallace, and Anderson, *Texas,* 169.

8. Riney L. Brown, *Riney Brown's Diary* (Boerne, Tex.: Toepperwein, n.d.), Feb. 16, 1875.

9. Ibid., July 25, 1876.

10. Ibid., Dec. 31, 1876.

11. Ibid., June 6, 1877.

12. Ibid., July 27, 1977.

13. Ibid., 1. His choice of Belton instead of the German-speaking settlements to the south of there, such as New Braunfels, indicates that Riney Brown was more comfortable in the English-speaking areas.

14. Seymore V. Connor, *Texas: A History* (New York: Cowell, 1971), 268–69.

15. Louise Brown Head, "History of the Brown and King Families," unpublished manuscript, Herman and George R. Brown Archives, to be housed at the Woodson Research Center, Rice University, Houston. Hereafter cited as HGRBA.

16. *Belton: Illustrated by the Journal Reporter, 1900,* Centennial Issue.

17. The expansion of Temple after the coming of the railroads is discussed in Tyler's *History of Bell County.* See 312–48.

18. Brown, *Diary,* foreword.

19. Ibid.

20. Ibid.

21. "The Brown Family," unpublished manuscript, HGRBA.

22. Temple High School Yearbook, Seniors, 1911, HGRBA.

23. "Meet George and Herman Brown," *Houston Post,* Dec. 31, 1942.

24. Tyler, *History of Bell County,* 346.

25. *The Belton Journal,* Centennial Edition, HGRBA.

26. Moorehead manuscript (undated version), HGRBA. This is a history of Brown & Root written and revised but never published.

27. Lady Bird Johnson interview.

28. Nancy Wellin, interviewed by Christopher J. Castaneda, New York City, Nov. 8, 1990.

29. "The Cotton Blossom, 1916," Temple High School. Copy in "High School" folder, HGRBA.

30. Fredericka Meiners, *A History of Rice University: The Institute Years, 1907–1963* (Houston: Rice University Press, 1982); Buenger and Pratt, *But Also Good Business,* 51, 82; Lipartito and Pratt, *Baker & Botts,* 53–61.

31. "George R. Brown," *Rice University Review* 3, no. 2 (Fall/Winter, 1968): 2–8.

32. H. Malcolm Lovett, interviewed by Christopher J. Castaneda, Houston, Sept. 25, 1990.

33. Hugh C. Welsh to the *Mirror,* Oct. 4, 1916; copy in HGRBA.

34. *Thresher,* May 4, 1977, Woodson Research Center (WRC), Rice University (RU).

35. Lovett interview.

36. Ibid.

37. Moorehead manuscript, undated, 23.

38. Lute Parkinson to Mr. G. Rufus Brown, E.M., Esq. Apr. 18, 1922, copy in HGRBA.

39. Colorado School of Mines, *Prospector 1922.*

40. *Congressional Record,* vol. 129, no. 3, Jan. 25, 1983, Eulogy of George R. Brown By Senator Lloyd Bentsen.

CHAPTER TWO

1. Newton Fuessle, "Pulling Main Street Out of the Mud," *Outlook,* Aug. 16, 1922, 640.

2. John D. Huddleston, "Good Roads for Texas: A History of the Texas Highway Department, 1917–1947" (Ph.D. diss., Texas A&M University, 1981), 19.

3. Ibid., 22.

4. Tyler, *History of Bell County,* 330.

5. Huddleston, "Good Roads," 24.

6. Ibid., 25.

7. Ibid., 26.

8. Ibid., 58.

9. Raymond Klempin, "The Odyssey of Brown & Root: How Skill, Luck, and Politics Built an Engineering Empire," part 1, *Houston Business Journal,* Aug. 30, 1982, 16.

10. Ibid.

11. Dan Dyess, "In the Beginning . . . ," unpublished reminiscences, 1–2, HGRBA.

12. Brown Booth, interviewed by Christopher J. Castaneda, Timpson, Tex., July 26, 1990.

13. Howard Counts, interviewed by Christopher J. Castaneda, Marble Falls, Tex., Oct. 9, 1990.

14. Dyess, "In the Beginning," 2.

15. Brown & Root, Inc., *Scoreboard* (Nov., 1962): 2.

16. Booth interview.

17. Moorehead manuscript, undated, 33.

18. Ibid.

19. Ibid., 32.

20. Ibid., 34.

21. Herman Brown, "Outline of Talk by Mr. Herman Brown to McGraw-Hill Editors' Meeting" (Shamrock Hotel, Houston, Nov. 21, 1952), 2. HGRBA.

22. Ibid.

23. Harry W. Nolan to Herman Brown, Aug. 8, 1961, "Brown, Herman: Personal Correspondence, 1961" folder, George R. Brown Records (GRBR), Houston. (The George R. Brown Records are to be housed in HGRBA. They consist of twenty storage boxes filled with folders culled from a much larger collection of records that were destroyed. They were stored at Brown & Root.)

24. Louisa Sarofim, interviewed by Christopher J. Castaneda, Houston, Jan. 15, 1991.

25. Moorehead manuscript, undated, 7.

26. Ibid., 10–21.

27. "Today: 25 Years Ago," *Taylor Daily Press,* n.d.

28. Klempin, "Odyssey of Brown & Root," part 1. Dyess, "In the Beginning," 6–7.

29. *Rice Review.*

30. Newspaper article, no date or title, ad announcement, HGRBA.

31. Herman Brown to George R. Brown (GRB), May 22, 1922, Folder: "Bio: GRB: Roadbuilding," HGRBA.

32. Moorehead manuscript, Apr., 1968, HGRBA, 24.

33. Dyess, "In the Beginning," 7.

34. Brown, "Outline of Talk," 2.

35. Moorehead manuscript, Apr., 1968, 41.

36. *Engineering News-Record,* Jan., 1926, 37.

37. *Wharton Spectator,* Jan. 29, 1926.

38. *Wharton Spectator,* Jan. 15, 1926.

39. *Wharton Spectator,* Jan. 29, 1926.

40. "Funeral Notice" of Minot Tully Pratt, *Mirror,* Temple, Tex., June 3, 1908, HGRBA.

41. Isabel Wilson, interviewed by Joseph A. Pratt, Houston, Aug. 16, 1984.

42. Frank "Posh" Oltorf, interviewed by Christopher J. Castaneda, Marlin, Tex, July 18, 1990.

43. Huddleston, "Good Roads," iv.

44. Ibid., 57, 76.

45. Dyess, "In the Beginning," 9.

46. See Frank M. Stewart, "Highway Administration in Texas: A Study of Administrative Methods and Financial Policies," *University of Texas Bulletin,* June 15, 1934.

47. Thomas MacDonald to Losh, Jan. 26, 1924, MacDonald Papers, Box 19, File 014, Texas A&M Archives.

48. Brown Booth to Christopher J. Castaneda, n.d., HGRBA.

49. "Brief for Appellants," Herman Brown, et al., Appellants, v. O. L. Neylands, Appellee, Court of Civil Appeals for the Third Supreme Judicial District of Texas at Austin, No. 9026, pp. 198–200, "Brief-(1940?) Herman Brown v. O. L. Neyland—Court of Civil Appeals" folder, GRBR.

50. See David McComb, *Houston: The Bayou City* (Austin: University of Texas Press, 1969).

51. Marilyn Sibley, *The Port of Houston* (Austin: University of Texas Press, 1968).

52. Marguerite Johnston, *Houston: The Unknown City, 1836–1946* (College Station: Texas A&M University Press, 1991).

53. Moorehead manuscript, Apr., 1968, 42

54. See River Oaks papers at the Houston Metropolitan Research Center.

55. Mrs. L. T. Bolin, interviewed by Christopher J. Castaneda, Houston, Aug. 30, 1990.

56. See Harris County Engineering Department, Houston, Boxes #1055 and #1053, Market Street Road Work.

57. Edgar W. Monteith, interviewed by Christopher J. Castaneda, Houston, Oct. 2, 1990.

58. See Timmons, *Jesse H. Jones,* for a description of R. M. Farrar.

59. Buenger and Pratt, *But Also Good Business,* 99–100.

60. Dugger, *Politician,* 448n 2.

61. Moorehead manuscript, Apr., 1968, 42–43.

62. Lady Bird Johnson interview.

63. Wellin interview.

64. Harry Austin, interviewed by Christopher J. Castaneda, Houston, Sept. 24, 1990.

CHAPTER THREE

1. "B&R and Associated Companies Employment Data, 1929–1979," HGRBA. This chart lists an estimated 500 workers in 1929, 545 in 1930, and 838 in 1935.

2. There is little formal history of the Public Works Administration or the Works Progress Administration in Texas. These two agencies had a profound impact on the state and the construction industry in the 1930s. Some of the story of "jobs programs" in the state, especially as they related to Lyndon Johnson, is told in Betty Lindley and Ernest Lindley, *A New Deal for Youth: The Story of the National Youth Administration* (New York, 1938). Also, Deborah Lynn Self, "The National Youth Administration in Texas" (M.A. thesis, Texas Tech University, 1974).

3. Mike Stude, interviewed by Christopher J. Castaneda, Houston, Oct. 31, 1990.

4. Nat Terence, "Does George Brown Run Houston," *Houstonian,* Mar. 1, 1941.

5. Ibid.

6. "Outline of Talk by Mr. Herman Brown," Nov. 23, 1952, Houston, HGRBA.

7. Lewis B. Walker, "Colorado-Harnessing Dreams Failed Three Times Before," *Austin American-Statesman,* July 7, 1946, 11–12. For a recent history of the events of the 1930s, see, John A. Adams, Jr., *Damming the Colorado: The Rise of the Lower Colorado River Authority, 1933–1939* (College Station: Texas A&M University Press, 1990).

8. Emmette S. Redford, ed., *Public Administration and Policy Formation: Studies in Oil, Gas, Banking, River Development, and Corporate Investigations* (New York: Greenwood, 1969), 201–204. Walker, "Colorado-Harnessing," 11.

9. Redford, *Public Administration,* 217–20.

10. James H. Banks and John E. Babcock, *Corralling the Colorado: The First Fifty Years of the Lower Colorado River Authority* (Austin: Eakin, 1988), 81–82.

11. Dugger, *Politician,* 275.

12. U.S. 50 Stat. 850.

13. For a review of the Thomas Valley Authority project see Thomas K. McCraw, *TVA and the Power Fight, 1933–1939* (Philadelphia: Lippincott, 1971).

14. Donald R. Whitnah, *Government Agencies* (Westport, Conn.: Greenwood, 1983), 36–40.

15. Moorehead manuscript, undated, 12.

16. Dugger, *Politician,* 448*n* 4.

17. Banks and Babcock, *Corralling the Colorado,* 95.

18. Moorehead manuscript, Apr., 1968, 47; *Roadbuilders with a Flair,* 100.

19. Redford, *Public Administration,* 212; Mary Rather to Lyndon B. Johnson (LBJ), Dec. 29, 1939, "Wirtz, A. J., 1939" folder, Box 36, Lyndon B. Johnson Archives Selected Names File (SN), Lyndon B. Johnson Library (LBJL), Austin; Dalleck, *Lone Star Rising,* 174–75.

20. Oltorf interview, July 18, 1990.

21. Verman Greer to C. McDonough, LBJL, 8–23.

22. Banks and Babcock, *Corralling the Colorado,* 95. See also Klempin, "Odyssey of Brown & Root," part 1, p. 4.

23. Caro, *Path to Power,* 379–80.

24. Ibid.

25. Ibid., 380–82.

26. Ibid., 383.

27. Lady Bird Johnson interview.

28. Banks, *Corralling the Colorado,* 101.

29. Dalleck, *Lone Star Rising,* 176; Caro, *Path to Power,* 461.

30. H. P. Bunger to C. E. Blakeway, July 13, 1937, "Brown, Herman," Box 13, SN, LBJL.

31. C. E. Blakeway to H. P. Bunger, July 26, 1937, "Brown, Herman," Box 13, SN, LBJL.

32. Herman Brown to LBJ, Aug. 3, 1937, "Brown, Herman," Box 13, SN, LBJL.

33. GRB to LBJ, Nov. 29, 1937, "Brown, Herman," Box 12, SN, LBJL.

34. A. J. McKenzie to LBJ, Jan. 24, 1938, "Brown, Herman," Box 13, SN, LBJL.

35. Herman Brown to LBJ, Jan. 6, 1939, "Brown, Herman," Box 13, SN, LBJL.

36. Redford, *Public Administration,* 221.

37. Alvin Wirtz, Undersecretary, U.S. Department of the Interior, memo concerning Marshall Ford Dam, "Wirtz–Marshall Ford Dam," Box 12, Alvin Wirtz Papers, LBJL.

38. Caro, *Path to Power,* 463–64; Banks and Babcock, *Corralling the Colorado,* 149.

39. Caro, *Path to Power,* 464; Banks and Babcock, *Corralling the Colorado,* 148.

40. Banks and Babcock, *Corralling the Colorado,* 149.

41. GRB to J. J. Mansfield, May 1, 1939, "Brown, George—2 of 2," Box 12, SN, LBJL.

42. LBJ to GRB, Aug. 11, 1939, "Brown, George," Box 12, SN, LBJL.

43. GRB to LBJ, Oct. 27, 1939, "Brown, George," Box 12, SN, LBJL.

44. Herman Brown to LBJ, Aug. 3, 1937, "Brown, George," Box 12, SN, LBJL.

45. Ross White to A. E. Moritz, Jan. 11, 1940, "LCRA Labor Contracts," Box 9, Alvin Wirtz Papers, LBJL.

46. LBJ to Harry Acreman, Jan. 27, 1940, "Marshall Ford Appropriation, 1940," Box 12, Alvin Wirtz Papers, LBJL.

47. C. McDonough to A. J. Wirtz, Jan. 29, 1940, "Wirtz–Marshall Ford Dam," Alvin Wirtz Papers, LBJL.

48. Alvin Wirtz to Harry Acreman, Feb. 3, 1940, "Marshall Ford Appropriation—1940," Box 12, Alvin Wirtz Papers, LBJL.

49. Ross White, "Marshall Ford Dam Involves Unique Construction Features," *Slide Rule* 2, no. 12 (Sept., 1944): 9–11. HGRBA.

50. Walker, "Colorado-Harnessing," 11.

51. Redford, *Public Administration,* 226.

52. Ibid., 220.

53. Banks and Babcock, *Corralling the Colorado,* 130.

54. Ibid.

55. Herman Brown to J. E. Van Hoose, Oct. 11, 1939, "Brown, Herman," Box 13, SN, LBJL.

56. Herman Brown to R. J. Beamish, Apr. 11, 1939, "Brown, Herman," Box 13, SN, LBJL.

57. Herman Brown to LBJ, Apr. 11, 1939, "Brown, Herman," Box 13, SN, LBJL.

58. See McCartney, *Friends in High Places.*

59. The most useful works, in order of publication, are Dugger, *Politician;* Caro, *Path to Power* and *Means of Ascent;* and Dallek, *Lone Star Rising.*

60. Lady Bird Johnson interview.

61. Ibid.

62. Sarofim interview.

63. Lady Bird Johnson interview.

64. Lyndon and Lady Bird Johnson to Herman Brown, Jan. 23, 1960, "Brown, Herman, 1960," Box 24, Senate Papers, LBJL.

65. Lady Bird Johnson interview.

66. Margarett Brown to Alice Brown, Aug., 1936, HGRBA.

CHAPTER FOUR

1. LBJ to GRB, Apr. 28, 1939, LBJL. Also see Caro, *Path to Power,* 581–82.

2. LBJ to GRB, May 16, 1939, LBJL.

3. In *Path to Power,* Robert Caro asserts that Thomas "was known to take Herman's orders unquestioningly" (583). This characterization demeans Thomas while also slighting the broader political context within which he functioned. Alvin Wirtz, close to Johnson and an attorney for Brown & Root, achieved a high-level status in Washington in early January, 1940, when Roosevelt appointed Wirtz as undersecretary of the interior. Another Texan, Everett L. Looney, became assistant U.S. attorney general. Looney, an associate of fellow Austin attorney Edward Clark, former Texas secretary of state, worked for Brown & Root on retainer. Thus, the interests of Texas were well represented on Washington's political scene.

4. Caro, *Path to Power,* 581–84.

5. W. E. Hilbert to W. L. Childs, Sept. 20, 1938, Box 2, "A," Albert Thomas Papers, WRC, RU. Also, see, *Corpus Christi Caller,* Sept. 26, 1938.

6. Pamphlet by Houston Chamber of Commerce, copy in Box 7, Albert Thomas Papers, WRC, RU.

7. Albert Thomas to Admiral A. J. Hepburn, Sept. 13, 1938; W. E. Hilbert to Albert Thomas, Sept. 20, 1938; "Miscellaneous Navy Department, 1937–40: Folder," Box 2, Albert Thomas Papers, WRC, RU.

8. Year Book, Corpus Christi Naval Air Station, 1941.

9. GRB to Roy Miller, May 1, 1939, LBJL.

10. GRB to LBJ, May 2, 1939, LBJL.

11. LBJ to GRB, May 4, 1939, LBJL.

12. *Houston Chronicle,* Jan. 24, 1940.

13. Albert Thomas to Charles Edison—Attn. Lewis Compton, May 16, 1940, "A," Box 2, Albert Thomas Papers, WRC, RU.

14. Lewis Compton to Albert Thomas, May 22, 1940, "A," Box 2, Albert Thomas Papers, WRC, RU.

15. Lewis Compton, "Memorandum for the President," Apr. 23, 1940; "A," Box 7, Albert Thomas Papers, WRC, RU. For related articles also see *Corpus Christi Caller,* Jan. 16, 1939; Jan. 10, 1939; Jan. 11, 1940; Jan. 24, 1940.

16. Caro, *Path to Power,* 584.

17. See *Corpus Christi Caller,* June 2, 1940; May 24, 1940; May 31, 1940; May 31, 1940; May 16, 1940; May 17, 1940; and May 23, 1940.

18. *Corpus Christi Caller,* June 7, 1940.

19. George R. Brown, interviewed by David G. McComb, Aug. 8, 1969, LBJL. *Corpus Christi Caller,* June 5, 1940; June 7, 1940; June 11, 1940; June 13, 1940; June 14, 1930; and Jan. 10, 1943.

20. George R. Brown, interviewed by David G. McComb, Aug. 6, 1969, LBJL.

21. *Corpus Christi Caller,* Feb. 27, 1941, and Feb. 14, 1941. See, also, Coleman McCampbell, *Texas Seaport: The Story of the Growth of the Coastal Bend Area* (New York: Exposition, 1952).

22. Moorehead manuscript, Apr., 1968, 56.

23. Foster, *Kaiser.*

24. Caro, *Path to Power,* 585, and Moorehead manuscript, Apr., 1968, 58.

25. A. J. Wirtz to GRB, Nov. 7, 1940, Personal Correspondence, Box 3, LBJL.

26. "In the Matter of Brown & Root, Inc., et al. and James F. Hoffman and Paul J. Bristol," *Decisions and Orders of the National Labor Relations Board,* vol. 51 (Washington, D.C.: GPO, 1944), 821. Hereafter cited as 51 NLRB 821.

27. McCampbell, *Texas Seaport.* Also, see, *Corpus Christi Caller,* Sept. 24, 1939; Feb. 14, 1941; Feb. 27, 1941; and Aug. 7, 1943.

28. "Navy Will Commission Air Training Station March 12," *Corpus Christi Caller,* Jan. 26, 1941.

29. *Corpus Christi Caller,* Jan. 10, 1943.

30. "Efficiency, Speed Win Army-Navy 'E' for Contractors—Naval Air Station was ready for Operation Eight Months after Start of Construction Despite Many Varying Problems," *Corpus Christi Caller,* Jan. 10, 1943.

31. "B," Defense Projects, 1941–42, Box 1, Albert Thomas Papers, WRC, RU.

32. Harry Austin, interviewed by Christopher J. Castaneda, Houston, Jan. 15, 1992.

33. Brown interview, Aug. 6, 1969.

34. Austin interview, Jan. 15, 1992.

35. "Shipyard Where Formerly the Buffalo Roamed," *Marine News,* Oct., 1943.

36. *The History of the Brown Shipbuilding Co., Inc.,* n.d., copy in Brown & Root, Inc., Archives (B&RA), Houston.

37. "Overcoming Scarcities: History of Brown Shipbuilding: No. 2," *Brown Victory Dispatch,* n.d., B&RA.

38. Austin interview, Jan. 15, 1992.

39. *Brownbuilder,* 13.

40. Ibid., 13.

41. "Platzer Shipuilding Co. to Build Subchasers," *Houston,* Oct., 1941. Also see Austin interview, Jan. 15, 1992.

42. "Wading to Work in Mud: History of Brown: No. 3," *Brown Victory Dispatch,* Nov. 9, 1945.

43. Ibid.

44. "Platzer Shipbuilding Co."

45. "No. 1 Industry," *Houston,* Apr., 1943.

46. "Full Crews Stay on Job, Pass Pickets, Build Ships," *Brown Victory Dispatch,* June 9, 1945.

47. For an interesting account of life at a wartime shipyard, see Katherine Archibald, *Wartime Shipyard* (Berkeley and Los Angeles: University of California Press, 1947).

48. Austin interview, Jan. 15, 1992.

49. Roy Grimes, "First of 8 Sub Chasers Launched Here," *Houston Post,* Feb. 28, 1942, and "Launching at Brown Shipbuilding Makes One More Confident of Victory," *Houston Post,* Feb. 28, 1942.

50. Grimes, "First of 8 Sub Chasers." *Brown Victory Dispatch,* Dec. 7, 1945.

51. Ed Kilman, "Third Subchaser Skids Down Way at Brown Shipyards," *Houston Post,* Apr. 12, 1942.

52. Ibid.

53. Ibid.

54. A. D. Baker III, *U.S. Naval Vessels, 1943* (Washington, D.C.: Naval Institute Press, 1943).

55. The two yards were consolidated in September, 1945.

56. "Records Fell Each Month," *Brown Victory Dispatch,* Nov. 16, 1945.

57. "Records Fell." Also see "Brown Shipyards Awarded Coveted Army-Navy 'E' Pennant," *Houston,* Jan., 1943. Also see "Bell County Brown Brothers Launch Seven Ships, Win 'E,' *Temple Daily Telegram,* Dec. 22, 1942. Also see "Seven Ships Launched and Navy 'E' Awarded at Brown," *Houston Chronicle,* Dec. 22, 1941; "Meet George and Herman Brown"; "Knox Lauds City Shipbuilders, 31,000 See Brown 'E' Ceremony," *Houston Press,* Dec. 21, 1942.

58. "Brown Shipyards to be Awarded Army-Navy 'E,'" *Houston Press,* Dec. 15, 1942.

59. "Brown Shipyard Awarded Army-Navy 'E' for Outstanding Work," *Houston Post,* Dec. 16, 1942.

60. "Texas Wonder Boys," *Time,* Jan. 11, 1943, 76. "Brown Yards Send 7 Ships Down Ways," *Houston Post,* Dec. 22, 1942.

61. "Brownship wins Star for 'E' Flag," *Brown Victory Dispatch,* July 23, 1943.

62. Ibid.

63. Brownship launched its twenty-fourth DE on schedule, and its fifty-fifth DE was delivered on January 1, 1944, seventy days early.

64. "Brown Gets Contract for 32 Destroyer Escorts," *Houston,* July, 1943.

65. *Brown Victory Dispatch,* Oct. 2, 1943. Also see *Houston Press* of late September and early October, 1943; "Brown to Build New Type of Naval Landing Craft," *Houston,* Oct., 1943; "Brown to Build Tank Landing Craft," *Brown Victory Dispatch,* Sept. 18, 1943.

66. "Urgent Need of Rocket Ships Causes 58-Hour Week at Brown," *Brown Victory*

Dispatch, June 16, 1945. Also see "1945 at Brown was Great Year," *Brown Victory Dispatch,* Jan. 4, 1946.

67. "Our Record to Date: 327 Warships in Action in All Parts of World," *Brown Victory Dispatch,* July 21, 1945.

68. "Your Future at Brown," *Brown Victory Dispatch,* n.d., copy in B&RA.

69. Albert Thomas to H. Harris Robson, Dec. 12, 1941, "Miscellaneous: War Production Board, 1941–43," Box 3, Albert Thomas Papers, WRC, RU.

70. George R. Brown to Harry D. Knowlton, Mar. 7, 1942, Misc. Defense Housing, 1941–43 "C," Box 1, Albert Thomas Papers, WRC, RU; G. T. Korink to Lt. Hudson, "Miscellaneous Defense Housing, 1941–43," Box 1, Albert Thomas Papers, WRC, RU.

71. "Labor Market Survey Report," War Manpower Commission, Feb. 1, 1943. AT papers, defense housing, 1942–44, Box 1. WRC, RU.

72. "Slugger or Slacker," *Brown Victory Dispatch,* Sept. 26, 1942.

73. "Labor Market Survey Report."

74. LBJ to GRB, Oct. 8, 1942, LBJL.

75. Robert M. McKinney, interviewed by Christopher J. Castaneda, Middleburg, Va., Mar. 9, 1991.

76. "Union Hasn't Played Ball According to Rules of Game; Attempted to Stop Vital War Job, Put Many Out of Job," *Brown Victory Dispatch,* June 9, 1945.

77. 57 NLRB 326–29. Also see 57 NLRB 1082 and 58 NLRB 138–43.

78. 59 NLRB 1352–53; 60 NLRB 196–99; 60 NLRB 733–35; 66 NLRB 978–82.

79. Labor folder citation of Herman Brown statement on unions.

80. "Union Hasn't Played Ball According to Rules of Game; Attempted to Stop Vital War Job, Put Many Out of Job," *Brown Victory Dispatch,* June 9, 1945.

81. Ibid., and "Brown Presses for Pay Raises for All Workers at Shipyard," *Brown Victory Dispatch,* Sept. 21, 1945.

82. Moorehead manuscript, Apr., 1968, 68.

83. Brown, "Outline of Talk," 2.

84. Ibid., 3.

85. "5 DEs from Brown Cited by President," *Brown Victory Dispatch,* May 12, 1945.

86. "Need by Navy Is Great," *Brown Victory Dispatch,* June 30, 1945.

87. "Samuel B. Roberts, Built at Brownship Meets Death of Hero in Philippines," *Brown Victory Dispatch,* Nov. 25, 1944.

88. "Brownship DE Rode Out Storm that Sank Three Destroyers," *Brown Victory Dispatch,* Feb. 10, 1945.

89. "Brownship PF Wins Last Atlantic Battle," *Brown Victory Dispatch,* June 2, 1945.

90. Sarofim interview.

91. "Brown Among 42 of 123 Wartime Yards to Be Retained," *Brown Victory Dispatch,* Feb. 8, 1946.

92. "Navy Lists Brownship as One of Its 'Post-War Activities,'" *Brown Victory Dispatch,* Sept. 28, 1945.

93. "They're Only One of Type Built," *Brown Victory Dispatch,* Apr. 12, 1946.

94. George J. Darneille, interviewed by Christopher J. Castaneda, Houston, Jan. 17, 1991.

95. *Brown Victory Dispatch,* Feb. 15, 1946. Also see "Brown Shipyards to Be Active in City's Future," *Houston,* Mar., 1946.

96. C. Bradford Mitchell, *Every Kind of Shipwork: A History of Todd Shipyards Corporation, 1916–1981* (New York, 1981), 175.

CHAPTER FIVE

1. See Jesse H. Jones, *Fifty Billion Dollars* (New York: MacMillan, 1951), 7.

2. 58 NLRB 293–99.

3. Walter Prescott Webb, *The Handbook of Texas,* vol. 1 (Austin: Texas State Historical Association, 1952), 622. Also, see, Don Hinga, "Heroes Lived Here," *Texas Parade,* Sept., 1949, and Caleb Pirtle III and Michael F. Cusak, *The Lonely Sentinel: Fort Clark on Texas's Western Frontier* (Austin: Eakin, 1985).

4. Pirtle and Cusak, *Lonely Sentinel,* 171.

5. Sarofim interview.

6. Darneille interview.

7. Ibid.

8. Ibid.

9. Ibid.

10. Ibid. Moorehead manuscript, Apr., 1968.

11. Darneille interview.

12. Moorehead manuscript, Apr., 1968.

13. Darneille interview.

14. Harry L. Tower to Curtis B. Dall, Dec. 8, 1942, Tenneco History Collection, Houston.

15. Clyde Alexander, interviewed by Alan Dabney, July 3, 1962, Tenneco History Collection, Houston.

16. Frank H. Love, "Construction Features of the Tennessee Gas and Transmission Company Pipe Line," *Petroleum Engineer* 16, no. 2 (Nov., 1944): 121–44. Also see Tenneco, Inc., "Tenneco's First 35 Years," (company publication, 1978).

17. "U-Boats Blast 2 More Tankers; Only Three Survivors Located," *Houston Post,* Feb.27, 1942.

18. Dana Blankenhorn, "The Brown Brothers: From Mules to Millions as Houston's Contracting and Energy Giants" *Houston Business Journal,* Mar. 19, 1979.

19. Christopher J. Castaneda, *Regulated Enterprise: Natural Gas Pipelines and Northeastern Markets, 1938–1954* (Columbus: Ohio State University Press, 1993), 66–93.

20. HR PR 1945, 3. See, also, Jones, *Fifty Billion Dollars,* 402.

21. U.S. Congress, *Special Committee Investigating Petroleum Resources,* "Hearings on War Emergency Pipe-Line Systems and Other Petroleum Facilities," Nov. 15, 16, and 17, 1945.

22. See Castaneda and Pratt, *From Texas to the East.* These events also are covered in Harold M. Hyman, *Craftsmanship and Character: A History of the Vinson and Elkins Law Firm of Houston, 1917–1997* (Athens: University of Georgia Press, 1998).

23. James W. Hargrove, interviewed by Christopher J. Castaneda and Louis Marchiafava, Mar. 31, 1988, Texas Eastern (TE) History Collection, WRC, RU. Also, Blankenhorn, "Brown Brothers," 1–4. Also, Castaneda and Pratt, *From Texas to the East.*

24. Charles I. Francis to Judge J. A. Elkins, June 14, 1946, TE History Collection, WRC, RU.

25. Ibid.

26. After Texas Eastern's successful bid, George Allen received substantial founder's stock from Frank Andrews, an original stockholder. See Castaneda and Pratt, *From Texas to the East.*

27. Oltorf interview, July 18, 1990.

28. War Assets Administration, *Government-Owned Pipelines: Report of the War Assets Administration to the Congress* (Washington, D.C.: GPO, 1946), 4, copy in TE History Collection, WRC, RU.

29. War Assets Administration, *Transcript of Proceedings: Proposals to Buy or Lease the Big and Little Big Inch Pipe Lines* (Washington, D.C.: GPO, 1946).

30. W. H. Leslie to GRB, Aug. 15, 1946, TE History Collection, WRC, RU.

31. Ibid.

32. GRB to Albert Thomas, Aug. 16, 1946; "Proposed Purchase or Lease of Big and Little Big Inch Lines—E. Holley Poe & Associates," Box 8, TE History Collection, WRC, RU.

33. "Former Big Names in Government Are Involved in Bids for Pipelines," *Washington Daily News,* Oct. 17, 1946.

34. Harold L. Ickes, "'Uncle Jesse's' Bid for Oil Pipelines Seen Loaded Two Ways for Monopoly," *Evening Star,* Oct. 2, 1946.

35. Harold I. Ickes to Charles I. Francis, Oct. 18, 1946, TE History Collection, WRC, RU. Four years after this furor, George Butler joined the board of Texas Eastern.

36. Marshall McNeil to GRB, Oct. 3, 1946; "Proposed purchase or lease of Big and Little Big Inch Lines—E. Holley Poe & Associates," Box 8, TE History Collection, WRC, RU.

37. War Assets Administration, Press Release, Oct. 3, 1946. TE History Collection, WRC, RU.

38. E. Holley Poe to L. Gray Marshall, Oct. 7, 1946, TE History Collection, WRC, RU.

39. See Elliot Taylor, "Thermally Thinking," *Gas,* Nov., 1946, 21–22. Also see Marshall McNeil "Former Big Names in Government Are Involved in Bids for Pipelines," *Washington Daily News,* Oct. 17, 1946.

40. Harold L. Ickes, "WAA Administrator Urged to Sell 'Inch' Pipe Lines Now, and for Cash" *Evening Star,* Oct. 28, 1946; Harold L. Ickes, "WAA Seen Yielding to Machinations of John L. Lewis on Natural Gas" *Evening Star,* Oct. 30, 1946; and "The New Lewis Threat," *News,* Oct. 24, 1946.

41. War Assets Administration, *Government-Owned Pipelines,* 7.

42. House, *Report of the Select Committee to Investigate Disposition of Surplus Property,* House Resolution 385, 79th Congress, 2nd Session, 1947, 14.

43. R. G. Rice to the Department of the Interior, Nov. 29, 1946; Tenneco History Collection, Houston.

44. War Assets Administration to Tennessee Gas in reference to PLE-1, n.d. (the company received the letter on December 3, 1946), Tenneco History Collection, Houston.

45. Charles Francis to General John J. O'Brien, Aug. 10, 1946, Charles I. Francis Correspondence File, TE Archives, WRC, RU.

46. Charles Francis to Jack Clarke, Feb. 24, 1955, Charles I. Francis Correspondence File, TE Archives, WRC, RU. Also see *Platt's Oilgram News,* Nov. 14, 1946.

47. Charles Francis to Robert Littlejohn, Dec. 16, 1946, Charles I. Francis Correspondence File, TE Archives, WRC, RU.

48. J. Ross Gamble, memo, Dec. 24, 1946, Folder #6, Box 70074-01, TE Archives, WRC, RU.

49. *New York Times,* Dec. 27, 1946, 10; *New York Times,* Dec. 28, 1946, 1.

50. War Assets Administration, *Government-Owned Pipelines,* 9.

51. August Belmont, interviewed by Christopher J. Castaneda and Joseph A. Pratt, June 14, 1988, TE History Collection, WRC, RU. See, also, Robert Sobel, *The Life and Times of Dillon Read* (New York: Truman Talley, 1991), 229–36.

52. Sobel, *Dillon Read,* 231.

53. ADG worksheet, Dec. 17, 1946.

54. Ibid. August Belmont to Christopher J. Castaneda, Feb. 6, 1989, copy in TE History Collection, WRC, RU.

55. Winsor H. Watson, *History of the Texas Eastern Transmission Corporation,* n.d., History Folder, Historian's Files, TE History Collection, WRC, RU. Also, see, Hargrove interview.

56. Belmont interview.

57. War Assets Administration, *Transcript of Proceedings: Proposals for Purchase of War Emergency Pipe Lines Commonly Known as Big and Little Big Inch Pipe Lines* (Washington, D.C.: GPO, 1947).

58. Booth interview.

59. Dugger, *Politician,* 282.

60. W. H. Leslie to George R. Brown, Feb. 10, 1947. TE History Collection. WRC, RU.

61. Belmont interview.

62. August Belmont to Chris Castaneda, Mar. 21, 1990; also, Clark Clifford to Chris Castaneda, July 5, 1991, letters in Castaneda's collection. Clifford stated that he no longer remembered events associated with the Inch Lines.

63. George E. Allen, *Presidents Who Have Known Me* (New York: Simon and Schuster, 1950), 197, 216–17.

64. "Allen Tells Cost of RFC Post to Him," *New York Times,* Feb. 8, 1946.

65. Charles Francis to Judge J. A. Elkins, June 14, 1946, TE History Collection, WRU, RU.

66. John W. Wilcker, "Fair Profit?" *Harvard Business Review* 26, no. 2 (Mar., 1948), 207.

67. Belmont interview.

68. W. H. Leslie to George R. Brown, n.d., "Texas Eastern Transmission Corp.—Second Bid," TE History Collection, WRC, RU.

69. Robert M. Poe, interviewed by Christopher J. Castaneda and John King, June 20, 1988, TE History Collection, WRC, RU.

70. Hargrove interview.

71. Oltorf interview, July 18, 1990.

72. Ibid. Jack Bixby, interviewed by Christopher J. Castaneda, Dec. 5, 1990, TE History Collection, WRC, RU.

73. Federal Power Commission, *Opinions and Decisions of the Federal Power Commission,* vol. 6. (Washington, D.C.: GPO, 1949), 148–75.

74. Testimony of Baxter Goodrich during FPC hearings on Docket G-880, certification of Texas Eastern Transmission Corporation, TE History Collection, WRC, RU.

75. Sobel, *Dillon Read,* 233–36.

76. Watson, *History of Texas Eastern Transmission Corporation,* 32–34.

77. Sobel, *Dillon Read,* 235.

78. *PM Daily,* Nov. 11, 1947.

79. Watson, *History of Texas Eastern Transmission Corporation,* 35.

80. *Harvard Business Review* 26 (Mar., 1948) and 26 (July, 1948).

81. Watson, *History of Texas Eastern Transmission Corporation,* 36.

82. Castaneda, *Regulated Enterprise.*

83. We have written about this history in greater detail in our book, *From Texas to the East.*

84. Bixby interview.

85. Belmont interview.

86. "Brown, Root, Inc., Buys Oil Firm, Other Properties," *Waco Paper,* Jan. 24, 1946.

87. Bixby interview.

88. Ralph O'Connor, interviewed by Christopher J. Castaneda, Houston, Apr. 8, 1992.

89. Ibid.

90. Orville S. Carpenter to George and Herman Brown, Mar. 25, 1957, "Highland Oil Company" folder, Box 7A, HGRBA.

CHAPTER SIX

1. "A $2.3 Million A-E Fee Can Look Small," *Engineering News-Record,* Nov. 15, 1962, 75.

2. "Roadbuilders with a Flair."

3. *Brownbuilder,* 15.

4. Delbert R. Ward, *From There to Here (With a Few Side Trips)* (Houston: Ganado, 1991), 203.

5. *Brownbuilder,* 19.

6. Oltorf interview, July 18, 1990.

7. Albert Maverick, interviewed by Christopher J. Castaneda, Houston, Oct. 21, 1990.

8. *Brownbuilder,* 19.

9. Ibid., 22.

10. Thomas J. Feehan, interviewed by Christopher J. Castaneda, Houston, Mar. 18, 1991.

11. "Outlook of Talk by Mr. Herman Brown to McGraw-Hill Editors' Meeting," Nov. 21, 1952, Houston, copy in HGRBA.

12. Feehan interview, Mar. 18, 1991.

13. Don E. Warfield, interviewed by Christopher J. Castaneda, Houston, Dec. 18, 1990.

14. Counts interview.

15. Bixby interview.

16. Ibid.

17. O'Connor interview.

18. Warfield interview.

19. Bixby interview.

20. Warfield interview.

21. Merritt A. Warner, interviewed by Christopher J. Castaneda, Houston, Oct. 10, 1990.

22. Brown, "Outline of Talk," 1–2.

23. "Biggest Contractors Set Volume Record; Profits Up but Margins Are Down," *Engineering News-Record,* Apr. 9, 1970, 45.

24. Ward Dennis, interviewed by Christopher J. Castaneda, New York City, Nov. 7, 1990.

25. Ibid.

26. Ibid.

27. Moorehead manuscript, Apr., 1968, 51.

28. Warner interview.

29. Moorehead manuscript, Apr., 1968, 49.

30. Ibid.

31. Warner interview.

32. Moorehead manuscript, Apr., 1968, 53.

33. Brown, "Outline of Talk," 1–2.

34. "The Freeway Just a Breeze to the Breeze," *Houston Press,* Aug. 1, 1951.

35. "New Orleans Expressway $30.7 Million Contract," *Engineering News-Record,* Nov. 11, 1954, 66.

36. "New Orleans Area," *Engineering News-Record,* Feb. 21, 1957, 100.

37. Warfield interview.

38. "A 24-Mile Bridge Across Louisiana's Lake Pontchartrain," *Engineering News-Record,* Aug. 30, 1956, 32.

39. "Roadbuilders with a Flair."

40. "Brown & Root, Inc., Engineers and Constructors," copy in HGRBA. Also see "$2.3 Million A-E Fee."

41. Brown, "Outline of Talk," 1–2.

42. Austin interview, Sept. 24, 1990.

43. Brown, "Outline of Talk."

44. *Lamp* 33 (June, 1951): 3–6.

45. *Brownbuilder,* 15–21.

46. We have written, with coauthor Tyler Priest, a detailed history of the evolution of the company's offshore work. See *Offshore Pioneers.*

47. Moorehead manuscript, Apr., 1968, 106.

48. Text from API awards ceremony for George R. Brown, 1982 Annual Meeting, copy in HGRBA. Also, "Ex Brown & Root Chairman to Receive Special Award," *Houston Post,* Nov. 8, 1982. For a detailed treatment of Brown & Root's earliest efforts in the Gulf of Mexico, see Pratt, Priest, and Castaneda, *Offshore Pioneers.*

49. Moorehead manuscript, Apr., 1968, 15 and 19.

50. Ibid., 107.

51. Ibid., 109.

52. *Brownbuilder,* 16–18.

53. "Roadbuilders with a Flair."

54. Brown, "Outline of Talk," 3.

55. Dan Dyess letter, *Austin American,* Feb. 22, 1950, 8.

56. Brown, "Outline of Talk," 3.

57. Andrew Carnegie, for example, had great difficulty adjusting to the new reality of giant steel mills, where "his boys" no longer knew and respected him. Ironically, the men who were most influential in transforming the traditional economy by revolutionizing the scale and technology of factories seem to have had the most difficulty understanding that they had destroyed the close personal ties between owner and worker in the process.

58. The traditional account of labor law in these years is Harry A. Mills and Emily Clark Brown, *From the Wagner Act to Taft-Hartley* (Chicago: University of Chicago Press, 1950).

59. See "Labor Costs and Wage Rates—Marshall Ford Dam" folder, GRBR.

60. Taft-Hartley presented tricky political problems for Lyndon Johnson. See Dalleck, *Lone Star Rising*, 303, 314–15.

61. *Texas Poll*, Sept. 25, 1946, cited in Thomas B. Brewer, "State Anti-Labor Legislation: Texas—A Case Study," *Labor History* 11, no. 1 (Winter 1970): 58–76.

62. Texas Federation of Labor, *Report of Officers*, Jan., 1947, as quoted in Brewer, "State Anti-Labor Legislation," 72.

63. John McCrea, *Texas Labor Laws: A Guide to Laws Affecting Employers, Employees, and Labor Organizations* (Austin: Bureau of Business Research, 1971). See also Green, *Establishment in Texas Politics*, 104–105.

64. Green, *Establishment in Texas Politics*, 107.

65. 77 NLRB 1137; 99 NLRB 1033.

66. Most of the information in this section is taken from NLRB hearings on the case and from "Ozark Dam Constructors vs. Labor Unions; Flippin Materials Co. vs. Labor Unions" folder, GRBR.

67. NLRB to Ozark Dam Constructors, Mar. 19, 1948; A. J. Wirtz to NLRB, Mar. 22, 1948; A.J. Wirtz to Herman Brown, Mar. 22, 1948; all from ODC folder, GRBR. See also "MRB Jurisdiction Viewed," *New York Times*, June 17, 1948; Dugger, *Politician*, 293.

68. A. J. Wirtz to Herman Brown, June 22, 1948, ODC folder, GRBR.

69. Tommy Topkins to Herman Brown, July 8, 1948, ODC folder, GRBR.

70. M. H. Slocum to Herman Brown, Aug. 8, 1948; Herman Brown to M. H. Slocum, July 30, 1948, ODC folder, GRBR.

71. A. J. Wirtz to T. J. Gentry, Aug. 27, 1948, ODC, GRBR.

72. John LeBus to Brown & Root, Oct. 1, 1948, ODC, GRBR; 99 NLRB 1076; 132 NLRB 486; 151 NLRB 241.

73. Moorehead manuscript, Apr., 1968, 48.

74. J. B. Bonny to Herman Brown, Mar. 16, 1949, ODC, GRBR.

75. Herman Brown to Jack B. Bonny, Mar. 25, 1949, ODC folder, GRBR.

76. Ben H. Powell, Jr., to All Joint Venturers, Ozark Dam Constructors, Sept. 30, 1949; Ben H. Powell, Jr., to Herman Brown, Dec. 4, 1949, ODC folder, GRBR.

77. 99 NLRB 1076; 132 NLRB 486; and 151 NLRB 241.

78. R. L. Ricketts to Herman Brown, Dec. 18, 1949, ODC folder, GRBR.

79. Herman Brown to R. L. Ricketts, Dec. 28, 1949, ODC folder, GRBR.

80. Slocum to Herman Brown, undated note, ODC folder, GRBR.

81. Booth interview.

82. "Pegler Says: A Recent Case in Beaumont," *San Antonio Light*, June 13, 1949, copy in "B & R vs. AFL—Beaumont" folder, GRBR (hereafter, Beaumont folder, GRBR).

83. Herman Brown to Homer Garrison, June 6, 1949; "Jester Considers Brown & Root Appeal for Aid," *Houston Chronicle*, June 3, 1949; Ben H. Powell, Jr., to Price Daniel, July 22, 1949, Beaumont folder, GRBR.

84. "Conference of Mr. Bell, Mr. Kelley of S.P. with Mr. Jones & Ditmer of H B & Trades C." undated memo regarding a Mar. 6, 1950, meeting between officials at the Southern Pacific Railroad and representatives of the Houston Building and Trades Council, Beaumont folder, GRBR.

85. "The Issues in the Brown & Root Case from the Standpoint of Management," Everett L. Looney, Address before the Twelfth Annual Meeting of the State Bar of Texas—Labor Law Section, July 6, 1951, copy in Beaumont folder, GRBR.

86. "Conference of Mr. Bell, Mr. Kelley of S.P. with Jones & Ditmer of HB & Trades," p. 4.

87. Herman Brown to Fred A. Hartley, Dec. 5, 1950, "B & R, Inc.-Labor Suit Against Unions, 1950–52" folder, GRBR.

88. A variety of newspaper clippings about the case are found in "B & R, Inc.-Labor Suit Against Unions, 1950–52" folder, GRBR.

CHAPTER SEVEN

1. "76 in U.S. Found to Have Fortunes above $75,000,000," *New York Times,* Oct. 28, 1957.

2. For two of the most recent treatments of Johnson's life before 1960, see Dalleck, *Lone Star Rising;* and Caro, *Path to Power* and *Means of Ascent.* See, also, Dugger, *Politician.*

3. For a useful overview of big business and political power in America, see David Vogel, *Fluctuating Fortunes: The Political Power of Business in America* (New York: Basic, 1989).

4. The literature on the 8-F crowd is itself crowded. For general discussion of 8-F, see Bert Swanson, "Discovering an Economic Clique in the Development of Houston," *Essays in Economic and Business History* 5 (1987): 101–14; Joe R. Feagin, *Free Enterprise City: Houston in Political and Economic Perspective* (New Brunswick, N.J.: Rutgers University Press, 1988); and Harry Hurt III, "The Most Powerful Texans," *Texas Monthly,* Apr., 1976. Green, *Establishment in Texas Politics,* places Houston in the context of Texas politics. For historical accounts of several of the men prominent in 8-F, see Dressman, *Gus Wortham;* Timmons, *Jesse Jones;* Leopold Meyer, *The Days of My Years: Autobiographical Reflections of Leopold L. Meyer* (Houston: Universal Printers, 1975); and Nicholson, *Mr. Jim.*

5. The fullest statement of this view is made in Feagin, *Free Enterprise City.*

6. Green, *Establishment in Texas Politics,* 17.

7. Hurt, "Most Powerful Texans," 79.

8. James Conway, *The Texans* (New York: Knopf, 1976), 102–104.

9. Feagin, *Free Enterprise City,* 107. See also Chandler Davidson, "Houston: The City Where the Business of Government Is Business," in *Public Policy in Texas,* edited by Wendell M. Bedichek and Neal Tannahill (New York: Scott, Foresman, 1982), 276–77.

10. Louie Welch, interviewed by Christopher J. Castaneda, Houston, Mar. 26, 1991.

11. Lipartito and Pratt, *Baker & Botts.*

12. Buenger and Pratt, *But Also Good Business,* provides a history of one of Jones's primary businesses, the National Bank of Commerce. For the standard biography of Jones, see Timmons, *Jesse Jones.* A broader perspective of Jones's role in the New Deal is included in Jordan Schwarz, *The New Dealers: Power Politics in the Age of Roosevelt* (New York: Knopf, 1993), 59–95.

13. Oltorf interview, July 18, 1990.

14. Naurice Cummings, interviewed by Joe Pratt, TE History Collection, WRC, RU.

15. Hyman, *Craftsmanship and Character;* James A. Elkins, Jr., interviewed by Christopher J. Castaneda, Houston, Jan. 17, 1991. See, also, Dressman, *Gus Wortham.*

16. Ben Love, interviewed by Christopher J. Castaneda, Houston, Dec. 20, 1990.

17. Castaneda and Pratt, *From Texas to the East.*

18. Nicholson, *Mr. Jim.*

19. Isabel Brown Wilson, interviewed by Michael Gillette, Houston, Feb. 19, 1988, LBJL.

20. Oveta Culp Hobby, interviewed by Christopher J. Castaneda, Houston, June 19, 1990; William P. Hobby, Jr., interviewed by Christopher J. Castaneda, Houston, Jan. 18, 1991.

21. O'Connor interview.

22. Nicholson, *Mr. Jim.*

23. Albert Thomas to GRB, Apr. 30, 1948, "Thomas, Albert, 1946–1961" folder, GRBR.

24. For a general discussion of money and Texas politics, see, Jimmy Banks, *Money, Marbles, and Chalk* (Austin: Texas Publishing Company, 1971).

25. Dressman, *Gus Wortham.*

26. Edgar Ray, *The Great Huckster* (Memphis: Memphis State University Press, 1980), 234.

27. Swanson, "Discovering an Economic Clique."

28. Lloyd Bentsen, interviewed by Christopher J. Castaneda, Washington, D.C., Mar. 7, 1991.

29. Of course, businessmen who control a giant construction company and one of the nation's major natural gas transmission companies would be expected to exercise considerable economic and political power even without close ties to a prominent national politician.

30. Belmont interview.

31. "Out of Six-Year Dream," *Houston Chronicle,* Feb. 16, 1964.

32. See, "Group May Buy Land for North Side Airport," unidentified newspaper clipping, Jan. 14, 1957, "Houston Intercontinental Airport" folder, Houston Metropolitan Research Center (HMRC), Houston.

33. Welch interview.

34. "Tract Bought to Provide Jet Airport," unidentified newspaper clipping, May 1, 1957, "Houston Intercontinental Airport" folder, HMRC.

35. Ibid. See, also, Welch interview. Also, see, "Council Backs Group's Plan for Jet Airport," unidentified newspaper clipping, May 2, 1957, and "Group to Hold Land for Airport Formed," unidentified newspaper clipping, "Houston Intercontinental Airport" folder, HMRC.

36. Welch interview.

37. Minutes of Special Meeting of Shareholders of Jetero Ranch Company, Mar. 24, 1958, "Jetero, 1957–61" folder, GRBR.

38. Ibid.

39. Lewis Cutrer to GRB, Dec. 1, 1959, "Jetero, 1957–61" folder, GRBR.

40. Welch interview.

41. Belmont interview.

42. Ronnie Dugger, "What Corrupted Texas," *Harper's,* Feb. 28, 1957, 68.

43. See Joe B. Frantz, *The Driskill Hotel* (Austin: Encino, 1973).

44. "'Mystery Rider' Keeps Brown & Root Building," *Texas Observer,* Feb. 19, 1957, 1–6.

45. Ibid.

46. Jack Robbins, interviewed by Christopher J. Castaneda, Houston, Feb. 22, 1991.

47. Searcy Bracewell, interviewed by Christopher J. Castaneda, Houston, Oct. 18, 1990.

48. Lyndon B. Johnson to Margarett and Herman Brown, Nov. 7, 1957, "Brown 32, 1957," Box 23, Master File index, U. S. Senate, 1949–1961, Papers of LBJ, LBJL.

49. Oltorf interview, July 18, 1990.

50. Ibid.

51. Ibid.

52. Ibid.

53. Stilwell, *Herman Brown Story*, 15.

54. Oltorf interview, July 18, 1990.

55. Lester Velie, "Do You Know Your State's Secret Boss?" *Reader's Digest*, Feb., 1953, 35–40.

56. Stilwell, *Herman Brown Story*, 1–2.

57. Dugger, *Politician*, 294.

58. See Brewer, "State Anti-Labor Legislation," 74–76.

59. Bracewell interview.

60. Oltorf interview, July 18, 1990.

61. Ibid.

62. Caro, *Means of Ascent*, 16; Dugger, *Politician*, 273–74.

63. George R. Brown, interviewed by Paul Bolton, Apr. 6, 1968, LBJL (copy in HGRBA).

64. Oltorf interview, July 18, 1990; Wilson interview, Aug. 16, 1994.

65. Frank Oltorf, interviewed by Joseph Pratt, July 7, 1995.

66. Wilson interview, Aug. 16, 1994.

67. Robert Sobel, *ITT: The Management of Opportunity* (New York: Truman Talley, 1982), 124–26, 146.

68. Charles Tillinghast, interviewed by Christopher J. Castaneda, Jan. 22, 1991.

69. "6 Board Candidates Proposed by T.W.A.," *New York Times*, Feb. 10, 1961.

70. Gus Wortham to Governor James Allred, Aug. 31, 1937; Weaver Moore to Governor James Allred, Aug. 31, 1937; Herman Brown to Governor James Allred, Aug. 31, 1937; and Governor James Allred to Herman Brown, Sept. 7, 1937, "1937, A–C," Box 297, Allred Collection, University of Houston.

71. "Truman Names 5 to Study Supplies," *New York Times*, Jan. 23, 1951.

72. The President's Materials Policy Commission, *Resources for Freedom* 1 (June, 1952).

73. "Congress Sets Up Panel to Spur Use of Atom in Peace," *New York Times*, Mar. 27, 1955.

74. Wilson interview, Aug. 16, 1994.

75. "Congress Sets Up Panel."

76. "Report Asks A.E.C. to Share Control of Atom in Peace," *New York Times*, Feb. 1, 1956.

77. Herbert Brownell, interviewed by Christopher J. Castaneda, New York City, Nov. 8, 1990.

78. "George Brown Heads River Study Panel," *Houston Press*, Dec. 18, 1958; and "G. R. Brown Named by Ike on Panel to Study Rivers," *Houston Post*, Dec. 18, 1958.

79. "Eisenhower Meets with Aid Leaders," *New York Times*, Jan. 28, 1958.

80. "U.S. Names Weeks a Business Advisor," *New York Times*, Jan. 29, 1959.

81. Meiners, *History of Rice University*, 211.

82. Lady Bird Johnson interview.

83. LBJ to Herman Brown, Dec. 12, 1957, "Brown 32, 1957," Box 23, Senate papers, LBJL.

84. LBJ to GRB, Nov. 25, 1959, "Brown, George—1959," Box 24, Senate papers, LBJL.

85. Brown interview, Apr. 6, 1968.

86. LBJ to Alice and George Brown, Sept. 28, 1958; LBJ to Margarett and Herman Brown, Sept. 28, 1958, "Brown 32, 1958," Senate papers, LBJL.

87. LBJ to GRB, Jan. 3, 1957, "Brown 32, 1957," Box 23, Senate papers, LBJL.

88. Naurice Cummings, interviewed by Christopher J. Castaneda, Houston, June 27, 1990.

89. Stude interview.

90. McKinney interview.

91. Ibid.

92. Belmont interview.

93. Bixby interview.

94. George R. Brown, interviewed by Michael Gillette, July 11, 1977, LBJL.

95. Caro's treatment of the Browns in the first two volumes of his biography of Johnson never attempts to place their campaign contributions in historical perspective by discussing common practices in this area in the years before the Watergate scandal produced reforms in campaign contribution laws. The revelations of widespread violations of the campaign contribution laws revealed in the Watergate scandal suggest that the Browns were not alone in aggressively seeking political influence. Indeed, the entire recorded history of corruption in American politics—from the much discussed accounts of late-nineteenth-century corruption to the Teapot Dome scandal of the 1920s to Watergate and even to more recent revelations of practices in the 1996 election—would seem to place the burden of proof on any historian who wants to argue that the Browns, or Boss Tweed or Warren Harding or Richard Nixon for that matter, moved corruption to a new level.

96. Dugger, *Politician*, 273.

97. Drew Pearson, "Lyndon–Brown & Root Connection Explored," *Texas Observer*, Mar. 28, 1956.

98. Robbins interview.

99. Lem Goodwin, interviewed by Christopher J. Castaneda, Houston, Dec. 4, 1990.

100. Herman Brown to LBJ, Feb. 16, 1953, LBJL.

101. Lyndon B. Johnson to GRB, June 8, 1956, "Brown 32, 1953," Box 23, "Bri-Brown" Master File Index, U.S. Senate, 1949–61 Collection, LBJL; GRB to Walter Jenkins, Oct. 6, 1955, "Brown, George 1 of 2," Box 12, LBJA Selected Names Collection, LBJL; Chas. I. Francis to Senator Lyndon B. Johnson, "Brown, George—1 of 2," Box 12, SN, LBJL.

102. Michael R. Beschloss, *Taking Charge: The Johnson White House Tapes, 1963–1964* (New York: Simon and Schuster, 1997).

103. Caro, *Means of Ascent*, 15.

104. The key work on this issue is Don E. Carleton, *Red Scare!: Right-wing Hysteria, Fifties Fanaticism, and Their Legacy in Texas* (Austin: Texas Monthly Press, 1985).

105. Oltorf interview, July 18, 1990.

106. Herman Brown to Committee for Constitutional Government, Inc., June 2, 1954, "Brown, Herman," Box 13, SN, LBJL.

107. See, for example, William H. Leslie to GRB, Apr. 26, 1950, "William Leslie," GRBR. In this letter, Leslie describes the progress of a variety of proposals for public works.

1. Meiners, *History of Rice University,* 44–69; Lipartito and Pratt, *Baker & Botts in the Development of Modern Houston,* 53–61.

2. "George R. Brown."

3. Buenger and Pratt, *But Also Good Business,* 82–83; Board Minutes, Rice University, Jan. 7, 1942, vol. 10, p. 38.

4. Felix A. Runion, "Prospects of a Million Dollars a Year," *Rice Owl* 6. no. 4 (Feb., 1944): 5, 16–18; Meiners, *History of Rice University,* 136–39.

5. Runion, "Prospects," 5.

6. The Rice Board spent much time examining this issue in the period from October through December of 1942, and their deliberations are summarized in the board minutes. Board Minutes, Rice University, Oct. 7, 1942, vol. 10, p. 190.

7. "George R. Brown," 5.

8. Board Minutes, Rice University, Oct. 7, 1942, vol. 10, pp. 192B–NN. The minutes include a copy of a portion of the court decision.

9. Ibid., Nov. 9, 1942, vol. 10, pp. 206, 211.

10. For the estimate as of the late 1960s, see "George R. Brown," 6.

11. Board Minutes, Rice University, Feb. 11, 1942, vol. 10, p. 59.

12. Meiners, *History of Rice University,* 141–47.

13. Board Minutes, Rice University, Jan. 20, 1943, vol. 10, p. 269; Mar. 31, 1944, vol. 11, p. 213.

14. Dressman, *Gus Wortham.*

15. Oveta Culp Hobby interview.

16. Kenneth Pitzer, interviewed by Joseph Pratt, Berkeley, Calif., Aug. 13, 1990.

17. Board Minutes, Rice University, July 30, 1945, vol. 11, p. 78; Meiners, *History of Rice University,* 140–41.

18. Board Minutes, Rice University, Dec. 28, 1945, vol. 12, p. 131.

19. Throughout the board minutes of this era, George Brown is recorded as serving on most committees to examine such issues as a postseason bowl game for Rice Stadium and the general affairs of the athletic program. For a short description of the Brown brothers' enthusiasm for Southwest Conference football, see, Everett Collier, "Those Amazing Brown Brothers," *Texas Parade,* Oct., 1951, 32.

20. Meiners, *History of Rice University,* 104, 132, 165–67.

21. Board Minutes, Rice University, Dec. 4, 1947, vol. 13, p. 7; Jan. 9, 1948, vol. 13, p. 26; Apr. 14, 1949, vol. 13, p. 166.

22. Ibid., Nov. 1, 1949, vol. 13, p. 207; Nov. 17, 1949, vol. 13, p. 215; "Leaders in Drive for New Stadium Talk Finances," *Houston Chronicle,* Apr. 6, 1949.

23. Board Minutes, Rice University, Nov. 17, 1949, vol. 13, pp. 216–18.

24. Dressman, *Gus Wortham,* 97–98; "Wortham," *Rice University Review,* Spring 1970, 6ff.

25. Board Minutes, Rice University, Nov. 17, 1949, vol. 13, p. 218.

26. Ibid., Dec. 30, 1949, vol. 13, p. 225; May 11, 1950, vol. 134, p. 264; "The Home Team Builds a Stadium," *Engineering News-Record,* Sept. 28, 1950, 42–43; Ward, *From There to Here,* 255–62; "The Brown Brothers: Houston's Contracting and Energy Giants," *Houston Business Journal,* Oct. 1, 1984.

27. Meiners, *History of Rice University*, 149–57.

28. Pitzer interview.

29. Board Minutes, Rice University, Aug. 5, 1949, vol. 13, pp. 94–95; Meiners, *History of Rice University*, 168–69.

30. Board Minutes, Rice University, Sept. 23, 1953, vol. 14, p. 171; Dec. 1, 1955, vol. 14, p. 328.

31. Pitzer interview.

32. "Rice Tuition Ban Blocks Funds, Trustee Testifies," *Houston Chronicle*, Feb. 14, 1964.

33. Pitzer interview.

34. Amilcar Shabazz, "The Opening of the Southern Mind" (Ph.D. diss., University of Houston, 1996). For an interesting account of desegregation in Houston, see Thomas R. Cole, *No Color Is My Kind: The Life of Eldreway Sterns and the Integration of Houston* (Austin: University of Texas Press, 1997).

35. Board Minutes, Rice University, Jan. 31, 1962, vol. 17, p. 153; May 23, 1962, vol. 17, pp. 195–96.

36. Ibid., Sept. 26, 1962, vol. 17, pp. 227–29.

37. Wilson interview, Aug. 16, 1994.

38. William D. Angel, Jr., "The Politics of Space: NASA's Decision to Locate the Manned Spacecraft Center in Houston," *Houston Review* 6, 2nd issue (1984): 63–81; Stephen Oates, "NASA's Manned Spacecraft Center at Houston, Texas," *Southwestern Historical Review* (Jan., 1964): 350–75.

39. Angel, "Politics of Space," 66–67.

40. Ibid., 70.

41. Brown interview, Aug. 6, 1969.

42. Board Minutes, Rice University, Aug. 23, 1961, vol. 17, p. 115; Oct. 25, 1961, vol. 17, p. 133; Carl E. Reistle, Jr., interviewed by Christopher J. Castaneda, Houston, Aug. 21, 1990.

43. James E. Webb to Vice President Lyndon Johnson, memo, May 23, 1961, "MSC Site Selection Correspondence, 1958–1962" folder, Box 10, JSC Files, Organization Series, MSC Site Selection, NASA History Archives, Clear Lake City, Tex.

44. Angel, "Politics of Space," 72–73.

45. James E. Webb to President John Kennedy, memo, Sept. 14, 1961, "Memorandum for the President-1961," Box 10, JSC Files, MSC Site Selection, NASA History Archives, Clear Lake City, Tex.

46. James Webb quoted in Angel, "Politics of Space," 63.

47. Pitzer interview.

48. Ralph Wood Jones, *Southwestern University, 1840–1961* (Austin: Jenkins, 1973).

49. Ibid., 167.

50. Claude Cody, Jr., to Herman Brown, July 11, 1939, "Brown, Herman," Box 13, SN, LBJL.

51. Jones, *Southwestern*, 297.

52. J. N. R. Score to Herman Brown, Apr. 6, 1946, "Brown, Herman" folder, Southwestern University Archives, Georgetown, Tex.

53. Herman Brown to J. N. R. Score, Dec. 1, 1948, "Brown, Herman" folder, Southwestern University Archives, Georgetown, Tex.

54. Jones, *Southwestern*, 313.

55. Minutes, Board of Trustees, Southwestern University, Nov. 12, 1953.

56. Jones, *Southwestern,* 306–307.

57. Grogan Lord, interviewed by Christopher J. Castaneda, Houston, Feb. 12, 1991.

58. Minutes, Board of Trustees, Southwestern University, Nov. 14, 1950, p. 132.

59. Ibid., Nov. 11, 1954, p. 85 (for pool); Nov. 15, 1955 (for pool); Apr. 17, 1952, p. 213 (for landscape work); and Apr. 4, 1954, p. 52 (for landscape work); Nov. 14, 1956, p. 217 (for exhibitions).

60. Dr. Durwood Fleming, interviewed by Christopher J. Castaneda, Dallas, July 12, 1990.

61. Minutes, Brown Foundation, July 2, 1951, and Jan. 2, 1952; Brown Foundation Grant Files, "Southwestern University."

62. Lord interview.

63. Minutes, Board of Trustees, Southwestern University, Apr. 17, 1951, p. 164.

64. "Houstonians Honored at Southwestern U.," *Houston Chronicle,* May 27, 1952.

65. Minutes, Board of Trustees, Southwestern University, Jan. 31, 1963, p. 116.

66. Remarks of President Durwood Fleming, Ceremony at the Naming of the Brown College of Arts and Sciences, Southwestern University, Apr. 18, 1975.

CHAPTER NINE

1. Darneille interview.

2. "Herman Brown Dies; Founded Great Firm," *Houston Press,* Nov. 16, 1962; "Brown & Root Founder Dies," *Houston Chronicle,* Nov. 16, 1962; "Herman Brown, Construction Firm's Founder," *New York Tribune,* Nov. 16, 1962; "Herman Brown, Builder, 70, Dies," *New York Times,* Nov. 16, 1962; "Herman Brown of Brown & Root Dead," *Houston Post,* Nov. 16, 1962.

3. Lady Bird Johnson interview.

4. Harold Geneen, interviewed by Christopher J. Castaneda, New York City, Nov. 8, 1990.

5. "Brown Is Eulogized by Vice-President," *Houston Chronicle,* Nov. 17, 1962.

6. "Herman Brown Mourned by 1,500 at Funeral," *Houston Post,* Nov. 18, 1962.

7. Lady Bird Johnson interview.

8. Fleming interview.

9. "Brown Foundation Gets Bulk of Estate," *Houston Chronicle,* Nov. 20, 1962.

10. "Brown Brothers: Houston's Contracting and Energy Giants."

11. Michael E. DeBakey, interviewed by Christopher J. Castaneda, Houston, Apr. 20, 1991. Also, see, Dr. E. L. Wagner, interviewed by Christopher J. Castaneda, Houston, Oct. 24, 1990.

12. John P. Harbin, interviewed by Christopher J. Castaneda, Dallas, Aug. 20, 1990. A nephew of Herman and George Brown, Harry Austin, was a member of the Brown & Root board of directors at the time of the sale.

13. "$2.3-Million A-E Fee."

14. Harbin interview.

15. Ibid. For a general history of Halliburton, including material on the history of Brown & Root before and after it was acquired by Halliburton, see Rodengen, *Legend of Halliburton.*

16. Harbin interview.

17. Ibid.

18. Ibid.

19. Raymond Klempin, "The Odyssey of Brown & Root: Today's Legacy of Power, Tradition," part 2, *Houston Business Journal,* Sept. 6, 1982, 4.

20. Harbin interview.

21. Raymond Klempin, "The Odyssey of Brown & Root: Today's Legacy of Power, Tradition," part 3, *Houston Business Journal,* Sept. 13, 1982.

22. D. Clayton Brown, *Rivers, Rockets and Readiness: Army Engineers in the Sunbelt: A History of the Fort Worth District, U. S. Army Corps of Engineers, 1950–1975* (Fort Worth, Tex.: Fort Worth District, U.S. Army Corps of Engineers, 1979), 58.

23. "$2.3 Million A-E," 70.

24. Ibid., 70.

25. Albert Sheppard, interviewed by Christopher J. Castaneda, Houston, Sept. 5, 1990. See also Tom Overton, "From This . . . to This . . . and Albert Sheppard Had a Hand in It," *Houston Post,* July 20, 1980, sec. BB, 1.

26. "$2.3 Million A-E Fee."

27. Ibid.

28. Meiners, *History of Rice University,* 212.

29. William R. Nelson, *The Politics of Science* (New York: Oxford University Press, 1968), 167.

30. See NAS Archives Divisions of the NRC Earth Sciences AMSOC Committee Mohole Project, 1957–1964, and Pratt, Priest, and Castaneda, *Offshore Pioneers,* 120–24.

31. Ibid.

32. See Willard Bascom, *A Hole in the Bottom of the Sea: The Story of the Mohole Project* (New York: Doubleday, 1961), for an in-depth review of the early phase of the project before Brown & Root became involved.

33. Ralph S. O'Connor memo, Feb. 5, 1992, regarding biography of Herman and George R. Brown, "Comments on Brown Biography Drafts" folder, HGRBA.

34. Herbert Solow, "How NSF Got Lost in Mohole," *Fortune,* May, 1963.

35. "$2.3-Million A-E Fee." Also see Darneille interview.

36. Pratt, Priest, and Castaneda, *Offshore Pioneers,* 123–34.

37. "$2.3 Million Fee," 86.

38. Pratt, Priest, and Castaneda, *Offshore Pioneers,* 123–34.

39. "Another Big Scientific Job Goes to Texas Firm," *Los Angeles Times,* Mar. 18, 1962.

40. Congressional Record (Senate), May 2, 1963, 7222. Also see "Mohole Contractor Scored by Senator," *New York Times,* May 3, 1963.

41. George H. W. Bush to the Editors, *Fortune,* May 2, 1963. In "National Science Foundation—Mohole—1963," B&RA.

42. Solow, "How the NSF Got Lost in Mohole."

43. Congressional Record (Senate), May 8, 1963, 7532.

44. Ibid., 7537.

45. E. J. Gerrity to H. S. Geneen, memo, May 9, 1963. In "National Science Foundation—Mohole—1963," B&RA, GRBR.

46. Brown Booth to Robert Six, May 16, 1963. In "National Science Foundation—Mohole—1963," B&RA, GRBR.

47. Robert F. Six to Joseph A. Thomas, May 29, 1963. In "National Science Foundation—Mohole—1963," B&RA, GRBR.

48. Robert C. Toth, "Mohole Concern Backed by Panel," *New York Times,* Aug. 7, 1963.

49. Leland J. Haworth to George R. Brown, Aug. 16, 1963. In "National Science Foundation—Mohole—1963," B&RA, GRBR.

50. "Mohole Lives," *Newsweek,* Jan. 25, 1965.

51. W. Henry Lambright, *Presidential Management of Science and Technology* (Austin: University of Texas, 1985), 160.

52. See HR 14921. Allott (R. Colo.): Amendment to delete $19.7 million in appropriations for the NSF for the continuation of Project Mohole, and to stipulate that no NFS funds may be used for Project Mohole. Rejected 37-46 (D 16-37; R 21-9), Aug. 10, 1966. A "nay" was a vote supporting the president's position. The bill was subsequently passed by a roll-call vote of 82-2. See, *New York Times,* Aug. 8, 1966.

53. "Congress Conferees Clash on Project Mohole Funds," *New York Times,* Aug. 18, 1966. Also, see, "House Rejects Fund for Project Mohole," *New York Times,* Aug. 19, 1966; also, "Big Spending Bill Passed by Senate," *New York Times,* Aug. 11, 1966.

54. "Mohole Contractor Is Lined to Gifts Made to Democrats," *New York Times,* Aug. 20, 1966.

55. Lambright, *Presidential Management of Science and Technology,* 160–61.

56. "Mohole Contractor."

57. "Johnson Denies Any Favoritism in the Award of U.S. Contracts," *New York Times,* Aug. 28, 1966.

58. "From the President's Office . . . Project Mohole," *Brownbuilder,* Jan.–Feb., 1967.

59. "Mohole Lives," *Newsweek,* Jan. 25, 1965.

60. Keith Wheeler and William Lambert, "The Man Who Is President," *Life,* Aug. 14, 1964, and Aug. 21, 1964.

61. LBJ and GRB, recorded telephone conversation, July 29, 1964, 9:10 A.M., Citation #4378, Recordings of Telephone Conversations—White House Series (RTC–WHS), LBJL.

62. Ibid.

63. Wheeler and Lambert, "Man Who Is President."

64. LBJ and Clark Clifford, recorded telephone conversation, July 29, 1964, 8:21 P.M., Citation #4409, RTC–WHS, LBJL.

65. LBJ and GRB, recorded telephone conversation, Feb. 14, 1964, 12:20 P.M., Citation #2082, RTC–WHS, LBJL.

66. Ibid.

67. Ibid.

68. Allen J. Matusow, *The Unraveling of America: A History of Liberalism in the 1960s* (New York: Harper and Row, 1984), 149–51.

69. "Construction Escalates in Vietnam," *Engineering News-Record,* Feb. 3, 1966, 11. Also see in the *Engineering News-Record,* "Vietnam Demand for A-Es Grows," July 21, 1966, 17–18; "Vietnam: On Schedule, but Slowing," Aug. 11, 1966, 61–64; "Air Force Civil Engineer-

ing Takes on Big Construction," Oct. 27, 1966, 26–28; "Vietnam Air Base Delivered Fast," Dec. 1, 1966, 17–18; "War or No, Mekong Projects Continue," Dec. 1, 1966, 18; "Vietnam Civilian Work Extended," Oct. 5, 1967, 21; "Vietnam Contractors Hit Again," May 30, 1968, 13; "U.S. Contractors in Vietnam Turn to Public Works Jobs," Sept. 5, 1968, 39–40; "Seabees Learn Fast Roadbuilding," Oct. 3, 1968, 43.

70. Richard William Tregaskis, *Building the Bases: The History of Construction in Southeast Asia* (Washington, D.C.: GPO, 1975), 183.

71. "Construction Escalates in Vietnam," *Engineering News-Record,* Feb. 3, 1966, 11.

72. H. W. Reeves, interviewed by Christopher J. Castaneda, Houston, Jan. 27, 1992.

73. Tregaskis, *Building the Bases,* 231.

74. Reeves interview; also, see, Tregaskis, *Building the Bases,* 231.

75. "Contractors Phase Out of Vietnam," *Engineering News-Record,* June 19, 1969, 75.

76. Tregaskis, *Building the Bases,* 437.

77. Klempin, "Odyssey of Brown & Root," part 2, p. 4.

78. "Brown and Root in Vietnam," *Texas Observer,* June 21, 1968.

79. "Tiger Cage Camp to Get New Cells," *New York Times,* Feb. 21, 1971.

80. "UT Alumnus Is Protested," *Austin American-Statesman,* Oct. 27, 1973.

81. Oltorf interview, July 18, 1990.

82. Ibid.

83. George R. Brown, interviewed by Michael L. Gillette, Houston, July 11, 1977, LBJL.

84. McCartney, *Friends in High Places.*

85. "B & R Inc. and Associated Companies Employment Data, 1929–1979," copy in HGRBA.

86. Klempin, "Odyssey of Brown & Root," part 3, p. 3.

87. Castaneda and Pratt, *From Texas to the East.*

88. Ibid., 179. This account is taken from chapter 8 on the Houston Center project in the Texas Eastern book, which, in turn, was written from internal records and interviews with the company's employees.

89. Ibid., 180.

90. Ibid., 181.

91. "Supplying Rigs Proves Toughest Operating Problem," *Oil and Gas Journal,* May 10, 1965, 154.

92. Millard Neptune, interviewed by Joseph Pratt, Oct. 31, 1988.

93. John E. Swearingen, interviewed by Christopher J. Castaneda, Chicago, Oct. 16, 1990. Also, see, Castaneda and Pratt, *From Texas to the East,* for a discussion of Texas Eastern's North Sea work.

94. Pratt, Priest, and Castaneda, *Offshore Pioneers,* has detailed discussion of Brown & Root's major projects offshore in the North Sea and around the world. See also "The North Sea . . . Oil's Biggest Gamble to Date," *Oil and Gas Journal,* May 10, 1965, 134.

95. "North Sea," Brown & Root publication, n.d., TE History Collection, WRC, RU.

96. Clive Callow, *Power from the Sea: The Search for North Sea Oil and Gas* (London: Golancz, 1973), 164.

97. Moorehead manuscript, Apr., 1968, 110.

98. Pratt, Priest, and Castaneda, *Offshore Pioneers.*

99. "Norway Claims Phillips Has a Big One," *Oil and Gas Journal,* Nov. 10, 1969, 130.

100. "North Sea."

101. "BP Borrows $936 Million to Develop Forties," *Oil and Gas Journal,* July 3, 1972, 25.

102. "BP Sets Sights on 400,000 b/d from Forties," *Oil and Gas Journal,* Dec. 27, 1971, 50; "The Gift Horse Gallops By," *Economist,* July 26, 1975, 21–24.

103. "Forties Field Development Planning Involves New Wrinkles," *Oil and Gas Journal,* Jan. 10, 1972, 78.

104. "First Two Platforms Set for Forties," *Oil and Gas Journal,* Jan. 24, 1972, 38.

105. Hugh R. Gordon, interviewed by Christopher J. Castaneda, Houston, Nov. 30, 1990, TE History Collection, WRC, RU.

106. "Grandjury Probes Brown & Root," *Engineering News-Record,* Dec. 1, 1977, 17.

107. "McDermott and Brown & Root Fined," *Engineering News-Record,* Dec. 21, 1978, 39.

108. "Offshore Contractors Sued in Civil Anti-trust Action," *Engineering News-Record,* Dec. 20, 1979, 46.

109. O'Connor interview.

110. Ibid.

111. James P. Sterba, "Plans to Drill for Oil in Park Creates Dispute in Houston," *New York Times,* Jan. 22, 1976.

CHAPTER TEN

1. "Brown Brothers: Houston's Contracting and Energy Giants," 1B. This article was originally written with George Brown's assistance in 1979 before being reprinted in 1984.

2. Bracewell interview.

3. Brown interview, Aug. 8, 1969.

4. "Brown Brothers: Houston's Contracting and Energy Giants," 1B.

5. Ibid.

6. Waldemar A. Nielsen, *The Golden Donors* (New York: Truman Talley, 1985), 6. Also, see, David F. Freeman, *The Handbook on Private Foundations* (Washington, D.C.: Seven Locks, 1981).

7. Minutes, Brown Foundation, July 2, 1951, and Nov. 15, 1952.

8. Ibid., Nov. 18, 1958.

9. Ibid., Nov. 26, 1962.

10. Ibid., Dec. 13, 1962, and Dec. 18, 1962.

11. Ibid., Jan. 29, 1963.

12. Ibid., Nov. 29, 1976.

13. Rice Minutes, Jan. 31, 1962.

14. "George R. Brown," *Rice University Review* (Fall/Winter 1968): 7–8; Board Minutes, Rice University, Jan. 31, 1962, vol. 17, p. 154; July 26, 1961, vol. 17, p. 113.

15. Minutes, Brown Foundation, Oct. 19, 1965.

16. Rice University News Release, Dec. 13, 1966, 1 and 2, copy in HGRBA.

17. Minutes, Brown Foundation, Apr. 7, 1965; Oct. 19, 1965; Mar. 22, 1966; May 25, 1967; "William Marsh Rice" printout, Brown Foundation.

18. "Minutes of Steering Committee Meetings" folder, "$33 Million Campaign" box, Development Office Papers, Rice University Collection, WRC, RU. Also, in same box, see, "Rice $33,000,000 Campaign News," vol. 1, no. 1, Jan., 1966.

19. "The Brown Challenge," 1976, copy in HGRBA.

20. "The Naming of the Brown College of Arts and Sciences," Southwestern University, Apr. 18, 1975, copy in the HGRBA.

21. "SU Benefactor Brown Dies at 84," *Bulletin of Southwestern University* 7, no. 3 (Feb., 1983): 1.

22. "Professorship and Medal at CSM Will Honor Alumnus George R. Brown," *Mines Magazine,* June, 1974.

23. "Dedication Ceremonies: George R. Brown Hall," Apr. 9, 1981, copy in "George Brown," HGRBA.

24. "Mines' Best," *Rocky Mountain News,* June 1, 1962, and "Engineers Told to Communicate," date and paper unknown.

25. "Brown, Darneille to Be Feted for 'Miracle' in Arkansas," date and paper unknown, copy in HGRBA.

26. Ann Holmes, "Fine Arts Museum Receives $1 Million," *Houston Chronicle,* Oct. 6, 1968, copy in "Brown Pavilion" folder, Museum of Fine Arts, Houston, Archives (MFAHA). Also, see, *Houston Post,* June 1, 1969, and June 2, 1973, MFAHA.

27. *Museum News,* Dec., 1974, 24, MFAHA.

28. See "Brown Pavilion" folder, MFAHA, which includes the following: Ann Holmes, "Fine Arts Museum Receives $1 Million," *Houston Chronicle,* Oct. 6, 1968; *Houston Chronicle,* Dec. 1, 1971; *Houston Magazine,* Jan., 1972; *ARTgallery,* Jan., 1974; and Charlotte Moser, "Abstract Elegance," *Houston Post,* Jan. 20, 1974.

29. *The Museum of Fine Arts, Houston: An Architectural History, 1924–1986: A Special Bulletin* (Houston: Museum of Fine Arts, Houston, n.d.), 110.

30. "Remarks by Frank Stanton," Museum of Fine Arts, Jan. 15, 1974, copy in HGRBA.

31. *Museum of Fine Arts, Houston,* 111.

32. "Brown Family" folder, MFAHA.

33. The Museum of Fine Arts, Houston, "The Brown Foundation, Inc. Ensuring the Museum's Excellence Past, Present, and Future," *Development Report,* Dec., 1989.

34. Ibid.

35. Wagner interview.

36. DeBakey interview; Wagner interview.

37. Ibid.

38. Dan Arnold, interviewed by Christopher J. Castaneda, Houston, Dec. 19, 1990.

39. DeBakey interview.

40. Ibid.

41. Arnold interview; DeBakey interview.

42. Lady Bird Johnson interview.

43. Ibid.

44. Brown interview, July 11, 1977.

45. Lady Bird Johnson interview.

46. Henry H. Fowler, interviewed by Christopher J. Castaneda, Alexandria, Va., Mar. 8, 1991.

47. Ibid.

48. Note in "Correspondence: LBJ School of Public Affairs" folder, copy in HGRBA.

49. "Lyndon B. Johnson School of Public Affairs," brochure, n.d., copy in HGRBA.

50. LBJ to George and Alice Brown, Oct. 2, 1970, "Correspondence, LBJ School of Public Affairs," copy in HGRBA.

51. Brown interview, Aug. 6, 1969.

52. Fowler interview.

53. Lady Bird Johnson interview.

CHAPTER ELEVEN

1. John Fritz Medal Program, "George R. Brown, Medalist for 1977," copy in HGRBA.

2. Pamphlet from API awards ceremony for George R. Brown, 1982 Annual Meeting, copy in HGRBA. See, also, "Ex Brown & Root Chairman to Receive Special Award," *Houston Post,* Nov. 8, 1982.

3. "City of Houston Resolution," Nov. 8, 1982, copy in HGRBA.

4. "Texas Business Hall of Fame Inducts Brown Brothers," *Brownbuilder,* Fall 1983, 16.

5. See, for example, Harry Hurt III, "The Most Powerful Texans," *Texas Monthly,* Apr., 1976, 73–76, 107–23.

6. Walter Mischer, interviewed by Christopher J. Castaneda, Houston, Oct. 30, 1990.

7. George Brown as quoted in J. Y. Smith, "George R. Brown, Supporter of Lyndon Johnson, Dies," *Washington Post,* Jan. 24, 1983.

8. Herbert E. Alexander, *Financing Politics: Money, Elections, and Political Reform,* 3rd ed. (Washington, D.C.: CQ Press, 1984).

9. *Houston Chronicle,* July 30, 1989, MFAHA.

10. John Bookout, interviewed by Christopher J. Castaneda, Houston, June 18, 1991.

11. Bracewell interview.

12. Oltorf interview, July 18, 1990.

13. Doris Johnson, interviewed by Christopher J. Castaneda, Houston, June 22, 1990.

14. Wagner interview.

15. George R. Brown's 1982 Calendar, HGRBA.

16. Smith, "George R. Brown, Supporter"; "George R. Brown, Industrialist, Dies," *New York Times,* Jan. 24, 1983. Copies of these and other clippings are in the HGRBA.

17. *Congressional Record,* 98th Congress, vol. 129, No. 3, Jan. 25, 1983.

18. Lady Bird Johnson as quoted in news clipping from *Houston Chronicle,* Jan. 24, 1983.

19. "George Rufus Brown, Honorary Member, ASCE, 1898–1983," 1983 ASCE Memoirs, copy in HGRBA.

20. In 1997, Duke Power acquired PanEnergy.

21. See Castaneda and Pratt, *From Texas to the East.*

22. "George Brown Statue Stands Up for Realism," *Houston Chronicle,* date unknown; "Brown Sculpture Gets Dedication," *Houston Chronicle,* Nov. 30, 1988.

INDEX

Pages containing illustrations or tables appear in italics.

Bell County, historical context, 5–7, 20
Bell County Engineering Department, 13,
    20–21
Bellows, W. S., 72
Belmont, August, IV: baseball team
    project, 167; on 8-F crowd, 170–71; on
    Herman's concerns about Texas Eastern,
    120–21; on Herman's reaction to LBJ
    VP nomination, 185; Inch lines bid,
    109–10, 111, 116–17
Belton, Texas, 5, 9
Bentsen, Lloyd, 98, 167, *168,* 276
Bhumiphol Dam, Thailand, 135
bidding process: and Mohole project, 231–
    33; for pipelines, 105–12; for road-
    building contracts, 36, *37;* for war
    surplus property, 96
Big Inch and Little Big Inch pipelines, 101–
    13, 140–41. *See also* Texas Eastern
    Transmission Company (TETCO)
Big Inch Gas Transmission Company, 106
Big Inch Oil, Inc., 105, 107
black Americans, 23, 85, 171, 189, 205–208,
    279
Blakeway, C. E., 52
Bloch, Adm. C. C., 81
Bolin, George R., 244
Bolin, L. T.: and Brown & Root sale, 224,
    225; and Dahlstrom Company, 97; at
    Herman's funeral, 222; promotion of,
    127; and shipbuilding operation, 74, 77,
    79
Bonny, J. S., 150
Bookout, John, 274
Booth, Brown, 112, 234
bootlegging, George's alleged, 17
bowl game for Rice, 200–201
BP (British Petroleum), 248
Bracewell, Searcy, 172, 176, 274
Brazos River Corporation and Reclamation
    District, 46
Breech, Ernest R., 181
bridge-building projects, 6, 27–29, 136, 243
British Gas Council, 246
British Petroleum (BP), 248

Brown, Alice (née Pratt) (wife of George):
    contributions to Brown Foundation,
    255; courtship and marriage of, 32–33;
    death of, 276; and Johnsons, 269, *269;*
    and Museum of Fine Art, 264, 265; at
    Rice, *220;* and Southwestern University,
    173, 211
Brown, Annie Otelia (née Ernst) (grand-
    mother), 7
Brown, Emma Lena (sister), 12
Brown, Fannie Maude (sister), 12
Brown, George Rufus: adjustment to
    Herman's death, 222; on aircraft salvage
    operation, 100; and airport campaign,
    169; Brown & Root hiring of, 28–29;
    Brown & Root post-Herman leader-
    ship, 225, *242,* 251–52; character of, xi–
    xii, 18, 128–29; college education, 14–17;
    courtship and marriage of, 32–33; death
    of, 276; education's importance to, 253–
    54, 273; family life, 64, *64;* health
    problems, 65, 275–76; and Herman, xii,
    13–14, 29, 96, 129–30, 215; and Houston
    garbage collection controversy, 44; and
    Jesse Jones, 163; and LBJ, *53, 57,* 63–64,
    68–69, 71, 85, 179, 183, 185, 188, 240–41,
    269; legacy to Brown & Root, 242–43,
    243; and Marshall Ford Dam project,
    50–51; military career in World War I,
    17; mining career, 17–18, 28; and
    Mohole project, 231, 233–34, 236; move
    to Houston, 37; national-level leader-
    ship of, 176–82, 268–69; and naval air
    station contract, 72; and North Sea
    projects, 246, 248; on offshore oil
    platforms, 140; philanthropic leader-
    ship, 217, 254–55, 262–63, 267–68, 270–
    71; on philanthropy, 253–54; and
    political fallout of LBJ association, 237–
    38; politics of, 274–75; public recogni-
    tion of, *82,* 272–73; recreational
    activities, *165,* 165–66, *168;* Rice
    University leadership, 193, 194, 195–96,
    197–211, *206, 217, 220,* 257–61; on road
    vs. dam construction, 48; sales talents

of, 14, 125; and shipbuilding operation, 75, 78, 84; style of leadership, 126–29; at Texas Eastern, 106, 109–10, 111, 113–14, 115, *116*, 243–45; Vietnam fiasco, 238–41; withdrawal from business activities, 241–42; youth of, 14, *15*

Brown, George (uncle), 8

Brown, Herman: and Albert Thomas, *70*; antiunionism of, 24–25, 58, 60, 86, 87, 142–43, 145, 146, 149–51, 154–55, 189; character of, xi, 18, 40, 90, 122, 128–29; on conflict of interest issue, 119; courtship and marriage of, 25–26, *27*; death of, 221–22; directorships of, 178; early road building career, 19, 20–21, 23–25, 26–28; and Ed Clark, 173; and Elkins, *163*; employee relations, 23–25, 30, 31, 81, 84–85, 129, 143, 150; enjoyment of politics, 175–76; estate gift to Brown Foundation, 223; family life, 89–90, *90*; frugality of, 45, 128; and George, xii, 13–14, 29, 96, 129–30, 215; and LBJ, 51, *53*, 57, 63–64, 179, 183, 184–85, 188; lobbying style of, 35–36, 174, 175; and McKenzie, 54; on Mohole project politics, 231; move to Houston, 125; and Oltorf, 173; politics of, 158, 188–91; recreational activities, 98, 164, 165–66; Southwestern University leadership, 211–17, *216*; style of leadership, 126–29, 150; Texas Eastern role, 115, *116*, 120–21; ties to home region, 137; youth, 12–13

Brown, Isabel (daughter of George), 64, 208, 222, 256

Brown, King (brother), 12

Brown, Louis (grandfather), 7

Brown, Louis King (brother), 11

Brown, Lucy Wilson (née King) (mother), 9, 28, 78, *80*, 215, 255

Brown, Maconda (daughter of George), 64, 121, 256

Brown, Margarett (née Root) (wife of Herman): as Brown Foundation board member, 255; courtship and marriage of, 25–26, *27*; as director of Brown & Root,

39; family life, 89–90, *90;* health problems, 98, 177, 222–23; on Herman's workaholism, 40, 65; intelligence of, 63; and Southwestern University, 211, 214, 217

Brown, Mary Louise (sister), 12

Brown, Nancy (daughter of George), 40, 64, 78, 256

Brown, Rhinehart ("Riney") (father), 5, 6, 7–12, *10*

Brown Building, Austin, 171–72

Brownco, Inc., 122, 250–51. *See also* Brown Securities Company

Brownell, Herbert, 114

Brown Engineering Company, 263

Brown Engineering Fund, 260

Brown Foundation: and Brown & Root sale, 224; Colorado School of Mines, 262–63; development of, 255–56; Herman's estate gift to, 223; Houston Medical Center, 266–68; LBJ School of Public Affairs, 269–71; matching grants summary, *266;* museum contributions, 264–66; as primary legacy of Browns, 276–77; status as charitable foundation, 254; as vehicle for university support, 193, 215. *See also* Rice University; Southwestern University

Brown Hardware Store, 11

Brown Pavilion, *259*, 265

Brown & Root: corporate culture, 30, 96, 126–30, 129, 142, 155; Depression-era survival of, 39–40, 44–45, 61; diversification, post–World War II, 124–31, 137, 142; early growth of, 33, 35; electric line installation project, 59–60; financial growth of, *125;* George's solo leadership of, 242–43, *243*, 251–52; headquarters of, *216;* and hiring of George, 28–29; incorporation of, 39; legacy of, 277–78; marketing practices investigation, 249–50; Mohole project, 230–36; origins of name, 27–28; and politics of influence, 48–49, 51, 67–68, 227–29, 236–38; predecessor of, 21–22;

Brown & Root: (*cont.*)
    reputation as efficient and effective, 33,
    34–35, 36, 73; sale of, 223–26; World
    War II growth of, 66. *See also* construc-
    tion projects; innovations; labor
    relations
Brown-Root Callahan Group, 102
Brown Securities Company, 39, 172, 250
Brown Shipbuilding Company, 66, 75. *See*
    also shipbuilding operation
Brown-Walsh-Raymond (BWR), 132
Buchanan, James P., 47, 49, 51
Buchanan Dam, 47
Building Trades Council of the United
    States, 150, 152
Bull Shoals dam project, 146–52, *147*
Bunger, H. P., 52
Bunker, Ellsworth, 239
Bureau of Public Roads, U.S., 36
Bureau of Reclamation, federal, 47, 48, 49,
    50
Burgess, W. Randolph, 178
Burkhart, Carl, 30, 127
Bush, George H. W., 233
Business Advisory Council, 182
business leaders, Browns vs. current, 278–
    79. *See also* 8-F crowd
Butler, George, 106, 113

Cadillac Fairview, 245
Cameron Iron Works, 161, 164
campaign contributions, 186–87, 273–74.
    See also lobbying, political
Canada, construction projects in, 142
Carmody, John, 60
Carnegie-Mellon Institute, 257
Caro, Robert, 185–86
Carpenter, Orville "Dick" S., 115
cars, and need for new roads, 19, 20
Carter, Amon G., 98
cattle drives, 5
Central Texas, 5–6, 46
challenge grants, initiation of, 261, 262, 265
charitable foundations, 196, 254. *See also*
    Brown Foundation

Charles Luckman Associates, 228
Childs, Marcus, 238
Church, Frank, 274
civic center for downtown Houston, 280
civic institutions, Browns' support of, 157,
    159, 167–70. *See also* Brown Foundation
civil rights movement, effect on Rice, 206
Civil War, 9
Clark, Ed, 155, 173, 211
Clark, Maj. John B., 97
Clear Lake, Texas, 209–10
Cleveland, A. S., 198
Clifford, Clark, 112, 237–38
Cody, Claude, Jr. (Herman's brother-in-
    law), 212
Cold War, as opportunity for military
    construction projects, 131, 133
colleges and universities: Baylor College of
    Medicine, 267–68; effect of Browns'
    leadership on, 192–93, 217–18, 256–57.
    See also Rice University; Southwestern
    University
Colorado River, and flood control in
    Central Texas, 46
Colorado School of Mines, 17–18, 28, 262–
    63
Columbia Construction Company, 71
communication, improvements in rural, 5–6
communism, Browns' stand on, 189–90
competition for contracts, and controversy,
    35–37. *See also* lobbying, political
Compton, Lewis, 69
conflict of interest issue for TETCO, 115,
    117, 118–19, *119*
Congress, U.S.: investigation into
    accounting in Vietnam projects, 240;
    and Marshall Ford Dam, 49–50, 52, 56;
    Mohole project debate, 232–34, 235–36;
    pipeline sale role, 107–108; war surplus
    disposition, 96; and World War II
    military construction projects, 67, 70–71
Connally, John, 98, *168,* 222, 240, 274
Conroe, Texas, Brown & Root road work
    in, 44–45
conservation commissions, 182

Parker, Foster, 222, 223
partnerships, in dam projects, 48. *See also* joint ventures
paternalism, in Herman's labor-management relations, 24–25, 81
patriotism, World War II, 79, 87
Patrol Craft (PC) subchaser program, 73–77
"paving paper," 39
PAW (Petroleum Administration for War), 103
PC boat (subchaser) program, 73–77
Pedernales Electric Cooperative (PEC), 59
Pegler, Westbrook, 153
Peligre Dam, Haiti, 133–35
Perry, E. B., 78
Perry, Jacqueline, 78
petrochemical plant construction, 137–38, *139*
Petroleum Administration for War (PAW), 103 ·
petroleum industry: construction projects for, 45, 102, 104–105, 115–16, 118–20, *119*, 137–42, *139*, 245–49; 8-F representation in, 161; George's questionable ventures in, 251; Houston as center of, 16, 27, 37, 101, 164; oil price collapse of 1980s, 277–78; Rice University's oil field acquisition, 194–96. *See also* natural gas industry
petroleum production company, Browns', 120, 121–22
philanthropy: decline of, 278–79. *See also* Brown Foundation
Phillips Petroleum, 248
physical education at Rice, 199–204
Pidot, George, 111
pipelines, natural gas/oil, 101–13, 140–41, 243–44. *See also* Texas Eastern Transmission Company (TETCO)
Pitzer, Kenneth, 198, 205, *206*, 209, 211
Platzer, Asa, 81
Platzer Boat Works of Houston, 74, 77
Poe, E. Holley, 103–104, 106–107, 110, 114, *116*
Poe group, 106–109

Political Action Committee of Texas, 273
politics: Austin network, 171–76; Browns' influence network (8-F group), 157–67; Brown & Root role in, 48–49, 51, 67–68, 227–29, 236–38; conservative pragmatism of Browns', 158, 188–91, 274–75; and corporate directorships, 178–80; and dangers of developing-nation work, 135; Fort Clark as recreational center for, 172–73; government commission participation by Browns, 180–82; and Mohole project, 231–36; mutual favors society, 85; Washington influence network, 176–78. *See also* Johnson, Lyndon Baines; lobbying, political
polyethylene plant construction, 138
population, Bell County growth in, 7, 11
Powell, Ben, 151, 222
Powell, Joseph, 178
Pratt, Alice (Mrs. G. R. Brown) (wife of George). *See* Brown, Alice (née Pratt) (wife of George)
Pratt, Bela L., 32–33
Pratt, Lillian Nelson (George's mother-in-law), 217
Pratt, Minot Tully (George's father-in-law), 32
presidential library, LBJ, 270
President's Materials Policy Commission, 180
price-fixing scandal, 249–50
private sector construction, 124, 137–42, 190
Project Mohole, 229–36
public scrutiny of Brown & Root, increases in, 227–29, 249–50. *See also* media coverage
Public Works Administration (PWA), 46–47
public works projects: dams, 45–48, *131*, 133–35, 146–52, *147*; expansion of, 95; lobbying as essential to obtaining, 35–37, 61, 72; military construction, 66–73, 130–33, *134*, 238–41; post–World War II expansion in, 124, 190; as source of